American Institute of Chemical Engineers

41st Loss Prevention Symposium 2007

Held at the 2007 AIChE Spring National Meeting

April 22-27, 2007
Houston, Texas, USA

Printed from e-media with permission by:

Curran Associates, Inc.
57 Morehouse Lane
Red Hook, NY 12571
www.proceedings.com

ISBN: 978-1-60423-353-7

ISBN: 978-1-60423-353-7

American Institute of Chemical Engineers
Proceedings
Three Park Avenue
New York, NY 10016-5991
Phone: 212-591-8100

www.aiche.org

American Institute of Chemical Engineers

41st Loss Prevention Symposium
2007

TABLE OF CONTENTS

Session 17 – Facility Siting and Building Design for Explosion Protection

A Practical (Semi Quantitative) Tool for The Determination of Dust Explosion Risks for Workers in Industry 1
A. Klein, M. De Maaijer, N. Versloot

The Employment of NFPA 59A in LNG Facility Siting 13
Anna Qiao, Caroline Deetjen, David Pertuz

Applying the Limitation of Effects ISP Strategy when Siting and Designing Facilities 23
David G. Clark

Determination of Pressure Levels for Siting Light Wood Trailers at Chemical Plants and Refineries 41
Michael A. Polcyn, Raymond H. Bennett, Quentin A. Baker

Session 25 – Preparing for Natural Disasters and Lessons Learned

Natural Hazards and Process Safety Management Elements: A Natural Mix. 53
Frank F. Liserio, Alan Svoboda

Damage to Petrochemical Facilities during Earthquakes 69
M. Cem Ozbey, Ali Sari

The Interaction of Earthquakes with Process Equipment in the Framework of Risk Assessment 81
Ernesto Salzano, Antonio Di Carluccio, Anita Garcia Agreda, Giovanni Fabbrocino,Gaetano Manfredi

Mitigating Hurricane Storm Surge Perils at the DuPont DeLisle Plant 99
Steven P. Harris, Duane O. Wilson

Preparing for Self-Sufficiency 119
Christina Francis

Lessons Learned and Mitigation Options for Process Facilities Following the 2005 Hurricane Season 121
James R. Bailey, Marc L. Levitan

Session 43 – Modeling in Fire and Explosion Protection

Why use Computational Fluid Dynamics for Explosion Studies? 135
David Herrmann

The Usefulness of Phenomenological Tools to Simulate the Consequences of Dust Explosions 151
Ch. Proust, E. Leprette, Daubech, I. Shochet

Simulation and Application of Blast Wave-Target Interaction 167
Jihui Geng, Kelly Thomas

The Design of Venting of Gas Explosions in the Presence of a Discharge Duct 177
Almerinda Di Benedetto, Paola Russo, Ernesto Salzano

CFD-based Risk Assessment for Hydrogen Applications 193
Olav Hansen, Prankul Middha

Application of CFD Models for the Onshore Process Industry 217
Gary Fitzgerald, Greg Knight, Anthony Gutierrez

Session 65 – Fire, Explosion and Reactive Hazards

The Relationship between Flash Point and LFL with Application to Hybrid Mixtures 237
Richard Prugh

Fire Hazards at Power Generation Facilities 249
John LeBlanc, Christopher Wieczorek

Fire and Explosion Framework Assessment on Oil and Gas Floating Production Storage Offloading (FPSO): an Effective Screening and Comparison Tool 261
Jaffee Suardin, Jeff McPhate, Anthony Sipkema, Matt Childs, Sam Mannan

Oleum and Chlorosulfonic Acid Spill Tests - Field Data for Model Validation 279
S. Dharmavaram

Estimating the Consequences of Deflagration to Detonation Transition (DDT) in Hydrogen Explosions 295
Prankul Middha, Olav Hansen

Use of Adiabatic Calorimetry and Aging Test for Safe Storage Study of Hydroxylamine Nitrate 315
Lijun Liu, Maria Papadaki, Yingchun Zhang, Yanjun Wang, William Rogers, Sam Mannan

Session 85 – Communicating Expert Knowledge to Technical Communities

Industry-Government Alliances Share Information to Help Prevent Workplace Injuries, Illnesses and Fatalities 325
Scott Berger, Lee Anne Jillings

Communicating RAGAGEP Responsibility Beyond PSM and RMP Coverage 331
Angela S. Blair

Education Materials for Universities and Industry 339
Joseph Louvar, Dennis Hendershot

CSB Safety Videos: A New and Effective Communication Tool for Accident Investigation Findings 349
Daniel Horowitz, Sandy Gilmour

New Tools and Approaches to Expand Global Process Safety Capabilities 355
Scott Berger, Karen Person

Session 104 – Case Histories and Lessons Learned

Explosion Hazards During Fuel Transition in Combustion Equipment 367
Scott Davis

What I Learned as an Investigator with the CSB 375
John Murphy

Recognizing In Situ Oxygen Analyzers as Ignition Sources in Fired Equipment 385
Sheila Van Geffen, Bernardo Corripio, Andy Weber

Case Study: Aluminum-Dust Explosion 389
Vahid Ebadat, Richard W. Prugh

Lessons Learned from an Unusual Hydrogen Reformer Furnace Failure 400
T. Mike Rogers

Vinyl Chloride Monomer Explosion 414
Lisa Long, James Lay, Katherine Leskin, Randy McClure, Allen Smith

Author Index

A Practical (Semi Quantitative) Tool for The Determination of Dust Explosion Risks for Workers in Industry

A.J.J. Klein
M. De Maaijer
N.H.A. Versloot

TNO Defense, Security and Safety
Lange Kleiweg 137,
P.O. Box 45, 2280 AA Rijswijk, The Netherlands
Phone: 281-721-6000
arjen.klein@tno.nl

Prepared for Presentation at
American Institute of Chemical Engineers
2007 Spring National Meeting
41st Annual Loss Prevention Symposium
Houston, Texas
April 22–26, 2007

UNPUBLISHED

A PRACTICAL (SEMI QUANTITATIVE) TOOL FOR THE DETERMINATION OF DUST EXPLOSION RISKS FOR WORKERS IN INDUSTRY

Authors: A.J.J. Klein, M. de Maaijer, N.H.A. Versloot

TNO Defence, Security and Safety, Lange Kleiweg 137,
P.O. Box 45, 2280 AA Rijswijk, The Netherlands
arjen.klein@tno.nl

Introduction

Many powders which are handled in industry are combustible (flammable). A dust explosion will occur if the concentration of the combustible dust suspended in air is sufficient to propagate a flame when the dust/air mixture (dust cloud) is ignited by a sufficiently energetic ignition source.

Employers in industry dealing with flammable dusts want to protect their workers against dust explosion hazard. Because of legislation and guidelines in this field, the risks have to be assessed so that adequate measures can be taken to reduce the risk to a tolerable level.

As a result of the implementation of the European ATEX directive 1999/92/EC [1] into the national legislation of the European Union member states, employers of many companies with dust explosion hazards need to assess and evaluate the risks in a systematic way. Furthermore these employers have to describe these activities in a mandatory explosion protection document and have to update this document continuously.

The need for instruments to assess and evaluate these risks in industry is increasing but simple and practical instruments are rare. Because of this, TNO develops a spreadsheet by which, on the basis of a limited amount of data, the risks in industry can be assessed in a semi-quantitative way as a first step to improve safety for the workers involved.

With this tool the risk related to specific equipment or process parts is determined. By doing this for each apparatus and process part the contribution to the total risk for the workers involved is made clear. Both technical and organizational preventive measures, but also explosion protective measures (mitigation) are considered.

Background

In compliance with article 4 of the ATEX directive 1999/92/EC [1] the employer shall assess the specific risks arising from explosive atmospheres, taking into account at least:

- the likelihood that explosive atmospheres will occur and their persistence;
- the likelihood that ignition sources will be present and become active and effective;
- the installations, substances used, processes, and their possible interactions;
- the scale of the anticipated explosion effects.

TNO develops a method to assess the risk by determination of risk levels. These risk levels are determined based on all above mentioned aspects. Harmful primary effects of dust explosions like the blast (pressure wave), flames and projectiles, but also secondary effects, like collapsing of constructions need to be translated to consequences for the workers involved. Obviously the consequence depends strongly on the presence of workers (exposure time) within the explosion effect distances.

Simple models to estimate the consequences of a dust explosion are still under development but at this moment a raw estimation is made based on the volume and the construction type of the process equipment.

Risk level determination method

The risk from dust explosions for process equipment depends on the probability of occurrence and the expected consequences. These consequences are based on the probability of people within harmful effect distances.
In many situations it is not possible to cover all the factors that effect risk, in particular those which contribute to the likelihood of a specified event occurring. Therefore risk is often expressed in a qualitative rather than a quantitative way.
A risk level determination system (risk matrix) is described in the RASE report [2]. This report from a European Union funded project describes different explosion risk assessment methodologies. For the development of the TNO risk level determination tool the risk matrix is used.

Severity can be expressed as defined levels, one or more of which can result from each hazardous event. Thus in terms of injuries or damage to health or system damage, severity can be expressed as shown in table 1.

Table 1: Severity levels description

severity levels description	mishap definition
Catastrophic	death or system loss
Major	severe injury, severe occupational illness or minor system damage
Minor	minor injury, minor occupational illness or minor system damage
Negligible	less than minor injury, occupational illness or system damage

In order to estimate the frequency of each severity level a screening technique can first be applied to determine the probability of each hazardous event in turn. The frequency of occurrence can be qualitatively expressed as shown in table 2.

Table 2: Frequency levels description

Frequency levels description	Specific individual item	Inventory
Frequent	likely to occur frequently	continuously experienced
Probable	will occur several times in life of an item	will occur frequently
Occasional	likely to occur sometime in life of an item	will occur several times
Remote	unlikely but possible to occur in life of an item	unlikely but reasonably be expected to occur
Improbable	so unlikely, it can be assumed occurrence may not be experienced	unlikely to occur, but possible

The link between severity levels on the one side and the frequency of their occurrence on the other, leads to the risk matrix as shown in table 3. The corresponding points in this matrix are allocated to the risk levels A, B, C, and D. The risk levels represent a ranking of the risks which enables an evaluation of further actions.

Thus:

- risk level A: very high
- risk level B: high
- risk level C: moderate
- risk level D: low

Table 3: Risk levels, linkage between severity and frequency level

Frequency of occurrence	Severity			
	Catastrophic	Major	Minor	Negligible
Frequent	A	A	A	C
Probable	A	A	B	D
Occasional	A	B	B	D
Remote	A	B	C	D
Improbable	B	C	C	D

As starting point for the risk level determination the dust explosion hazardous areas have to be classified. This classification, also called zoning, provides information about the likelihood that explosive atmospheres will occur and their persistence. Performing a hazardous area classification is mandatory with respect to the European directive 1999/92/EC [1].

Classification of dust explosion hazardous areas (zoning)

The classification of hazardous areas (zoning) for dust explosion hazards is described in the European Standard EN 61241-10 (IEC 61242-10:2004) [3]. As a practical guideline, with respect to zoning, the document 'The Dutch Practical Guideline NPR 7910-2' [4] is used in the Netherlands. This document describes a 3 step approach:

Step 1: assess necessity of classification;
Step 2: assess class of hazardous area (zone);
Step 3: assess the dimensions of a classified area.

The dust explosion hazardous area classification depends on the frequency and time of occurrence of an explosive atmosphere. A three-zone system is utilized for dust/air mixtures as well as for gas/air mixtures. For dust explosions these are zones 20, 21 and 22. The inside of equipment as well as the work places is classified according to this system.

Zone 20: An area in which an explosive atmosphere is present permanently or for long periods of time (more than 1000 hours/year). Examples are silo's, mixers or mills.

Zone 21: An area in which in normal operation an explosive atmosphere is sometimes present (dust cloud 10 -1000 hours/year and/or dust deposits more than 1000 hours/year). Examples are in the direct vicinity of pneumatic transport systems or bag filling installations.

Zone 22: An area in which an explosive atmosphere under normal conditions is not likely present, and if it is, only for short periods of time. (dust cloud less than 10 hours/year; dust deposits less than 1000 hours/year). Examples are leakages from flexible connections and leakages from bags.

When dust explosion atmospheres are not likely under any conditions this area will be defined as a 'non classified area' (NCA). When a dust explosion hazard is present but also one or more effective ignition sources which are directly related to the process, this area is called a deviant area (DA).

Figure 1 and 2 shows flowcharts taken from the NPR 7910-2 which visualize how dust explosion hazardous areas (zoning) are classified based on the existence of dust clouds and dust layers.

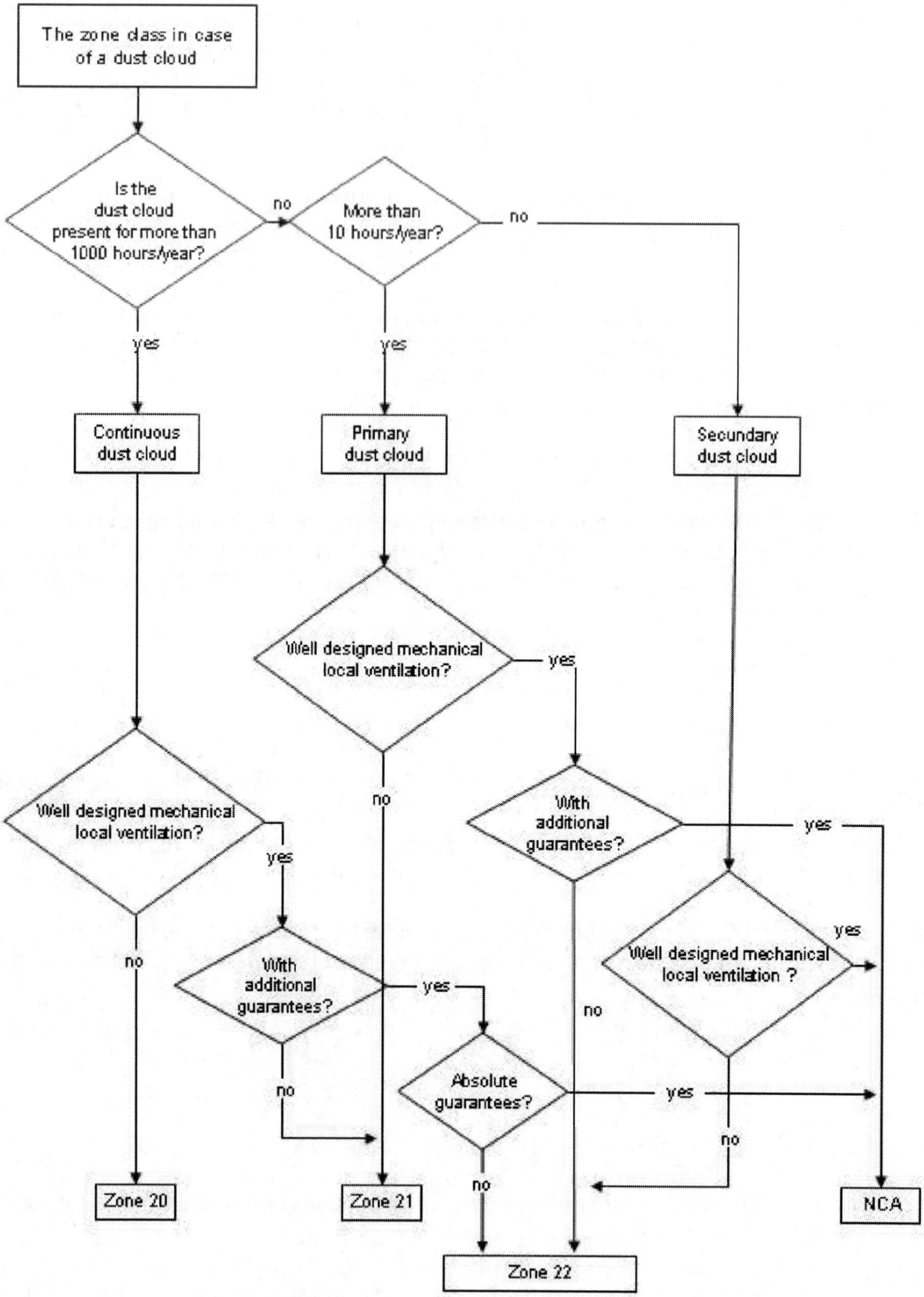

Figure 1: Flow chart for the zone class in case of a dust cloud (NPR7910-2).

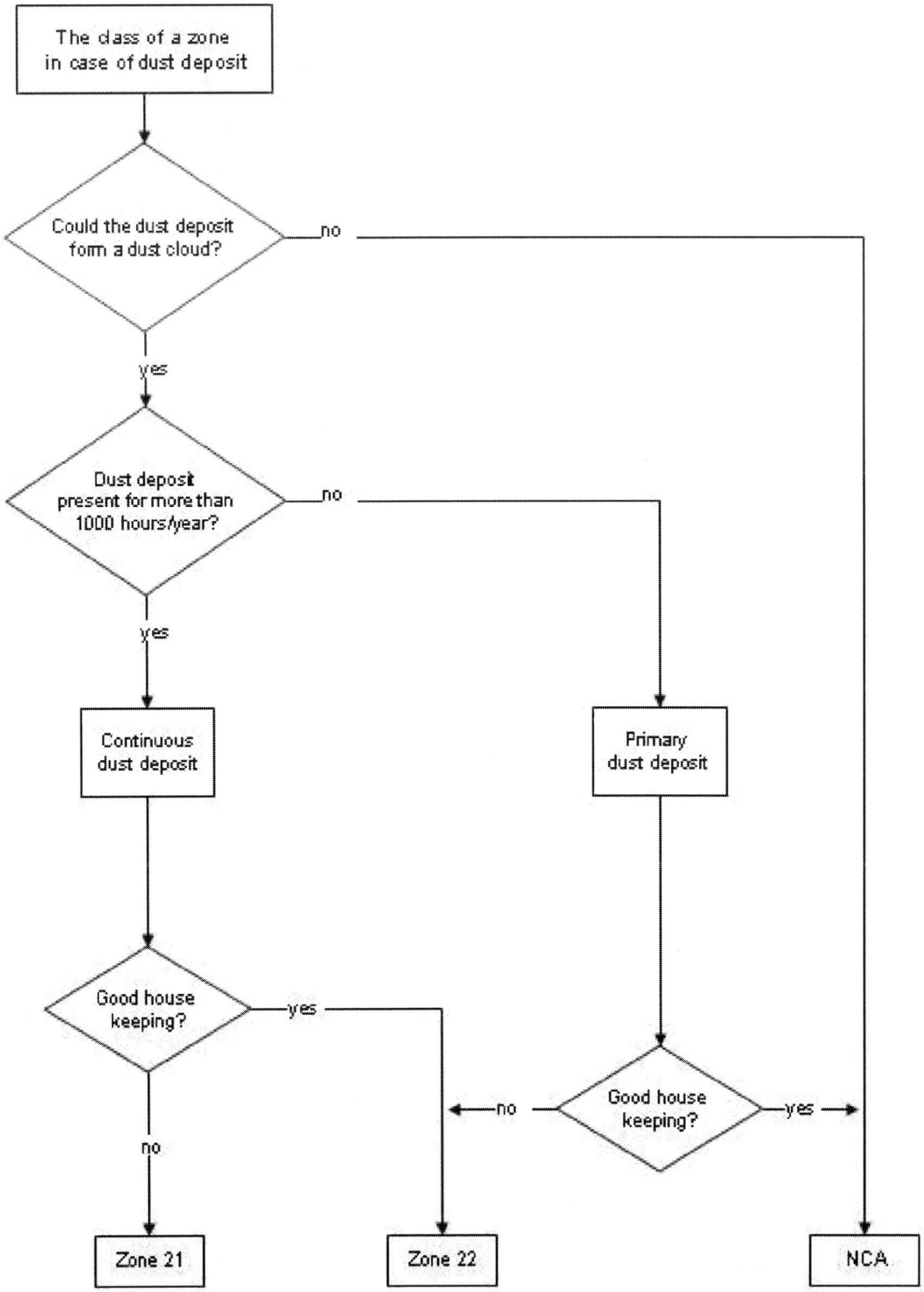

Figure 2: Flow chart for the zone class in case of dust deposits (NPR7910-2).

Frequency level determination

The probability of occurrence (frequency level) for an explosion in each of the process parts is estimated based on the hazardous area classification (zoning), an ignition risk assessment and the preventive measures already taken by the owner of the installation.

For the ignition risk assessment the dust explosion ignition properties of the product and the process conditions are taken into account. Required data is collected by a plant visits, interviews, and evaluation of technical documentation like piping and instrumentation diagrams (P&ID).

For the ignition risk assessment all relevant ignition sources are considered like:

- Hot surfaces
- Flames, hot gasses
- Mechanical sparks
- Static electricity
- Electrical equipment
- Exothermic reactions
- Ignition by primary explosions

For each process equipment, for example bucket elevators, rotary valves, cyclones, filter units and spray dryers it is estimated on basis of case history and expert opinion how often an ignition will occur if an explosive dust/air mixture would be continuously in contact with the inside of this equipment. The frequency (ignition probability) is expressed as a theoretical value as described below.

0 = very rare
1 = rare
2 = expected
3 = normal

Based on the hazardous area classification (zoning) the ignition probability for each process part is corrected as follows, but the resulting value is always ≥ 0.

Zone 20 = 0
Zone 21 = -1
Zone 22 = -2

When for specific equipment preventive measures are already taken the ignition probability is corrected as follows, but the resulting value is always ≥ 0.

ignition probability considerably lower than default value	= -2
ignition probability some lower than default value	= -1
default ignition probability	= 0
ignition probability higher than default value	= +1

It will be clear that each correction has to be documented well so it becomes clear on basis of what corrections are made. The total score for the frequency level is expressed as the following values.

0 = improbable
1 = remote
2 = occasional
3 = probable
4 = frequent

Severity level determination

To determine the severity level for a dust explosion, each process part (equipment) harmful primary effects of dust explosions like the blast (pressure wave), flames and projectiles, but also secondary effects, like collapsing of constructions should be estimated and translated to consequences for the workers involved. Simple models to estimate the consequences of a dust explosion are still under development. At this moment the explosion effects (severity) are estimated based on the internal volume of the process part and the type of construction.

Effects (severity)
0 = small volume and/or weak construction (or explosion resistant)
1 = moderate volume and/or moderate robust construction
2 = large volume and/or robust construction (but not explosion resistant)

It is obvious that mitigation like explosion venting devices or suppression systems has to be taken into account if applicable. The following reductions are then applied.

0 = no effect reduction measures
-3 = adequate effect reduction measures

Obviously the consequence depends also strongly on the presence of workers (exposure time) within the explosion effect distances. For example a process part in the vicinity of a control room will give rise to an increase of the consequence level.

0 = not often people present and/or only during a short times
+1 = often people present and/or during longer time

The total score for the severity level is expressed as the following values.

0 = negligible
1 = marginal
2 = critical
3 = catastrophic

Example; determining risk levels for a spray dryer installation

For demonstration the explosion risk assessment tool is applied for a spray dryer installation. A description of this installation is partly taken from the RASE report [2].

Determination of intended use

Spray Drying is the widely used industrial process involving particle formation and drying. It is suited for the continuous production of dry solids in either powder, granulate or agglomerate form from liquid feed stocks such as emulsions and suspension.
Spray Drying involves the atomization of the liquid feed stocks into a spray of droplets and contacting the droplets with hot air in a drying chamber. The sprays are produced by either rotary or nozzle atomizers. All systems can be provided with post-treatment equipment, for example: fluid bed dryer/cooler, agglomerator, de-duster and conveyor.

Description of the system

Atomization plays a central role in the process. The formation of sprays having the required droplet size distribution is vital to both the operation and the explosive atmospheres in the form of a cloud of combustible dust occurring. Exhaust air is subject to a cleaning process in cyclones, bag filters, and occasionally wet scrubbers.

The remaining part of this application example deals with a specific milk powder spray dryer. All equipment and its components have stainless steel housing or are mounted in a stainless steel casing. Only the spray dryer chamber is equipped with explosion relief panels.

Figure 3 shows a schematic diagram of the spray dryer installation.

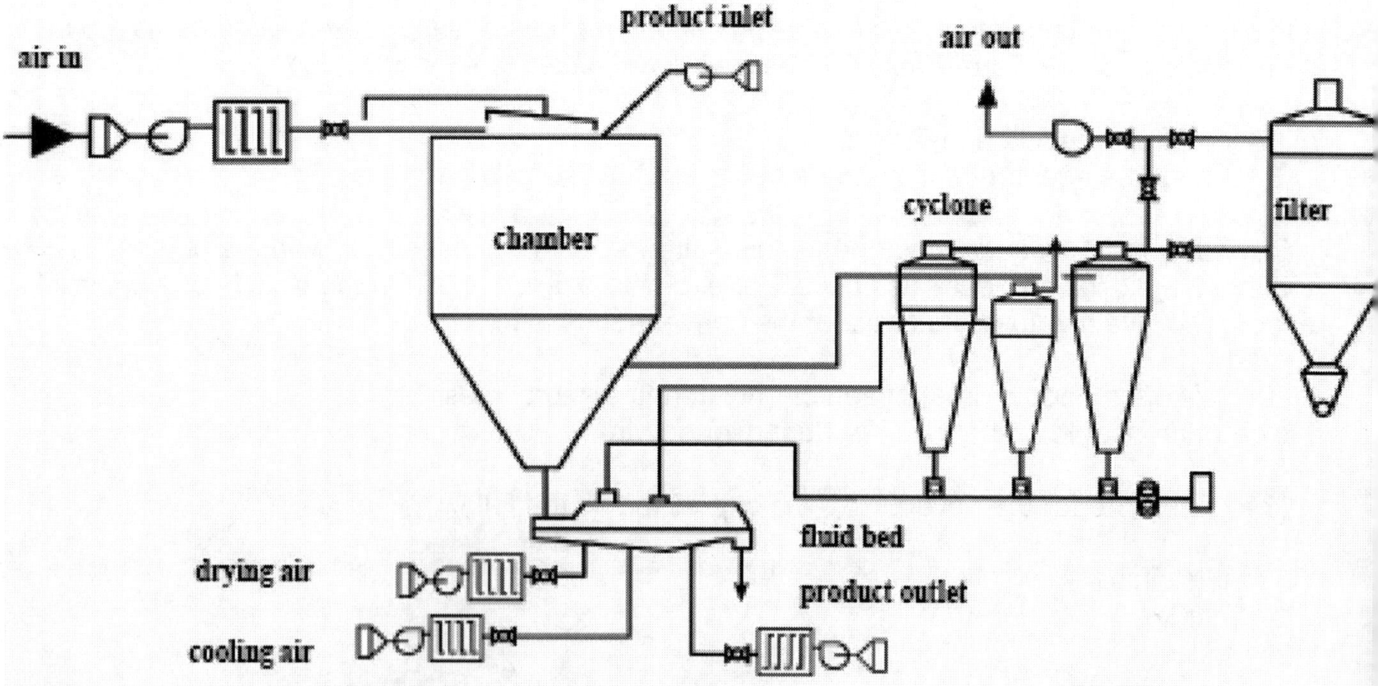

Figure 3: Schematic diagram spray dryer installation

Below the risk assessment calculation sheet is shown. For each process part (rows) the columns displays the hazardous area classification, the frequency level, the severity level and the values on basis these levels are calculated. The last column shows the calculated risks.

Table 4: Determination of risk levels spray dryer installation

Process item	**Area Classification**	Ignition Value	Ignition Reduction	**Frequency Level**	Effects	Effects reduction	Persons Exposure	**Severity level**	**Risk Level**
Dryer chamber	**20**	3	0	**3**	2	-3	1	**0**	**D**
Fluid bed	**21**	1	-1	**0**	0	0	1	**1**	**C**
Cyclone 1	**21**	1	0	**1**	0	0	1	**1**	**C**
Cyclone 2	**21**	1	0	**1**	0	0	1	**1**	**C**
Cyclone 3	**21**	1	0	**1**	0	0	1	**1**	**C**
Rotary valve 1	**22**	1	-1	**0**	0	0	1	**1**	**C**
Rotary valve 2	**22**	1	-1	**0**	0	0	1	**1**	**C**
Rotary valve 3	**22**	1	-1	**0**	0	0	1	**1**	**C**
Filter	**20**	2	0	**2**	2	0	1	**3**	**A**

Based on the risk levels shown in table 4 it can be concluded that the filter is responsible for the highest risk. For example mounting explosion relief panels or replacing the dry filter by wet scrubbers will reduce the risk significantly. To further reduce the risk levels for the cyclones and the rotary valves organizational measures like clear maintenance procedures and hot work permits have to be taken.

Conclusions

With this tool it is possible to get insight how the risk related to a specific item can be reduced and to what extent. This instrument will contribute to efficiently assessing and evaluating dust explosion risks as a first step and can help employers to take the right priorities for technical or organizational measures in order to reduce the risks for the workers most effectively.

At this moment a raw estimation of the consequences is made based on the volume and the construction type of the process equipment. Simple models to improve the estimation of the consequences of a dust explosion are still under development.

References

[1] Directive 1999/92/EC of the European parliament and of the council of 16 December 1999 on minimum requirements for improving the safety and health protection of workers potentially at risk from explosive atmospheres.

[2] RASE report 'Methodology for the Risk Assessment of Unit Operations and Equipment for Use in Potentially Explosive Atmospheres, EU Project No: SMT4-CT97-2169, 17th March 2000.

[3] European Standard EN 61241-10 (IEC 61242-10:2004), 'Electrical apparatus for use in the presence of combustible dust', Part 10: Classification of areas where combustible dusts are or may be present (this standard supersedes EN 50281-3:2002).

[4] NPR7910-2 (Dutch), Dutch Practical Guideline - Part 2 'Hazardous Area Classification, regarding dust explosion hazard', Dutch Standardization Institute NEN, July 2001.

LOSS PREVENTION SYMPOSIUM

The Employment of NFPA 59A in LNG Facility Siting

Anna Qiao
Caroline Deetjen and David Pertuz
DNV
Houston, TX 77450 U.S.A.
Phone: 281-721-6000
Anna.Qiao@dnv.com

Prepared for Presentation at
American Institute of Chemical Engineers
2007 Spring National Meeting
41st Annual Loss Prevention Symposium
Houston, Texas
April 22–26, 2007

UNPUBLISHED

The Employment of NFPA 59A in LNG Facility Siting

Anna Qiao
Caroline Deetjen and David Pertuz
DNV
Houston, TX 77450 U.S.A.
Phone: 281-721-6000
Anna.Qiao@dnv.com

Abstract

This paper presents the application of NFPA 59A, National Fire Protection Association publication 59A *Standard for the Production, Storage, and Handling of Liquefied Natural Gas (LNG) 2006 Edition*, in the siting of LNG facility. NFPA 59A is an industry consensus standard that covers general LNG facility considerations. To assist the development of LNG facility plan layouts, a series of studies must be conducted in accordance with NFPA 59A to determine the feasibility of the LNG facility in terms of the surrounding community safety.

According to NFPA 59A, the siting evaluation is conducted by assessing the consequence of LNG spills from the facility, and calculating the hazard zones for LNG fire and flammable vapor dispersion. PHAST, the most popular consequence analysis software in oil and gas industry, is employed in determining the safe distance from the facility to the property line.

This study also compares NFPA 59A with Canadian LNG standard and the Code of Federal Regulation on LNG facility. Recommendations are made on the model selections and parameter determinations in applying NFPA 59A for LNG facility siting.

1. Introduction

The National Fire Protection Association (NFPA) publication 59A: Standard for the Production, Storage, and Handling of Liquefied Natural Gas (LNG) 2006 Edition [1] is an industry consensus standard that covers general LNG facility considerations. The standard is used worldwide as the final word for site selection, design, construction, and fire protection of LNG facility.

For the construction of an LNG regasification terminal, the project involves the marine transportation, offloading, regasification and send-out of natural gas. NFPA 59A defines the standards to follow from the design of the facility, like the size of the storage tank, to the handling of LNG, like the fire protection. This paper specifically addressed the employment of NFPA 59A in the siting of the facility, including the determination of the property line, to define the distance between different storage tanks, and to define the locations of other equipments in the facility.

NFPA 59A Section 5.2.3 "Impounding Area Siting" provides guidelines for determining the required minimum safe distance from the impounding area to the property line in the event of an LNG spill. Section 5.2.4 "Container Spacing" defines the separation distance between two containers. The calculations provide a basis for the siting of the storage tank and the spacing of two LNG storage tanks. The criterion also applies to other equipments in the facility which could cause potential LNG release.

The evaluation is conducted in a case study by assessing the thermal radiation flux from a fire in the facility and calculating safe distances in accordance with the parameters defined in NFPA 59A. The safe distance to the property line and the safe distance between storage tanks can be determined based on the calculation in accordance to NFPA 59A.

2. Study Basis

The siting of the LNG facility has to base on the assessment of consequences associated with all of the activities related to the proposed operating facility. This covers:

- LNG unloading operations
- Operation of LNG terminal storage tanks
- Operation of LNG re-gasification and send out by pipeline
- Export pipeline from metering to facility boundary line

The consequence analysis need to be performed under the conditions defined in NFPA 59A. The facility background and conditions for consequence analysis are detailed in this section.

2.1 LNG Storage Tank and Impounding Area

In the case study, it is proposed that two LNG Full Containment Tanks (FCT) with equal dimensions are located in the facility. The summary of the information is as follows:

- Net capacity of the inner LNG storage tank is 160,000 m3
- There is no penetration below the liquid level in the tank
- The secondary container is constructed of reinforced concrete with both inner and outer tanks designed to be capable to hold the gross volume of cryogenic liquid without loss of containment
- The secondary container roof and sidewalls are capable of withstanding fire exposure and heat radiation from a relief valve discharge fire with the relief valves discharging at the maximum relieving rate.
- The secondary container roof is spherical in shape and made of reinforced concrete
- For the FCT, the outer tank diameter was considered as the impounding area for the purpose of the calculations.

2.2 Atmospheric Conditions

2.2.1 Parameters Defining Distance of Facility to the Property Line

The following atmospheric conditions are based on defining thermal radiation flux limits to property lines and occupancies in NFPA 59A Section 5.2.3.2: zero wind speed, 70 °F temperature, and 50% percent relative humidity.

In this study, the wind speed is selected as 0.1 m/s instead of 0 m/s, because it is almost impossible that the atmosphere can have a zero wind speed. Calculations cannot be validated for the 0 m/s wind speed condition. A more realistic assumption is 0.1 m/s wind speed rather than 0 m/s wind speed. Pasquill atmospheric stability is not defined for assessment in NFPA 59A Section 5.2.3.2. According to the Pasquill stability classification system [2], the Pasquill stability could be A, A/B or B when the wind speed is less than 2 m/s. All three stabilities will be considered in this study. The atmospheric conditions for thermal radiation flux calculation are summarized as:

- Wind speed: 0.1 m/s
- Pasquill atmospheric stability: A, A/B, B
- Temperature: 70 °F
- Relative humidity: 50%

For the dispersion consequence, according to NFPA 59A Section 5.2.3.4, the dispersion distance is computed based on the actual liquid characteristics and the maximum vapor outflow rate from the vapor containment volume. The wind speed and Pasquill-Gifford atmospheric stability shall be either that resulting in "the longest predictable downwind dispersion distance that is exceeded less than 10 percent of the time", or the combination of Pasquill-Gifford stability Category F with wind speed 2 m/s. In this study, the former condition is employed in the dispersion consequence calculation. It is assumed in the case study that the following condition can resulting in "the longest predictable downwind dispersion distance that is exceeded less than 10 percent of the time."

- Wind speed: 6 m/s
- Pasquill atmospheric stability: C
- Temperature: 85 °F
- Relative humidity: 70%

2.2.2 Parameters Defining Separation Distance of Two FCTs

In accordance with NFPA 59A Section 5.2.4.2 (A), the ambient conditions for thermal radiation flux shall be based on the conditions within the range expected for the site, calculated to produce the highest temperature in the secondary container structure. Therefore, historical data on the weather condition should be analyzed for the site of LNG facility. The case study employs the same condition for dispersion calculation presented above.

2.3 End Point Criterion

Per NFPA 59A Section 5.2.3.2, the endpoints for thermal radiation flux calculation are selected to be 5000 W/m2, 9000 W/m2, and 30000 W/m2, and the final thermal exclusion zone is defined by 30000 W/m2, which can be considered as the safe distance to the property line.

For the dispersion calculation, as outlined in NFPA 59A, an average concentration of methane in air of 50 percent of Lower Flammable Limit (LFL) should not extend beyond the property line. Therefore, the LFL, which is 4.4 percent of methane, will be considered for the purpose of evaluating the hazard zone.

The final exclusion zone is determined by thermal exclusion zone and dispersion exclusion zone, whichever is greater.

3. Methodology

3.1 Hazard Identification

Defining scenarios was a critical step in the Quantitative Risk Assessment (QRA) carried out in assessing the consequence, and the following outlines the approach used. First, generic industry failures were postulated (e.g. equipment failure leads to release of flammable material). These generic failures were inclusive of a range of loss of containment event types, with the segregation of event types made at the greatest level of detail for which there were significant data available to support an assessment of frequency (e.g. leak or rupture). These generic incident cases were supplemented by detailed cases that were unique to the LNG facilities and operations.

For this study, the spill and duration for incidents were determined in accordance with Table 5.2.3.5 in NFPA 59A. For each release, the largest flow from any single line was considered. Thus only the rupture scenario was considered for the release from each segment, and the release rate was the normal flow rate in the line. For the release from unloading arm, the release rate is based on the rupture of all the unloading arms (4 unloading arms are proposed for the case study). The duration for the release was modeled to be 10 minutes per the guideline in NFPA 59A Table 5.2.3.5.

3.2 Consequence Analysis

The calculations in this report where conducted using the Process Hazard Analysis Safety Tool (PHAST) Version 6.4.2. PHAST is a widely used consequence package in the oil and gas industry.

PHAST Version 6.4.2 meets the modeling requirements listed in NFPA 59A Sections 5.2.3.3(2) and 5.2.3.4(2).

The following figure shows the block diagram for the LNG facility for case study. Based on the operating process, each category has been grouped into smaller isolatable segments where, in each segment, the release characteristics would be similar. The equipment could have only one segment for the release scenario. However, the unloading arm and piping and process equipment have several isolatable segments where the release characteristics would be different.

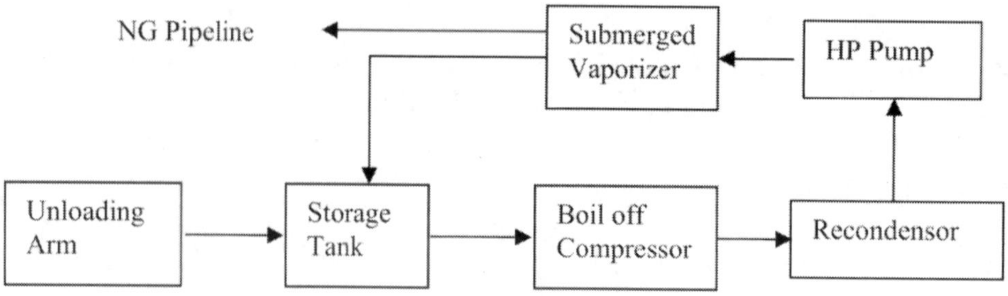

Figure 1. Block Diagram for LNG Terminal Facility

Based on the process analysis, the scenarios for consequence analysis using PHAST are presented in Table 1.

Table 1. Emission Severity Rating Scale

Segment Group	Phase	Release Direction	Temp (C)	Press (bar gauge)	Normal Flow Rate (kg/s)
Rupture of loading arms while ship is unloading due to shifting of ship or failure of loading arm	Liquid	Horizontal	-164	4.0	1.49E+03
Unloading line (Jetty)	Liquid	Horizontal	-159	3.8	1.49E+03
Unloading line (Shore)	Liquid	Vertical	-159	3.8	1.49E+03
Sendout line (unloading)	Liquid	Vertical	-158	10.8	2.33E+02
Vapor return (shore)	Gas	Vertical	-120	0.2	6.10E+00
Vapor return (jetty)	Gas	Horizontal	-120	0.2	6.10E+00
Rupture of vapor return arm while ship is unloading due to shifting of ship or failure of loading arm	Gas	Horizontal	-120	0.2	6.10E+00
Sendout line (holding) - shore	Liquid	Vertical	-159	10.8	1.00E+01
Sendout line (holding) - jetty	Liquid	Horizontal	-159	10.8	1.00E+01
BOG Recondenser	Liquid	Horizontal	-156	7.0	2.34E+02
Sendout pumps	Liquid	Horizontal	-159	103.0	2.34E+02
Vaporizers	Gas	Vertical	1	82.7	2.73E+04
Metering	Gas	Horizontal	5	82.7	2.73E+04
Compressors (unloading)	Gas	Horizontal	-36	7.6	2.00E+00
Compressors (holding)	Gas	Horizontal	-36	7.6	6.60E-01
Natural Gas pipeline	Gas	Impinged	5	82.7	2.73E+04

3.3 Siting Resutls

3.3.1 Siting on LNG Facility Property Line

Based on the PHAST calculation, the consequence modeling results are shown in the following table. The facilities in jetty will decide the exclusion zone in the water, while the process facilities will decide the distance to the property line on the land. As shown in the results, upon a release from the LNG facility, it is usually the vapor dispersion results in a longer impact distance than that caused by the thermal radiation. The highest distance is associated with the release from natural gas line, i.e., the distance of the natural gas line has to be away from the property line up to 1971 m.

Table 2. Determination of Distance to Property Line

Segment Group	Scenario ID	Distance to 30 kW/m^2 (m)	Scenario ID	Distance to ½ LFL (2.2% Methane) (m)
Rupture of loading arms while ship is unloading due to shifting of ship or failure of loading arm	UL01RLF-H	32	UL01RLF-H	616
Unloading line (Jetty)	UL02RLF-H	33	UL02RLF-H	765
Unloading line (Shore)	UL03RLF-V	95	UL03RLF-I	516
Sendout line (unloading)	UL04RLF-V	61	UL04RLF-I	399
Vapor return (shore)	UL05RGF-V	-	UL05RGF-I	330
Vapor return (jetty)	UL06RGF-H	73	UL06RGF-H	192
Rupture of vapor return arm while ship is unloading due to shifting of ship or failure of loading arm	UL07RGF-H	69	UL07RGF-H	169
BOG Recondenser	PR01RLF-H	34	PR01RLF-H	407
Sendout pumps	PR02RLF-H	35	PR02RLF-H	138
Vaporizers	PR03RGF-V	-	PR03RGF-V	304
Metering	PR04RGF-H	873	PR04RGF-H	857
Natural Gas pipeline	PL01RGF-I	490	PL01RGF-I	1971

3.3.2 Siting on Two Adjacent LNG Tank

PHAST is employed to calculate the radiation distance from the FCT to the adjacent container Table 2 presents the results from PHAST from the fire within the secondary tank of an FCT to the adjacent FCT.

Table 3. Thermal Radiation Distance between Two FCTs

Radiation level (W/m2)	Radiation Distance (m)
30,000	128

Besides the radiation distance to the adjacent tank, two criteria are also applicable in determining the separation distance from one FCT to another:

- One-half (½) of the diameter of the largest tank as defined in Section 5.2.4.2 (B) of NFPA 59A
- One-fourth (¼) of the sum of the diameters of adjacent containers as defined in Table 5.2.4.1 of NFPA 59A

As presented in Table 3, the 30,000 W/m2 radiation distance is 128 meters. It is assumed that the two FCTs have the same diameter (86.8 meters); thus, one-half of the diameter of the largest tank and one-fourth of the sum of the diameters of adjacent containers are both 43.4 meters. The longest distance will decide the allowable separation distance between two adjacent tanks to meet NFPA 59A criteria. Therefore, 128 meters being the longest distance is the safe separation distance between two FCTs.

4. Discussions

4.1 *Feasibility of NFPA 59A Employment if LNG facility Siting*

NFPA 59A provides the standards for the determination of the distance to the property line, the distance between two storage tanks, and other siting issue. It provides the guidelines on the calculation needed to determine the minimum safe distance from the LNG facility to the property line in the event of an LNG spill. It will provide a basis for the siting layout of the storage tank, impounding area, and other facilities and to secure the necessary permits from the authorities having jurisdiction.

4.2 *Comparisons of NFPA59A to CFR and CSA*

Canadian Standard Association (CSA) Standard Z276-01, Liquefied Natural Gas – Production, Storage, and Handling [3] and the Title 49 of Code of Federal Regulations (49 CFR) [4] also set the standards/regulations for LNG facility siting. Difference exists in these three standards/codes.

4.2.1 Definition on Distance to Property Line

49 CFR Section 193.2057, NFPA 59A Section 5.2.3.2, and CSA Z276-01 suggest using the same thermal radiation levels as endpoints for thermal radiation flux calculation. i. e., 5000 W/m2, 9000 W/m2, and 30000 W/m2, and the final thermal exclusion zone is defined by 30000 W/m2.

Different criterion are given on the end point for dispersion calculation. 49 CFR Section 193.2059 defines that an average gas concentration in air of 2.5 percent shall be used in computing dispersion distances, which will be the gas concentration at the boundary of dispersion exclusion zone. NFPA 59A suggests using ½ of the LFL for the end point for dispersion calculation, which is similar as the regulation in 49 CFR since the LFL for methane is 4.4%. CSA Z276-01 does not define the concentration limit at the boundary of the property line.

4.2.2 Parameters Used to Determine Distance to Property Line

The parameters, especially the parameters relating to meteorological conditions, used to model thermal radiation flux and vapor dispersion are selected based on the regulations and guidelines in CSA Z276-01 and 49 CFR. The parameters used in CSA and CFA calculations were simplified as follows:

The temperature, humidity, wind speed and stability, Elevation of output contour in dispersion calculation is 0.5 for CFR, Surface roughness for CFR is 0.001 m.

4.2.3 Update suggestions on NFPA 59A

NFPA 59A Section 5.2.3.2 defines that the following atmospheric conditions should be employed in defining the thermal radiation flux limits to property lines and occupancies: zero wind speed, 70 °F temperature, and 50% relative humidity. However, to define the distance between two adjacent tanks, In accordance with NFPA 59A Section 5.2.4.2 (A), the ambient conditions for thermal radiation flux shall be based on the conditions within the range expected for the site, calculated to produce the highest temperature in the secondary container structure.

Zero wind speed usually results in shorter radiation distance than a higher wind speed if the release scenarios are the same. The wind speed on the site of the LNG facility could be much higher than 0 m/s. Therefore, it is not conservative enough to use zero wind speed in defining the distance to the property line. It also creates conflicts in calculating the distance from the facilty to the property line and the distance between two storage containers. It is suggested that the conditions to define the distance of the storage tank to property line should be defined according to the same conditions as defined in NFPA 59A for two tanks distance determination, or the same as that defined as CFR 42.

6. References

[1] NFPA 59A Standard for the Production. Storage, and Handling of Liquefied natural Gas (LNG) 2006 Edition with Erratas

[2] Pasquill, F. and Smith, F. B., Atmospheric Diffusion: A Study of the Dispersion of Windborne Material from Industrial and Other Sources. 3rd Edition. Chichester, West Sussex, England: Ellis Horwood Limited, 1983

[3] Canadian Standard Association (CSA) Standard Z276-01, Liquefied Natural Gas (LNG) – Production, Storage, and Handling

[4] Title 49 – Transportation, Code of Federal Regulations (49 CFR)

LOSS PREVENTION SYMPOSIUM

Applying the Limitation of Effects ISP Strategy when Siting and Designing Facilities

by

David G. Clark
Process Safety and Fire Protection Engineering
E. I. du Pont de Nemours & Co., Inc.
1007 Market Street
Wilmington, DE 19898
Voice: (302) 774-8044
Fax: (302) 773-6508
E-mail: david.g.clark@usa.dupont.com

Prepared for Presentation at
American Institute of Chemical Engineers
2007 Spring National Meeting
3rd Global Congress on Process Safety
41st Loss Prevention Symposium
Houston, TX
April 22–26, 2007

UNPUBLISHED

Applying the Limitation of Effects ISP Strategy when Siting and Designing Facilities

by

David G. Clark
Process Safety and Fire Protection Engineering
E. I. du Pont de Nemours & Co., Inc.
1007 Market Street
Wilmington, DE 19898
Voice: (302) 774-8044
Fax: (302) 773-6508
E-mail: david.g.clark@usa.dupont.com

Abstract

Minimize, Substitute, Moderate, and Simplify are four primary Inherently Safer Processing (ISP) strategies. The Moderate strategy is primarily Attenuation, a strategy that emphasizes the use of materials under less hazardous conditions. Limitation of Effects is also an ISP term bantered about as being included within the purview of this Moderate strategy, although it is at times presented as a separate, unique ISP strategy. Regardless of the taxonomy, the application of such a strategy can significantly lower overall process risk. This paper discusses the author's understanding of the meaning of the term Limitation of Effects and presents examples of the application of this strategy to the siting of plants, processes, and buildings.

Key words: Inherently Safer Processing; ISP; Inherently Safer Design; Limitation of Effects

Introduction

The asteroid is 933 km in diameter. And has a mass of 7.6E20 kg. And is traveling at a speed, relative to earth, of 17 km/s. It is headed our way and spotted 100 years in advance. A direct hit is predicted. The delivery of the asteroid to the earth is a batch process to be run only once. The consequence analysis indicates that the outcome is less-than-satisfactory: the destruction of life on earth, including, especially, us. The hazard is the 1.098E29 joules of kinetic energy to be delivered in about 61 seconds, the difference between the time that the asteroid's leading edge hits the outer reaches of the earth's 120 km-thick atmosphere and the time its trailing edge impacts the earth's surface. Nothing draws uncommon enemies together like a common enemy. So the political leaders get together. The economists get together. The physicists and other scientists get together. And the PHA gurus get together. And from them a small Inherently Safer Processing (ISP) group breaks off to tackle the problem. In so doing they recognize that this accident in-the-making has some parallels to other (though less onerous) accidents. As highlighted in 'Inherently Safer Chemical Processes: A Life-Cycle Approach' [1], ISP could/should/can/may be applied to any of the 3 steps which most accidents follow. These steps are Initiation (Generation), Propagation (Transportation), and Termination (Reception).

Something starts it. Something (pressure, temperature, molecules,, etc.) is transported elsewhere. And someone is impacted by what is transported. With this in mind the ISP team seeks to answer the question: Can we make this process inherently safer?

Had they been there and had they been consulted at the time of the creation or formation of the asteroid and at the time of its placement in its current trajectory, they may have suggested that it not be created or be formed at all, thereby not only saving all life as we know it but also, importantly, providing a most memorable example of Elimination, a sub-set of the ISP strategy Minimization.

Or they may have suggested that it still be one that strikes the earth but that it be one that strikes with less mass, a little asteroid of say 7.6E19 kg, or with a little less speed, a slow one at say 5670 m/s. With the smaller mass or lower speed the energy is a mere 1.098E28 J. While the lights may go out, it is absorbable with an acceptably minimum (by some measure) damage and fatality consequence. And it would have provided a notable, primary example of the ISP strategy Minimization if the weight is reduced or an example of 'Moderation' of process variables if the speed is reduced.

Or how about exchanging the rock (having a density of 1800 kg/m3) with cotton candy (having a density of 58 kg/m3). Saving the world and providing an example of the ISP strategy Substitution at the same time.

[[The ISP team might have drawn a blank on how to simplify the process After all it is just a rock at one place and time with a given speed and weight. And so are we. It can't get much simpler than that.]]

All of the above are only-if-we-were-there-then solutions. We aren't there then. We are here now.

What about the sub-set of the Moderation strategy that is named Limitation of Effects? We see the 'stroid coming. Can we do something about, can we limit, the effects? Four billion of us pack our bags and move to where 2 billion of us already are (India and China). The earth's position moves ever so slightly from where it would have been had the population shift not occurred. The very gentle shift in the center of earth, relative to where the center of the earth was before the emigration, makes all the difference. The cumulative effect over 100 years of the wobble induced by that relocation causes the asteroid to pass us rather than hit us and, notwithstanding the chaotic real estate business in the post-pass era, we and all other life forms are saved. Such is the value of this Limitation of Effects ISP-Moderation strategy. (Note: The author does not believe that the resulting ~0.1 μm shift in the position of the earth's center would actually do much. But, it is a good illustration of Limitation of Effects assuming that it does.)

In the following discussion, examples of Limitation of Effects ISP solutions are presented for several different types of hazards to proactively and inherently limit the effects of accidents. The value of separation distance is emphasized.

Limitation of Effects ... relative to a BLEVE Fireball

Assume that there is a fireball that results from a BLEVE (Boiling Liquid Expanding Vapor Explosion) of a tank truck containing 25,000 kg of flammable liquid. The igniter is assumed to be a fire that also caused the BLEVE.

The duration (sec) of the fireball is a function of the fuel mass (kg), $t \propto m_f^{1/3}$. The radiation intensity q (W/m2) is a function of the atmospheric transmissivity, τ (-), the radiative fraction of the heat of combustion, R_f (-), the heat of combustion, H (J/kg), the fuel mass (kg), and the distance from the center of the fireball to the receptor, L (m): $q \propto \tau R H m_f^{2/3} / L^2$. In turn L is a function not only of the lateral, horizontal distance, L_h (m), to the recipient target but also a function of the vertical distance to the center of the fireball, L_v (m): $L = \left(L_h^2 + L_v^2\right)^{1/2}$. In turn, L_v is typically assumed to be the fireball radius and is a function of the fuel mass, $L_v \propto m_f^{1/3}$. If the receptor is a person and radiation burns are the hazardous injury of interest, the degree of injury is a function of the radiation intensity and duration. A rendition of a diagram to that effect and found in "Thermal Radiation: Physiological and Pathological Effects" [2] is included below:

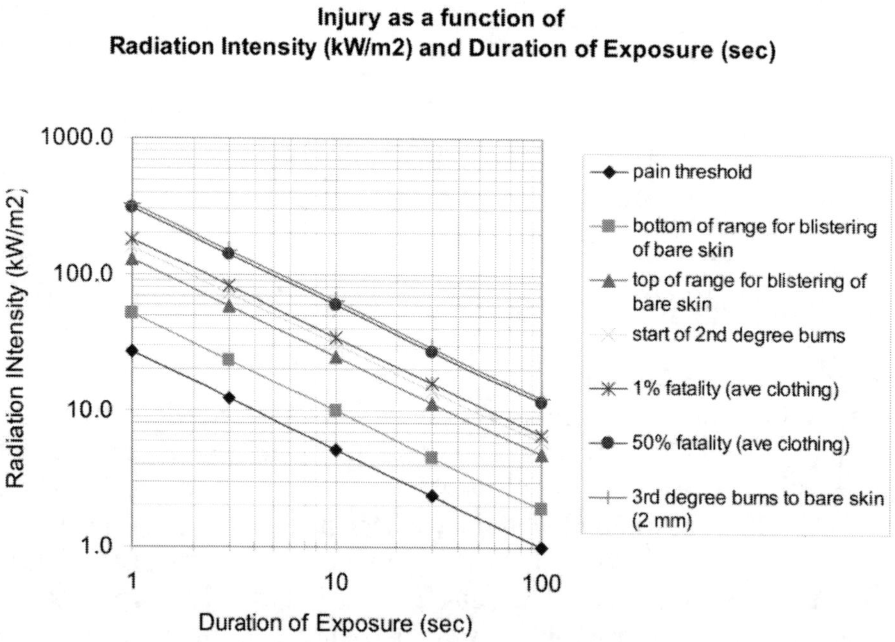

Exposure duration is either the fireball duration (if it is assumed that the recipient cannot remove himself from harm's way within the time frame of the fireball) or the escape time (if it is assumed that the recipient can and does get out of harm's way). Injury is a function of this duration as well as the degree of personal protection that the recipient has or can apply.

Anything done to impact the transport (propagation) of radiation from the fireball to the recipient, or to impact the duration of exposure, or to impact the absorption of radiation that reaches him would be an example of Limitation of Effects. And, if such is an inherent thing (an imbedded, integral, ever-present, and inseparable means of doing so), such is an example of the ISP Limitation of Effects strategy.

For a 10 second fireball, the top of the range for skin blistering is a dose of 750 (kW/m2)^$^{4/3}$-sec. Assuming that $L_h >> L_v$, doubling the distance reduces the dose to 118 (kW/m2)^$^{4/3}$-sec, a dose that is only slightly greater than the pain threshold. Even a 64% increase in distance is an incremental cushion that lowers the dose to the bottom of the range for skin blistering. Such is the power of the power of 2. Increasing the 'L' in the transport term is an example of the application of Limitation of Effects.

Limitation of Effects ... relative to a Toxic Cloud

A toxic cloud will result from the accidental release of toxic fumes for a given duration and at some average rate. As an example, assume that there is a release of a toxic gas that has an 'equivalent harm' equation of the form $K = C^n t$, with C=concentration (ppm), n=2, and t=time (sec) and assume that the probit equation is of the form $PROBIT = a + b * LN(K)$, with a=-35.9 and b=1.85.

Population and Weather

Assume also that the cloud dissipates in a Gaussian way, with the form of the transport function being $c_L \propto c_0 Q / (\sigma_x \sigma_y u)$ with c_L=concentration (ppm) on the centerline at distance L(m), Q=the leak flow rate (m3/s), σ_x & σ_y=the downwind & crosswind standard deviations, and u=wind speed (m/s). The standard deviations are a function of distance $\sigma_L \propto L^n$. For example, n~=.8 for close distances and E atmospheric stability and thus c_L varies with the inverse of $L^{1.6}$. Doubling the distance reduces the centerline concentration by a factor of 3. All other things being equal, including the exposure duration, if the probit equation indicates a 99% fatality rate at some distance, it is only 4% at 2x the distance. A 40% increase in distance results in a drop from 99% fatality to 10% fatality. Such is the power of the power of 1.6.

Further, supposing that a plant needs to be planted somewhere and ignoring all other considerations, with the wind speed in the denominator of the dispersion equation it would behoove us to plant the plant in the windiest place available. Weather is an imbedded, ever-present, stable-within-a-tolerable-range 'thing', an inherent property, despite its constant variability. The annual average relationships between the wind speed, the wind direction, and weather stability is an inherent property of a place. Planting a plant in one place with a wind speed of 4 m/s in deference to another with a wind speed of 2 m/s will decrease the expected concentration at any given distance of any chemical released, on average, by 50%. A factor of 2. For free. (Caution: The local airport is likely to have a wind rose for the airport. Due to topography and other factors, it may not represent what it is at a local site.)

The local population density distribution can be important in reducing risk when planting a plant as well.

Consider the four situations presented in the following figure:

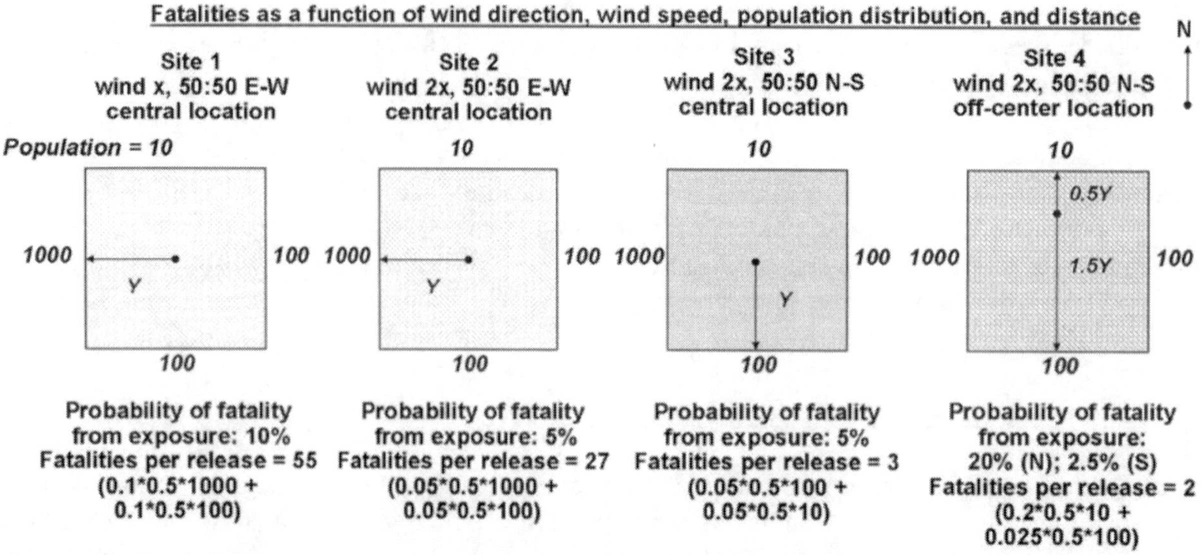

The 4 sites are identical relative to the surrounding population: 10 people north, 100 east, 100 south, and 1000 west. Sites 1 and 2 have a wind blowing east 50% of the time, west 50%; Sites 3 and 4 have a wind blowing north 50% of the time, south 50%. The wind speed is x at Site 1 and 2x at Sites 2, 3, and 4. Site 4 is different from Site 3 only in the location of the plant on the site (an off-center location vs. a central location). On sites 1, 2, and 3, the process is y distance from the property line in all directions. On site 4, it is 0.5y distance from the north line, 1.5y from the south line, and y from the east and west lines.

Which is the best site/plant location for a process with a potential toxic gas release? The quantitative risk analysis indicates that, with a center-of-plant location, given a release, 55 fatalities are likely for Site 1, 27 for Site 2, and 3 for Site 3. It also indicates that if the most probable release point in the process is not arbitrarily placed centrally but is judiciously placed as on Site 4, only 2 fatalities are likely. Such is the power of the power of 1.6 on the standard deviations, the power of the wind speed in the denominator of the dispersion equation, and the power of paying attention to where people are. Likely, the people are not as inherent as the wind, nor the wind as inherent as the distance, but they are all somewhat inherent properties of the system and taking them into account bestows an inherent advantage. In this illustration that advantage is a factor of 27.

Receptor Modification

A 'safe haven' is sometimes provided with the intent that it be a safe place to be in the event of a toxic release. It is necessary to design the haven such that the toxic gas accumulation in the haven does not reach toxic levels. This can be controlled to some degree by limiting the air changes per hour (ACPH) to less than some maximum value. If this ACPH is maintained under all operating conditions and is not 'turned on' only in response to a plant alarm, then the limited ACPH provided is inherent in the process.

A continuous, steady-state accidental toxic gas release from a limited-volume source will rapidly create a steady-state condition downwind for the cloud passage time. As the cloud passes a building the concentration outside of the building will rise to some steady state value and decline only after the cloud has passed. The concentration within the building will rise. The rate of rise will depend on the driving force between the concentration outside and the concentration inside and will depend on the building in-leakage rate from both openings and from deliberate ventilation. After the cloud passes, the inside concentration will fall. The mass balance differential equation for the build-up (accumulation = input-output) is: $dC_{inside} = ACPH * C_{outside} dt - ACPH * C_{inside}$, with C_{inside} being the instantaneous concentration inside the building at time *t*, and $C_{outside}$ being the steady-state cloud concentration outside, and *ACPH* being the air changes per hr (from ventilation and from leakage through openings). A rough but reasonable average C_{inside} can be estimated and compared to a specified toxic concentration, taking into account not only the cloud passage time but also time period to purge the toxic fumes from the building subsequent to the cloud passing. The lower the ACPH, the slower the inside concentration approaches the outside concentration and the longer the tolerable cloud passage time will be. A graph such as that below can be used to determine the ACPH needed to keep an indoor concentration lower than some toxic limit:

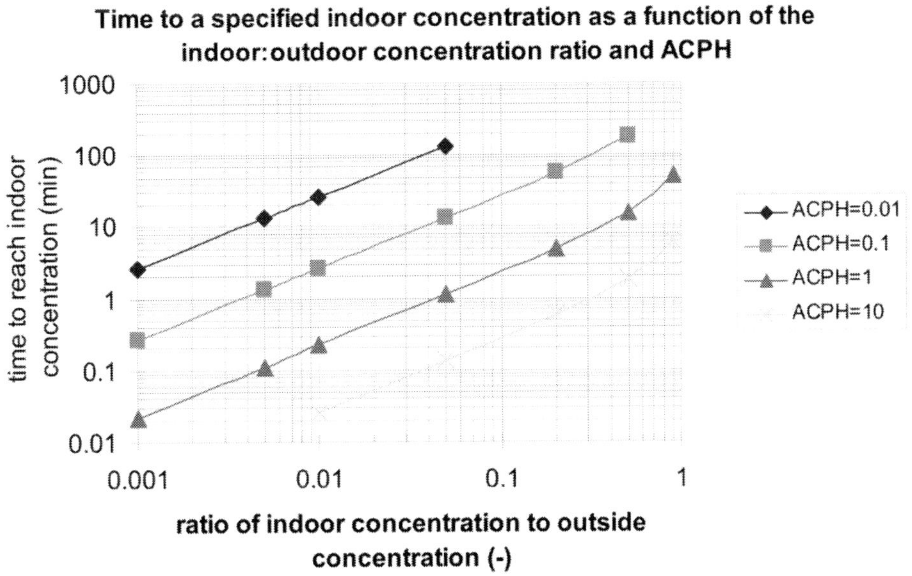

For example, assume that the cloud passage time is 5 minutes and that a concentration deemed toxic for a duration of exposure (2x the passage time; 10 min) is 10% of the outside concentration at the building location at the time the cloud passes the building. As illustrated in the figure, the building and its ventilation system would have to be sufficiently well designed such that there are no more than approximately 0.5 ACPH so that the occupants would not be exposed to a toxic dose. The above is somewhat simplified and should not be used in an actual evaluation. However it illustrates that the ACPH is a significant factor in any similar evaluation. A design that includes continuous ventilation at or below some minimum can make the building more inherently safe, at least relative to this particular hazard.

Limitation of Effects ... relative to Explosions

Structurally (and ignoring injuries from flying glass from window breakage) most buildings of most types of construction will be relatively safe for their occupants if an explosion blast wave delivers less than 10 kPa (1.5 psig). And most buildings of most types of construction will be structurally unsafe for their occupants if an explosion blast wave delivers more than 25 kPa (3.5 psig). That is a fairly narrow range.

The generator (initiator) term for a condensed phase explosive is: $E = me_s$, where E=energy released (j), m=mass (kg) and e_s=specific explosion energy (j/kg) for the compound. For the bursting of a vessel containing compressed gas, the (Brode; corrected for surface burst) equation is $E = 2(P - P_o)V/(\lambda - 1)$, where P=burst pressure (N/m2; pas; absolute), P_o=initial pressure (N/m2; pas; absolute), V=volume (m3), and λ=the heat capacity ratio of the gas in the vessel at the temperature at the time of rupture.

The principal variables in the transport term for pressure vs. distance are R_bar and P_s. The definition of R_bar is: $R\_bar = L(P_o E)^{1/3}$ and of P_s is: $P\_s = P/P_o - 1$. Their relationship is found in graphs such as in the "Guidelines for Evaluating the Characteristics of Vapor Cloud Explosions, Fires, and BLEVEs" [3].

The values of P_s for the above pressures, 10 kPa and 25 kPa, are 2.3 and 1.2, respectively. The distance (m) to the 10 kPa pressure limit is: $L \cong 0.050E^{.333}$. The distance (m) to the 25 kPa pressure limit is: $L \cong 0.025E^{.333}$. To place a building far enough away such that the building damage will cause no more than a low percentage of fatalities (<1%) in the building, use the explosion energy and the former equation. To determine the radius within which an explosion will cause major structural damage, with major being defined as sufficient to cause a high % fatalities, determine the explosion energy and use the latter equation to calculate the radius. Or use a graph such as that below:

Separation distance between an explosion origin and a typical building as a function of energy and damage level

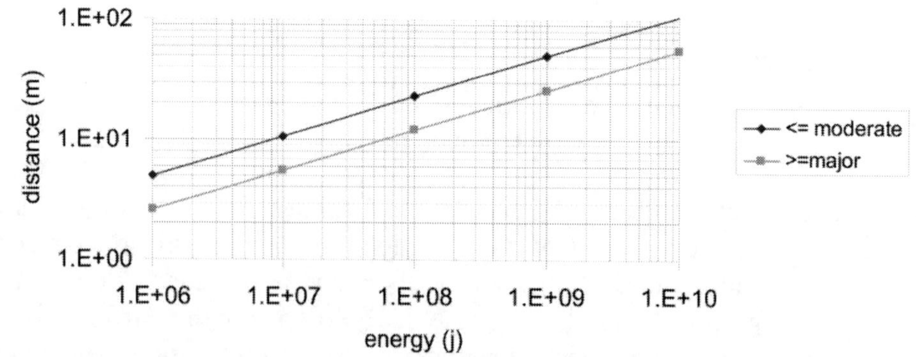

A figure such as the above is specific for TNT-type explosions. And it utilizes only the static pressure for determining damage potential. And implies a somewhat arbitrary definition of high and low damage that is not intended to be design guidance. It does indicate however that distance is a very good friend. Moving a vulnerable building 2x the distance away from an explosion center may reduce the damage and injury potential by as much as an order of magnitude.

The resistance of buildings and building components to blast damage from an explosion can be represented by pressure-impulse (p-i) diagrams such as the one below in which a brick building is assumed and 3 levels of damage are defined (Note: The low-medium-high damage level data is from the 'Guidelines for Evaluating the Characteristics of Vapor Cloud Explosions ... ' [1], Figure B-1, pg. 348). Superimposed on the figure are the p-i combinations for various explosion energies from TNT-type of explosions:

Explosion Damage for a brick building as a function of pressure and impulse with the pressure-impulse curves for 4 orders of magnitude of explosion energy superimposed

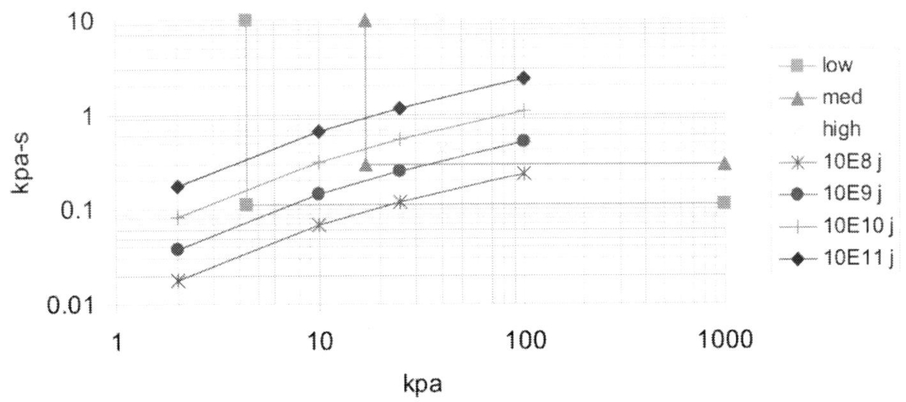

As evident from the figure, there is typically, for any damage level, a pressure below which, regardless of impulse, the defined damage will not occur and an impulse below which, regardless of pressure, the defined damage will not occur.

TNT-type explosions yield unique sets of pressure and impulse. These sets vary with explosion energy but for any energy and p (or i) there is a corresponding i (or p). Superimposing these p&i sets for explosions of different energy indicates that damage from higher energy explosions are more likely to be controlled by pressure than by impulse and the damage from lower energy explosions are more likely controlled by impulse than by pressure. The figure below presents the same information presented above but does so differently:

Distance to specifed damage level as a function of TNT-type explosion energy for a brick house

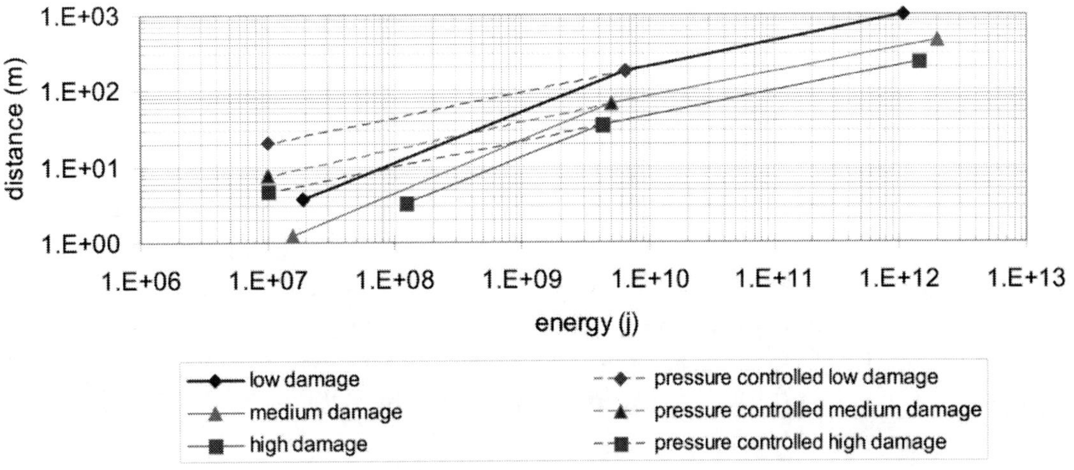

This type of figure can be used in the following way: Given an explosion energy of 1.E+08 j, if only the pressure criteria is considered, preventing a brick building from sustaining 'more than low damage' would seem to imply that it be no less than 60 m away. And so a plant design team may consider moving the location. Or if this is too far a stand-off distance for practical purposes they may be moved to add safety features or change the design or accept higher risk and not rely on the inherent advantage of distance. But the real damage curve, for that energy and that damage level, is governed by the impulse. A 20 m distance is all that is required. The building can be located more closely to the process than some generic pressure-damage table may infer and yet be inherently safe. Process cost is in some way proportional to process area. Being limited in what can be placed within a radius no less than 20 m rather than being limited in what can be placed within a radius no less than 60 m is a 90% reduction in process area within which the design need be limited.

For lower energy explosions the above means that the distance to a specified damage level may not be as large as would be estimated from the pressure only. Or, for the purposes of this paper, a separation distance chosen for a specific type of building and chosen considering explosion pressure only may actually be more inherently safe than one chosen based on the pressure-impulse-damage level curve.

To some degree for vessel bursts and to a very significant degree for vapor cloud explosions, the above relationship is somewhat more complicated. For vessel bursts the pressure and impulse combinations are a function also of the initial burst pressure. While the energy may be then same, a vessel burst with a high P and a low V yields different p vs. distance and i vs. distance curves than a vessel burst with a high V and a low P. For vapor cloud explosions, the pressure and impulse combinations are a function of energy also of the flame speed, which are, in turn, functions of confinement, congestion, fuel reactivity, and geometry. It is nonetheless true that there are energy-burst pressure combinations for vessel bursts and energy-flame speed combinations for vapor cloud explosions for which the impulse is the important variable relative to damage and combinations for which the explosion pressure is the important variable relative to damage.

Limitation of Effects ... relative to Vapor Cloud Explosions

Vapor cloud explosions are complicated. They demand 1) a release 2) the formation of a cloud, 3) the collection of the cloud in a confined and/or congested area, and 4) ignition.

If the release is actually in the confined and congested area and does not have to be driven by wind into it, then the distance between the confined and congested area and the release point is zero and moot. With an open processing structure housing a confined and congested processing area, if it is known in advance that the release would likely be from a particular unit operation within the overall process, locating that operation on the downwind side of the process area will be inherently safer, relative to vapor cloud explosion effects, than locating it elsewhere in the process. The frequency of the catastrophic event (ignition of a cloud within the process area) will be less since at least a significant portion of the time the wind will help to avoid the worst case. In this sense, effects are limited by the expected and anticipated (inherent) atmospheric conditions in a similar manner as above for toxic releases, though the distance scale is substantially less.

The dispersion of released flammable gases is typically not Gaussian in the near field. But wind direction, wind speed and to some degree weather stability are still key factors in the dispersing of the vapors. If the release is in one area and a proximate area is a confined and congested area, then there are weather conditions that will, and weather conditions that will not, be conducive to forming an explosive cloud in an latter area from a leak in the former. The cloud has to be driven by wind into the confined and congested area such that the concentration is still ignitable when it gets there. Thus, the distance between the leak point and the confined and congested area is important, as are the probability of ignition, the wind direction, and, to some degree, the wind speed.

Orientation of equipment within a process can also limit the effects of a vapor cloud explosion. The 6 process areas in **Figure 1** are rectangular in shape, with roofs and with one long side being of easily-blown-away, damage-limiting construction. They are all identical in shape and size and seemingly in congestion. However in the 3 to the left of the figure, the equipment is arranged perpendicular to the weak long wall and in the 3 to the right of the figure the equipment is arranged parallel to the weak long wall. In the top-most processes, there are no interior walls. In the vertically-central processes there is an interior wall isolating one side of the process area from the other and creating relatively cubic process areas. In the bottom-most processes there are interior walls isolating the process area into 4 equal parts such that the depth of the area is 2x the width of the external easily-blown-away wall that is associated with that portion of the process. A vapor cloud is assumed to be generated in the process and to be ignited in the top-left corner of the process area.

The interior walls create confinement but also limit the extent of flammable mixtures. The real estate covered by damaging pressures is proportional to the cloud size. Peak flame speed and peak pressures are generally lower, all other things being equal, when the equipment orientation is perpendicular with the weak wall than when the equipment orientation is parallel to it.

The interior walls create confinement but also limit the extent of flammable mixtures. The real estate covered by damaging pressures is proportional to the cloud size. Peak flame speed and peak pressures are generally lower, all other things being equal, when the equipment orientation is perpendicular with the weak wall than when the equipment orientation is parallel to it.

The peak pressure is lowest when the process area is relatively cubic. When the process area has a relatively long wall that is an easily blown away wall, it takes time after ignition before the pressure opens the vent area. By then, the flame has developed fully and is progressing parallel with the long wall toward the opposite end of the area, creating significant reflected pressures in the far corner.

When the process area has an easily blown away wall that has a width shorter than the process area's depth, the process area is 'pipe-like'. With ignition at a far end of the process area (at the closed end of the 'pipe') the flame progresses toward the weak wall, but accelerates so quickly

Figure 1: Explosion pressure (bar) as a function of equipment orientation and departmentalization (Pressure snapshots at elevation = 1 m and at the time the peak pressure is observed)

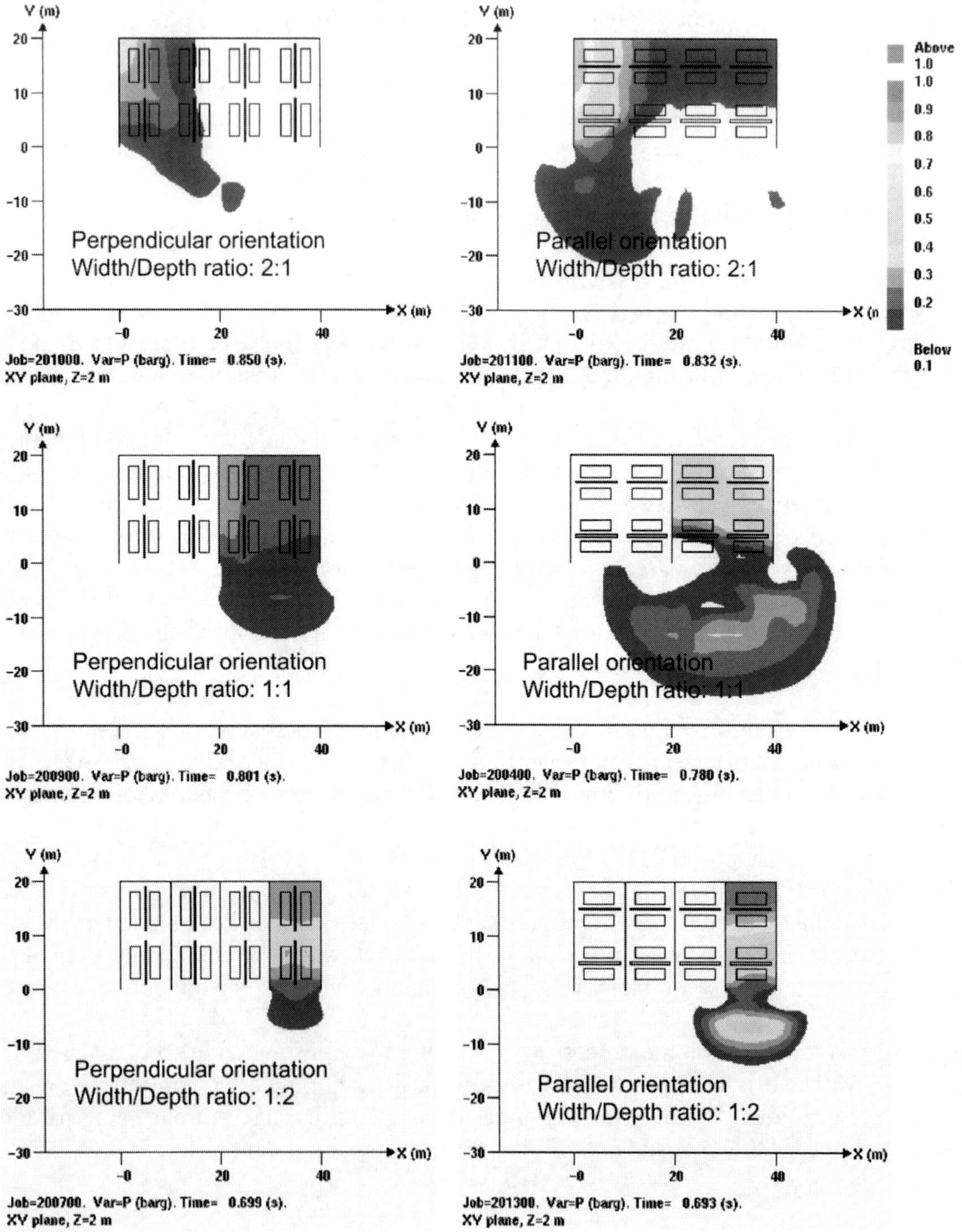

Note: Ignition in top right corner

that the peak pressures experienced are higher than that experienced when the process area is cubic (with the characteristic depth about equal to a characteristic vent wall width).

From this example it can be seen that process area shape and equipment orientation can have a significant bearing on the extent and magnitude of a vapor cloud explosion hazard. Appropriate design and orientation can lead to an inherently safer process area because it limits the turbulence, minimizes the distance from the ignition point to the free field, and limits the cloud size.

Limitation of Effects ... relative to Explosions in Barricades

Explosion barricades and missile walls are sometimes used to mitigate the effects of explosions. A single-side-open barricade (with 3-walls and a roof) with a free standing missile wall spaced apart from and parallel to the open side is one such common design. The intent is to direct the blast wave that may occur from an explosion within the barricade in a preferential direction yet simultaneously prevent missiles, pressure, and impulse from straying too far from the area. The barricade and missile wall could be considered an inherently safer Limitation of Effects solution. The barricade mitigates the pressures that would otherwise be observed in the directions that the barricade is walled. But in doing so it intensifies the pressure in the open-side direction between the barricade and the missile wall. Also, there is considerable wrap-around pressure. Even with the gentle forcing of the blast wave toward the open side, the wave expands around the sides and over the top. This is illustrated in the **Figure 2** below, in which 4 barricade-missile wall designs are compared (no barricade and no wall; barricade and no wall; barricade and a partially open missile wall; barricade and an entirely closed missile wall). An explosion of a stoichiometric acetylene/air mixture in a congested volume inside a barricade is assumed.

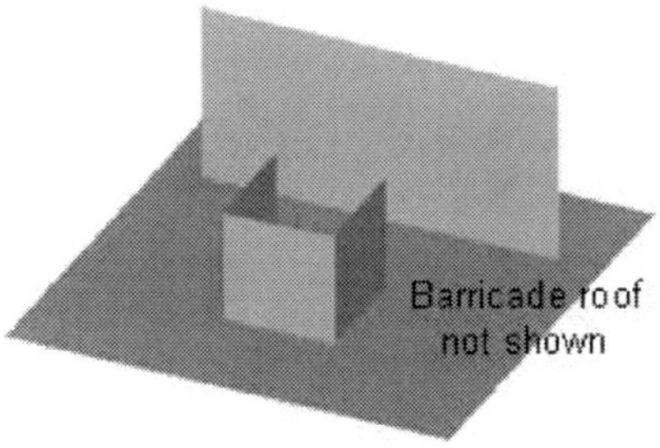

While the barricade's free-standing missile wall does mitigate the passage of pressure in the open-wall direction, it does so by reflecting some of that blast wave energy back toward the process being protected. Thus the barricade provides passive pressure protection in some (walled) directions, though not necessarily eliminating higher-than-desired pressures in those directions, and imposes a higher hazard in the open wall direction (at least in the area between the barricade and the wall). And although the free-standing missile wall stops missiles (at least some of them; some will go over and around) and attenuates pressure in one direction, the wall strengthens the explosion pressure in the reverse (back toward the process) direction. Therefore, by means of the barricade and the missile wall, a limitation of effects (pressure and impulse; missiles) occurs at some points, but at the expense of imposing additional effects at other points. A barricade may be passive and inherent. But the effects in the walled directions may still be substantial due to wrap-around issues.

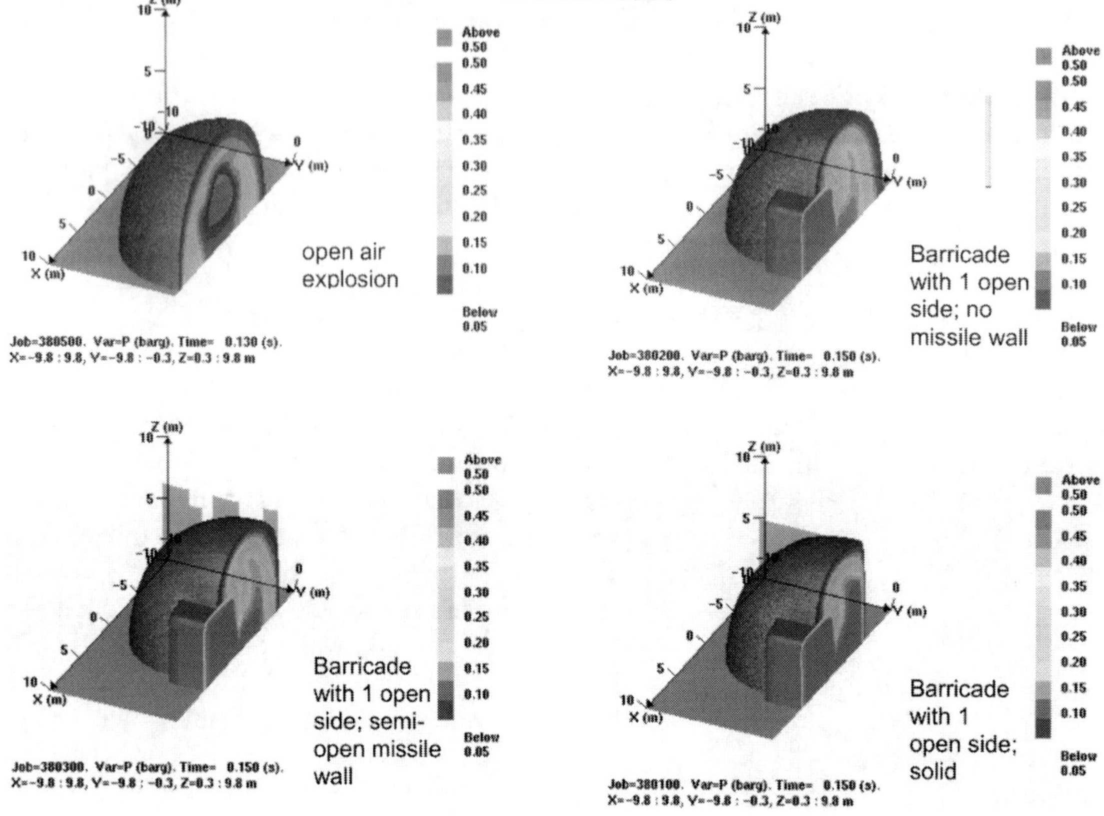

Figure 2: Explosion Pressure (bar) from an explosion in a barricade

Conclusion

All accidents can be described as having 3 components: initiation (generation), propagation (transportation), and termination (reception). Whenever the process is made inherently safer by means of impacting the initiator (generator), it is a form of hazard minimization, substitution, or moderation of process quality. However, whenever the impact of a safety strategy lessens the accident impact by affecting either the propagation (transport) of ... or the termination (reception) of ... the consequences of the hazard, and whenever the safety strategy is integral to, ever-present in, and not-easily-separable from the process, it is an example of a Limitation of Effects ISP Moderation strategy. Opportunities for evaluating and implementing Limitation of Effects strategies can be identified by inspection of the propagation (transportation), and termination (reception) algorithms used to predict the hazardous consequences of initiating (generating) events.

A note on ISP semantics

In this author's view there is only one ISP strategy: Minimization (Intensification). Elimination is Minimization as a hazard approaches the zero limit. Substitution is Elimination followed by addition with the end result being a net loss in overall hazard. It is Minimization, although not perhaps to zero. Moderation of process variables (pressure, temperature, etc.) such that the hazard is lessened is also Minimization, though one is not minimizing a quantity to reduce a hazard but one is minimizing primarily a process quality (pressure, temperature, velocity, etc.) to do so. It is still Minimization. And Simplification is a means to minimize either risk or hazard. It is still an example of Trevor Kletz's 'What you don't have can't leak.'

Limitation of Effects is not an exception. It is also Minimization and could actually be re-named Minimization of Effects. An condensed-phase explosion hazard exists, say from an explosion of 50 metric tons TNT. An operator is 100 meters away in a block building with a reinforced concrete frame and unreinforced masonry walls. The building will sustain heavy damage, with heavy being defined as that which has a greater than 10% chance of fatality. But beef the building up and make it a block building with a reinforced concrete frame and reinforced masonry walls and the building will sustain, upon exposure to the pressure and impulse from the explosion, light-moderate damage, with light-moderate being defined as that which has a less-than-1% chance of fatality. One has *not* inherently minimized the explosion *hazard.* But a characteristic quality of the building, its reinforcement, is integral, seemingly ever-present, and not-easily-separable from the building and the building is an integral, seemingly ever-present and not-easily-separable part of the overall process. Thus the reinforcement is inherent in the process or system, ever-vigilant to destroy the pressure and impulse enemies when and if they appear.

But it is just semantics. What bucket the 'thing that is done to make a plant inherently safer' is put in is not materially important relative to what it actually, inherently (by being integral, ever-present, and not-easily-separable) does.

One of the things that makes the understanding ISP somewhat difficult is that is there is some residual insistence to emphasize that it is the process hazard that must necessarily be inherently minimized if a particular strategy is to truly be an ISP strategy. Then we trip over ourselves attempting to come up with good examples ... and we find a lot ... but they all really don't fit our own narrow definition. The soon-to-be published new CCPS book on ISP, an expanded version of reference [1] will utilize the narrow definition but emphasize the rewards that can come from applying the ISP strategies to sub-systems (rather than a process or plant as a whole) and to risk (and therefore to frequency, to protection systems, and to consequences). By emphasizing that it is the process or system that must necessarily be inherently safer and by emphasizing that it is the risk that must be inherently lower if a particular strategy is to be an ISP strategy, a lot of the good examples do fit. The risk, being a function of the type of hazard, the frequency of an undesired initiating event, the unavailability of the protection layers/loops, and the consequence, can be reduced by reducing the hazard, lowering the likelihood (frequency), improving the unavailability of the protection layers/loops, and/or reducing the magnitude of the consequences. The risk is inherently reduced inherently if any of these reductions are inherent (integral, ever-present, and not-easily-separable) to the process. Thus, Limitation of Effects, which seems even by its name not to imply a removal of hazards but rather to imply that hazards still exist and that only their impact will be limited, does not feel like it fits into the narrow definition of ISP. But it easily does so under the broader definition.

A note on how E. I. du Pont de Nemours & Co., Inc implements ISP in its PSM Program

The DuPont company implements ISP as an essential element within, and an integral part of, the corporate Process Safety Management (PSM) program and does not implement ISP as a distinctly separate initiative.

When any Process Hazards Analysis (PHA) is done, an ISP review, guided by a corporately-developed Q&A checklist structured around the principal ISP strategies, is included within that review.

For DuPont projects, a PHA is actually several PHAs completed in succession (one at the project conceptualization and screening stage; one at the project authorization stage; one at the detailed design stage). The depth and detail of the PHA at each of these stages is proportional to the depth and detail of the process design available at each stage. The depth and detail of the ISP review within each PHA is proportional as well. However, PHA teams recognize the greater value (and lower cost) of adopting inherently safe processing concepts earlier rather than later in project design. Therefore experience indicates that ISP can more strongly impact process design at the 'screening' PHA stage than at later stages. ISP is given greater emphasis, relative to the emphasis given to other hazard control and risk reduction strategies, at the screening stage than at later project development stages.

For existing hazardous processes, PHAs are completed cyclically, with the preferred method being an 'evergreen' method. This entails the creation of as-if-new PHA documentation at each cycle. Much of the prior cycle's PHA documentation is usable and used but re-creation drives a fresh look-see. The re-creation allows for 1) updates in our internal PHA guidance to be applied, 2) a review of process modifications that have been made and 3) a review of incidents that have occurred. Improvements in ISP guidance are reflected in changes in the ISP checklist. Thus, the use of the most recent ISP checklist is a means to ensure that the ISP review and its associated documentation is up-to-date and 'evergreen'.

Research and Development (R&D) projects are more numerous, smaller, less expensive, more transient, more hands-on, and much less steady-state. The magnitude of a hazard inherent in an R&D experiment and/or pilot plant setup is typically lower than the full-scale process it mirrors. However, chemists being what chemists are, they are an order of magnitude closer to the hazard than an operator is to the full-scale plant. And they are by nature more adventurous in their approach to their beakers than operators are to their vessels. Nonetheless, experiments are conceptualized, designed, and built in a manner similar to projects. The chemists are required to do PHAs on their experiments, primarily to protect themselves and their technicians. These PHAs are not as complicated or codified as PHAs on projects and existing processes but all of the applicable elements of the DuPont PSM program are addressed. Therefore, ISP, being one of the essential PSM elements, is considered from the standpoint of making the experiment/pilot plant run as inherently safely as possible.

The company recognizes the higher value of applying ISP at the R&D life cycle stage than at later process life cycle stages. In addition to making all experiments and pilot plants as inherently safe as possible .. for the chemists and technicians ... DuPont desires to develop full-scale, profitable processes that are as inherently safe as possible. Taking a process out of the lab and turning it into a full-scale process thus demands that several corporately-erected ISP barriers or hurtles be overcome along the way. When a process is developed in the lab and R&D proposes that it be moved to a higher level of development and/or to a larger size there is a requirement that this proposal be accompanied by the identification of other potential routes to the same/similar end. Each of the routes, including the one preferred by R&D, is scored using a simple, semi-quantitative system that takes into account how hazardous the chemicals are, how severe the operating conditions are, and how large the inventories would be in the full-scale process that would be developed from the potential route. The end result is a number for each process route that is indicative of the degree of inherent safety in the process and that is usable for comparison to the other numbers generated for the other, alternate processes. Each is reviewed based on their respective ISP score as well as other factors such as profitability, capital cost, and technological risk before it is passed along to the next level of development, or passed back to the lab for further development, review, and/or experimentation.

ISP is highlighted within the corporate PHA and PSM education courses. In addition, an ISP-specific instruction course has been developed and instructors trained. The CCPS ISP book is a mainstay of the course. Special emphasis has been placed on providing instruction and guidance to R&D personnel.

References:

[1] Bollinger, Robert E., et al, "Inherently Safer Chemical Processes: A Life Cycle Approach," Center for Chemical Process Safety of the American Institute of Chemical Engineers (CCPS/AIChE), New York, NY, 1996.

[2] Hymes, Ian, et al, " Thermal Radiation: Physiological and Pathological Effects," Institution of Chemical Engineers (IChE), Rugby, Warwickshire, 1996.

[3] Center for Chemical Process Safety of the American Institute of Chemical Engineers, Guidelines for Evaluating the Characteristics of Vapor Cloud Explosions, Flash Fires, and BLEVEs", CCPS/AIChE, New York, NY, 1994.

LOSS
PREVENTION
SYMPOSIUM

Determination of Pressure Levels for Siting Light Wood Trailers at Chemical Plants and Refineries

Michael A. Polcyn
Raymond H. Bennett, P.E., Ph.D.
Quentin A. Baker, P.E.
Baker Engineering and Risk Consultants, Inc.
3330 Oakwell Court, Suite 100
San Antonio, TX 78218 USA
Phone: 210-824-5960
MPolcyn@BakerRisk.com
RBennett@BakerRisk.com
QBaker@BakerRisk.com

Prepared for Presentation at
American Institute of Chemical Engineers
2007 Spring National Meeting
41st Annual Loss Prevention Symposium
Houston, TX
April 22–26, 2007

UNPUBLISHED

Determination of Pressure Levels for Siting Light Wood Trailers at Chemical Plants and Refineries

Michael A. Polcyn
Raymond H. Bennett, P.E., Ph.D.
Quentin A. Baker, P.E.
Baker Engineering and Risk Consultants, Inc.
3330 Oakwell Court, Suite 100
San Antonio, TX 78218 USA
Phone: 210-824-5960
MPolcyn@BakerRisk.com
RBennett@BakerRisk.com
QBaker@BakerRisk.com

Abstract

The Explosion Research Cooperative (ERC), a joint industry research program, was started by Baker Engineering and Risk Consultants, Inc. (BakerRisk) in 1993 and is now completing its thirteenth year. The ERC is funded by chemical, petrochemical, refining, and insurance companies interested in understanding and predicting explosion hazards and their consequences on buildings, process equipment and people, as well as preventing or mitigating explosions. Since January 2006, the ERC has been conducting a study on the blast capacity of portable buildings at refineries and chemical plants. The work summarized in this paper was performed by Baker Engineering and Risk Consultants, Inc. (BakerRisk). The work presented should not be considered an official guidance document or standard of the ERC or its member companies.

The ERC tasked BakerRisk to develop a set of consistent damage curves (Pressure-impulse or P-i curves) for a variety of portable buildings including light wood trailers so that ERC participants could site portable buildings while considering occupant safety. The ERC agreed to make the pressure asymptotes for the damage curves for light wood trailers available to the American Petroleum Institute for use in API RP 753 – *Management of Hazards Associated with Location of Portable Process Plant Buildings*. The trailer properties selected are representative of the weakest trailer construction types that are likely to be encountered in the continental United States.

The asymptotes are presented in terms of the free-field overpressure and account for reflections and clearing effects. Asymptotes are provided for Building Damage Levels 2A and 2B, as discussed below.

1. Background

Categories of Building Damage Levels (BDL) have been used to provide a standardized means of rating building damage[1]. BakerRisk expanded the categories defined in Reference 1 to five categories, summarized in Table 1, which cover the range of damage from cosmetic to collapse[2].

All analyses performed in this study utilized BakerRisk's BDL categories. In addition, building damage observed in explosion incidents was rated according to the BDLs observed. Use of BDLs also facilitates risk studies by correlating building damage levels to estimates of occupant vulnerability from life threatening injuries, as was done in a previous study conducted by BakerRisk for the U.S. Government[2].

Table 1. Building Damage Level Categories Defined by BakerRisk

Building Damage Level	Damage Description
1	Onset of visible damage. Repairs only needed for cosmetic reasons. Building is reusable after an explosion.
2A	Localized building damage. Building performs function and can be used; however, major repairs are required to restore integrity of structural envelope. Total costs of repairs moderate.
2B	Widespread building damage. Building cannot be used until major repairs are completed. Total cost of repairs is significant, approaching replacement costs of the building.
3	Building has lost structural integrity and may collapse due to environmental conditions (i.e. snow or wind). Total repair cost exceeds replacement cost of building.
4	Building fails completely. Repair is not feasible.

2. Analysis Methods

A sensitivity study was performed on the response of wood trailers using single-degree-of-freedom (SDOF) and two-degree-of-freedom (2DOF) models. Parameters evaluated included the type of exterior sheathing, the quality (type of wood and grade) and geometry of wall studs (notched or un-notched), and strength of connections.

The dynamic responses of the structural components (wall sheathing, walls studs, and roof) of the trailers were calculated in accordance with the procedures in the ASCE manual *Design of Blast Resistant Buildings for Petrochemical Facilities*[3] and Department of the Army Technical Manual TM 5-1300 *Structures to Resist the Effects of Accidental Explosions*[4]. In the analysis, components were initially modeled as an equivalent single degree-of-freedom (SDOF) system consisting of a spring-mass system (see Figure 1). The equivalent SDOF system properties were determined such that the deflection of the equivalent spring-mass system would be equal to the maximum deflection of the component using procedures in the previously stated technical manuals.

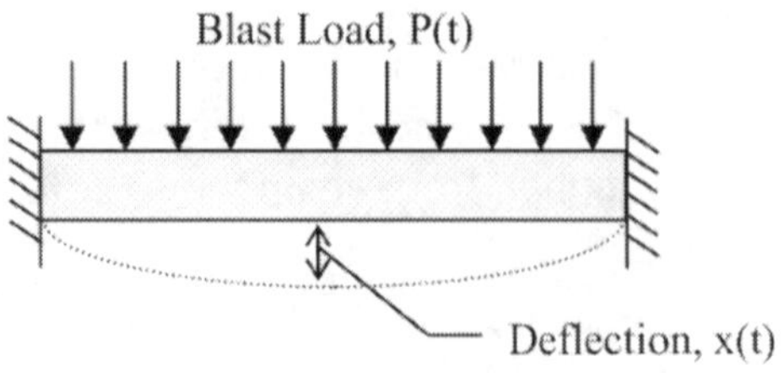

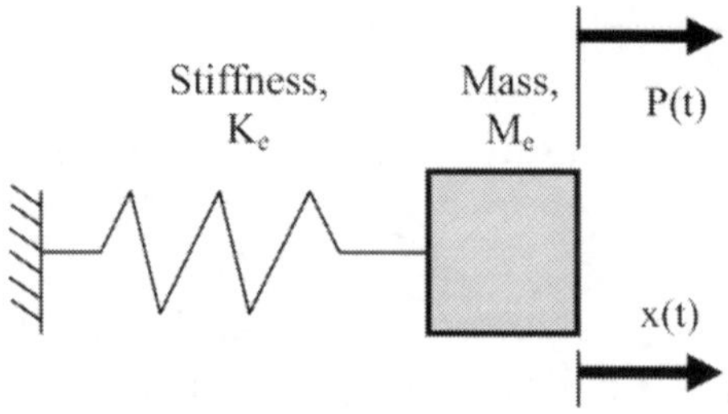

Figure 1. Equivalent Spring-Mass SDOF System

The stiffness and strength of the equivalent SDOF system springs were based on the material properties, geometry, and boundary conditions of each member. The masses of the equivalent SDOF systems were based on the weights of the components and their load-mass factors. The 2DOF analysis method follows a similar approach except that the two springs and masses (representing the sheathing and the wall studs) are modeled as two springs in series.

The maximum deflections of the equivalent SDOF and 2DOF systems representing each structural component were calculated using BakerRisk's computer program ISADS, which was developed for the ERC. The ISADS program calculates the response-time history of a SDOF or 2DOF system by solving for the acceleration, velocity, and displacement at each time step based on the equation of motion. The methodology is discussed in the text by Biggs[5]. The value x_m is the maximum deflection of the component over the span, L. The ductility ratio for each component was determined by dividing x_m by x_y, the deflection at yield.

There is an infinite suite of pressure and impulse combinations that will produce the same level of damage. The combinations can be represented by a Pressure-impulse (P-i) curve or iso-damage curve as shown in Figure 2. There will be a different P-i curve that corresponds to each BDL shown in Table 1 for each variation of wood trailer construction. The member companies of the ERC agreed to release the pressure asymptotes for the P-i curves for weak wood trailer configurations (i.e., those that divide BDL 2A from BDL 2B, and BDL 2B from BDL 3).

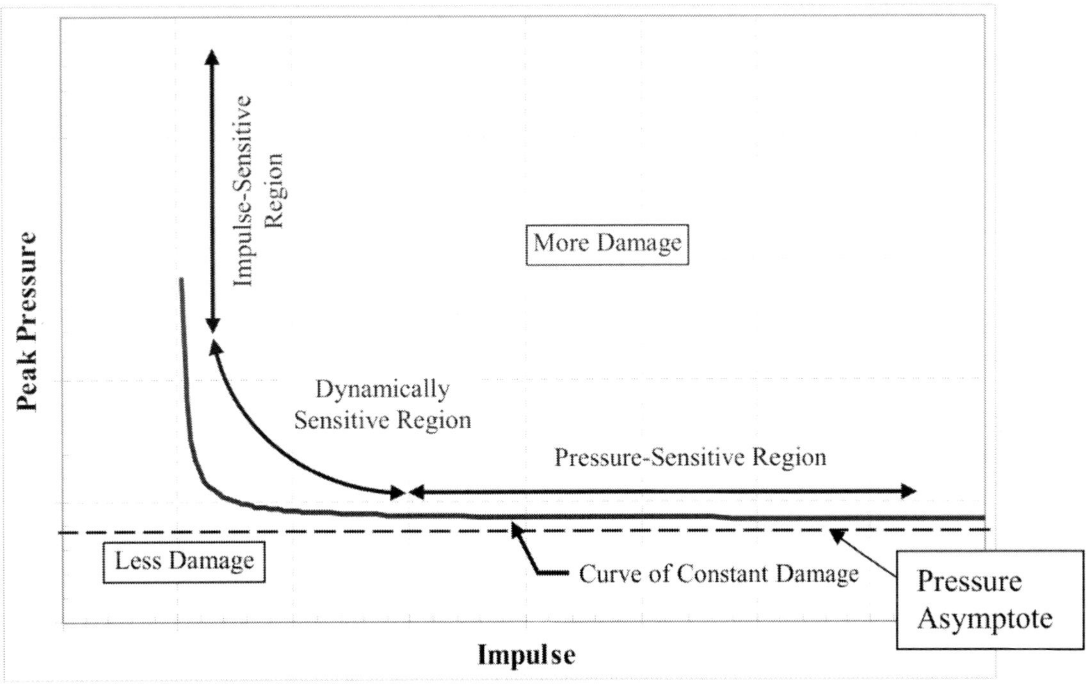

Figure 2. P- i Curve

The pressure asymptotes are a conservative representation of the P-i curve. The pressure asymptotes represent the lowest pressure at which a certain level of damage would occur, which would be associated with a high impulse (long duration) blast. Shorter duration blasts may be of higher pressure while causing the same level of damage if the impulse is less and if the combined pressure and impulse fall onto either of the portions of the curve labeled "Dynamically Sensitive Region" or "Impulse-Sensitive Region" in Figure 2.

Based on the findings of the sensitivity study, a trailer with a wall design consisting of 2x4 studs with a thin aluminum outer skin was selected as a weak type of construction likely to be encountered in wood trailers located in U.S. plants. This type of wood trailer construction was also the most common type present at the March 23, 2005 Texas City explosion.

2.1 Shape of Assumed Blast Loads

The blast load history is assumed to have the shape of a right triangle, as shown in Figure 3, which causes greater structural damage than a centrally peaked shape. Figure 3 also shows that the impulse is the area under the pressure history.

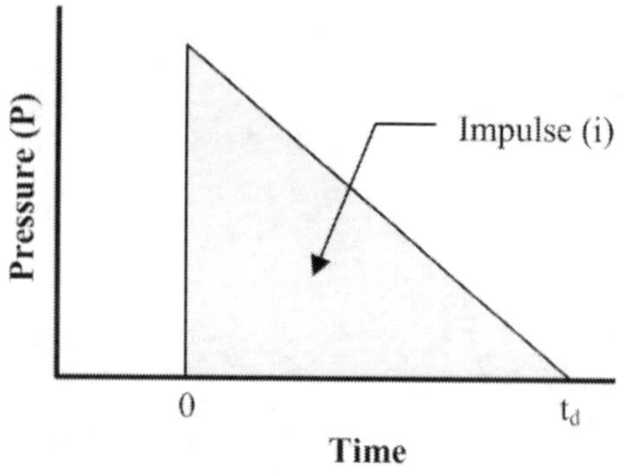

Figure 3. Assumed Shape for Incident Blast Pressure in Blast Analysis

The effects of clearing were included in developing the loading for the reflected face (the side of the building facing the explosion) of the trailer. The effects of reflection and clearing are shown in Figure 4 for a suddenly applied shock load. The pressure on the blastward trailer wall increases (essentially instantaneously) to the reflected pressure, and the pressure on the wall decays to nearly the incident pressure waveform at a time referred to as the clearing time. Reflected pressure is at least twice incident pressure, with the actual multiplier depending on the magnitude of the incident pressure. For a typical trailer, the clearing time is approximately 20 ms.

For the reflected wall of a trailer, the free-field or incident blast pressure was modified to the form shown in Figure 4, and this modified waveform was then used to evaluate the structural response. The results, however, are reported in terms of free-field or incident pressure. When the duration of the blast is less than the clearing time, then the full reflected pressure is indicated by the dashed line in Figure 4.

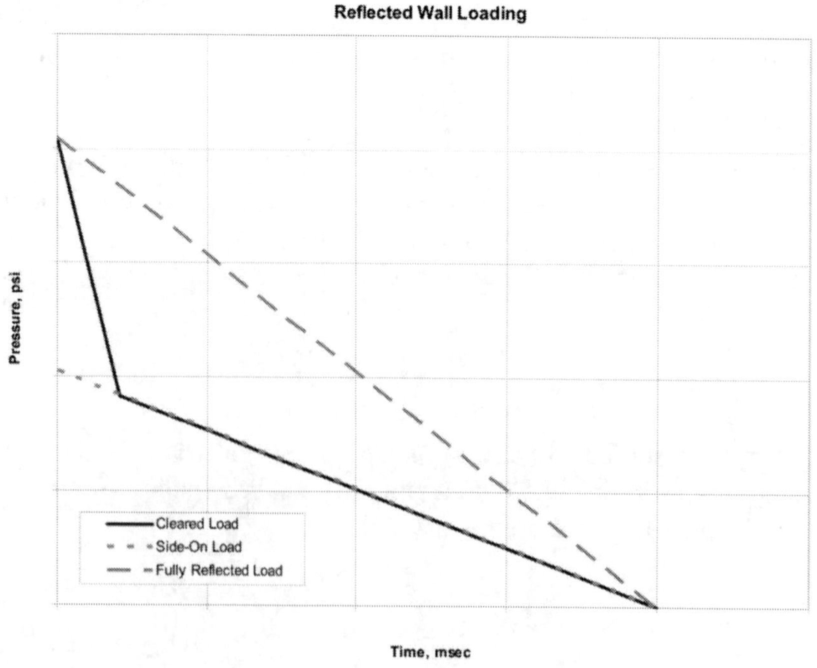

Figure 4. Example of the Effects of Reflection and Clearing on Reflected Face Loading

3. Selected Portable Building Construction

3.1 Typical Light Wood Trailer Construction

The values presented in this paper are based on what is expected to be the weakest wood trailer or portable building typically encountered at plants inside the continental United States. The trailer walls consist of a relatively thin metal exterior skin supported by 2x4 wood studs that span 8 feet with nominal 1x2 wood furring strips that are notched into the studs. The effects of these furring strips are significant in that the notches reduce the section modulus of the studs and provide a point of stress concentration.

3.2 Material Properties

The static and dynamic material properties used in the analysis for the light wood trailers are shown in Table 2.

Table 2. Material Properties Used in Analyses

Material Property	Wood Studs	Aluminum Sheathing
Allowable static bending stress (psi)	875	N/A
Bending stress at rupture (psi)	4375	N/A
Static yield stress (ksi)	N/A	34
Dynamic yield stress (ksi)	N/A	48
Modulus of elasticity (ksi)	1500	10,000
Weight	30 pcf	0.42 psf

4. Results

4.1 Pressure Asymptotes

The pressure asymptotes determined by this study for light wood trailers are shown in Table 3. These asymptotes are expected to provide a reasonably conservative estimate of the pressures at which the stated damage may occur. The values in the table should not be extrapolated upward for trailers with stronger construction without the user performing an independent structural analysis for the specific construction type in consideration. The trailer construction evaluated for the values in Table 3 was assumed to respond flexurally such that the component strength was not limited by the strength of the connections at the ends of the studs or by the performance of the roof.

Table 3. Pressure Asymptotes

BDL	Pressure (psi)	Basis
Boundary between 2A and 2B	0.6	Response of reflected wall reaches ductility ratio of 2
Boundary between 2B and 3	0.9	Response of incident wall reaches ductility ratio of 2

4.2 Comparison to Accident Data

To evaluate the reasonableness of the asymptotes presented in Table 3, the asymptotes were compared to available accident data. The wood trailers involved in accidents were assessed based on the observed damage and the estimated pressure and impulse that the trailers had experienced in the accident. As can be seen in Figure 5, the asymptotes conservatively bound the data. That is to say, no incidence of BDL 2B was found below the 2A/2B asymptote, and no incidence of BDL 3 was found below the BDL 2B/3 asymptote. Photographs showing post-incident damage levels BDL 2A and 2B are shown in Figure 6 and Figure 7.

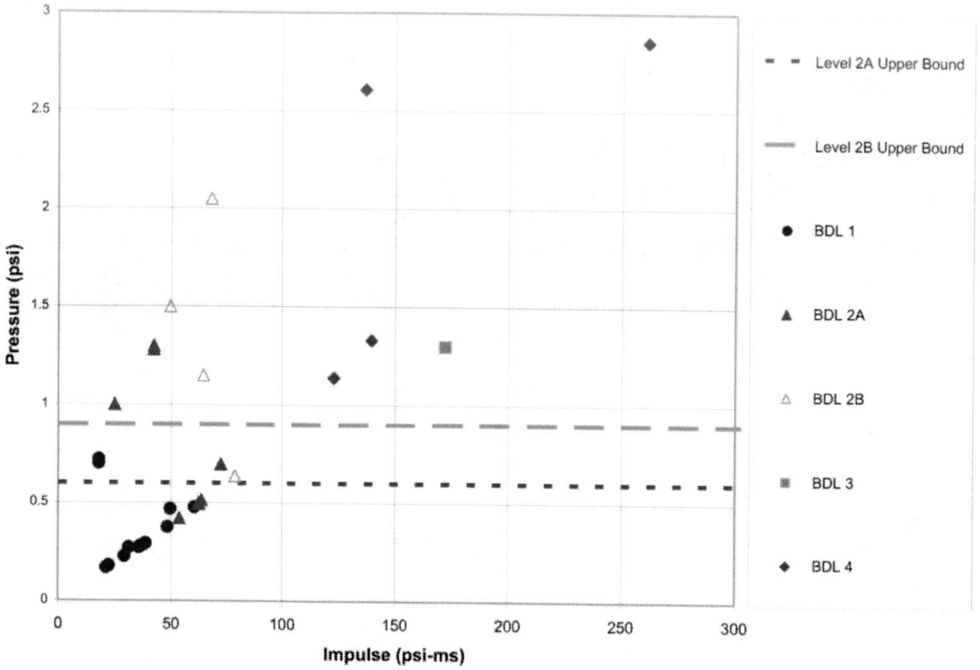

Figure 5. Comparison of Recommended P-i Curve Asymptotes to Test Data

Figure 6. Building Damage Level 2A – Incident Pressure ~0.5 psi

a) front wall

b) back wall

Figure 7. Building Damage Level 2B – Incident Pressure ~ 0.65 psi

5. Summary and Other Work Underway

5.1 Summary

The results of a study performed by BakerRisk and funded by the ERC have been presented in this paper to assist industry in the siting of wood trailers. BakerRisk expects the results from this study to be conservative for the following reasons:

- The trailer construction selected represents the weakest type of wood trailer likely to be encountered on industrial plants,
- The material properties selected were those for the weakest grade of lumber available, and
- The analysis did not account for the effects of blast pressure leaking into a trailer. Sensitivity analyses performed during the study indicated that the percent of fenestration can decrease the amount of damage.

5.2 Other Work Underway

The ERC continues to fund research to improve safety related to potential explosions. Upcoming studies include development of P-i curves for a variety of portable building types and shock tube testing of wood trailer walls.

6. References

[1] Center for Chemical Process Safety of the American Institute of Chemical Engineers, *Guidelines for Evaluating Process Plant Buildings For External Explosions and Fires*, New York, 1996.

[2] Charles J. Oswald and Quentin A. Baker, "Vulnerability Model for Occupants of Blast Damaged Buildings," presented at the 34th Annual Loss Prevention Symposium, March 6-8, 2000.

[3] American Society of Civil Engineers, "*Design of Blast Resistant Buildings in Petrochemical Facilities*," ASCE Task Committee on Blast Resistant Design, New York, NY, 1997.

[4] Departments of the Army, the Navy and the Air Force, *Structures to Resist the Effects of Accidental Explosions*, Army TM5-1300/Navy NAVFAC P-397/Air Force AFR-8-22, Washington, D.C., Rev. 1, November 1990.

[5] John M. Biggs, Introduction to Structural Dynamics, McGraw-Hill Book Company, 1964.

LOSS PREVENTION SYMPOSIUM

Natural Hazards and Process Safety Management Elements. A Natural Mix.

Frank F. Liserio, Jr. and Alan Svoboda
FM Global
Plano, TX 75024 U.S.A.
Phone: 972-377-4808
frank.liserio@fmglobal.com
alan.svoboda@fmglobal.com

Prepared for Presentation at
American Institute of Chemical Engineers
2007 Spring National Meeting
41st Annual Loss Prevention Symposium
Houston, TX
April 22–26, 2007

UNPUBLISHED

Natural Hazards and Process Safety Management Elements. A Natural Mix.

Frank F. Liserio, Jr. and Alan Svoboda
FM Global
Plano, TX 75024 U.S.A.
Phone: 972-377-4808
frank.liserio@fmglobal.com
alan.svoboda@fmglobal.com

Abstract

Natural hazards are real and likely risks to chemical and industrial sites. A complete understanding of all process variables and safety knowledge that influence the natural hazard exposure is often unknown and under recognized. Furthermore, it is rarely contained under existing Process Safety Management (PSM) programs. This is a recommended component of an exposed facility to minimize or prevent episodic releases or events that can cause significant property damage and business interruption. This paper explores the need to consider natural hazard events in existing PSM elements. It also provides event considerations for process hazard analysis and event response. Adopting natural hazards into an integrated PSM program can effectively identify and manage these hazards and resulting risks.

1. Introduction

It is generally understood, through site selection and market constraints, that natural hazards pose various risk levels to chemical and industrial facilities. Their proximity to coastal or unique geological areas needed for raw material acquisition, distribution, and finished goods usage makes them prone to damage and chemical releases during such events. The common expectation throughout the industry is that through specific site selection, design and emergency response, natural hazards will be mitigated. However, this may not be the case as the true potential for loss, even under proper site selection and design, is misunderstood.

The tremendous power exerted by various natural hazard events on chemical and industrial facilities should not be underestimated. Past events have demonstrated incomplete comprehension and hazard conception associated with natural hazards. The enormous energy possessed within hurricanes, floods, earthquakes, freeze, and other natural hazard events is often difficult to comprehend when experience is limited. It is not commonly recognized that a facility located in a 100-yr flood boundary, subjected to 100-yr wind speeds, or significant shaking intensities associated with ground fault ruptures during a 100-yr event will experience significant property damage and business interruption.

Furthermore, natural hazard events, as rare and infrequent as perceived, are not part of a normal process operating condition and as such, may not be identified in Process Safety Management (PSM) programs. Design specifications during construction are intended to address these hazards. Compliance with these design specifications is expected to limit and prevent any episodic releases that may lead to health hazards to employees, contamination or significant property damage and resulting business interruption. Normal construction design standards address some level of abnormal event but not the desired abnormal level due to the abnormal hazard of the process. They also are based upon a perceived liklehood of the hazard occurring without a full appreciation of the consequences of higher intensity less liklehood events and resulting risk.

PSM, as established, has been ingrained as a regulation that must be complied with to prevent accidental release of highly hazardous chemicals that may be toxic, reactive, flammable, or explosive, or may exhibit a combination of these properties. As listed in Title 29 of the Code of Federal Regulations, Part 1910.119, elements are outlined that must be managed to prevent such episodic releases with resulting property damage and business interruption. PSM is the most comprehensive management plan established at such facilities for management of process hazards.

In the application of an effective PSM program, effort is made to establish comprehensive sets of policies, procedures, and practices designed to ensure that barriers to episodic incidents are in place, in use, and effective. Major elements of a PSM program include accountability and responsibility, process safety knowledge, process hazard analysis, process risk management, management of change, process and equipment integrity, incident investigation, training and performance, human factors, standards, codes, laws, audits/corrective actions and emergency response planning.

The following sections explore the need to consider natural hazards in these PSM elements. While site specific conditions may eliminate the introduction of natural hazards into all PSM elements, consideration should be given to understanding the true hazards that would be present by utilizing process hazard analysis or other process safety knowledge into the these events. Design specifications alone may not minimize or prevent episodic releases or events that can cause significant property damage and business interruption. Emergency response can be ineffective if the true hazards posed during natural hazard events are not properly identified. PSM elements are integral in establishing a safe, productive site that successfully manages these natural hazards to the best of its abilities. Consideration of natural hazard events in application of PSM elements will also assist in the identification of sites hazards, vulnerabilities and mitigation opportunities to these events. It is a complete understanding of all the perils related to site specific natural hazards that will influence the natural hazard exposure and resulting risk. PSM programs can naturally assist in this understanding.

2.0 Background

There is a common understanding within the chemical industry and broader industrial classes that through proper site selection and design guidelines, potential damage imposed on structures, buildings and processes during a natural hazard event will be mitigated. However, through site audits, research, code reviews and loss experience, this perception may not be correct.

2.1 Design Specifications

Design specifications are intended to address the minimum load requirements on process structures and buildings based upon the perceived probability that a likely natural hazard event will impact a respective geographical area. These design specifications ensure that structures or buildings satisfy strength limit states in which members are proportioned to carry the design loads to resist buckling, yielding, fracture, etc. that can lead to episodic releases.

It is expected that a factor of safety is incorporated into these specifications based upon the level of strength needed for each process component. These safety factors are meant to address unknowns or undefined loads which might be placed on a process structure or building. Most often, the safety factor is incorporated directly into the allowable loadings, such that the designer automatically uses them, by simply using the appropriate design criteria. Practical engineering design protocol precludes the use of safety factors to satisfy known design considerations, such as wind loads, flood loads or earthquake loads.

In addition to strength, process structures and buildings should also satisfy serviceability limit states that define functional performance and behavior under load. Strength limit states have traditionally been specified in building codes because they control the safety of the structure. Serviceability limit states, on the other hand, are usually perceived as noncatastrophic (deflection and vibration), define a level of quality of the process structure, and are matters of judgment as to their application. The serviceability limit states for the most part are not included within the building codes. Exceeding a serviceability limit in a process structure usually means that its function is impaired. In a natural hazard event, this can result in damage or deterioration that potentially can lead to an episodic release under repetitive loads. Serviceability limit states involve the perceptions and expectations of the designer and are often subjective and difficult to define or quantify. The user may not be involved in determining the serviceability limit. This is highly dependent upon the nature of the process and involves two different disciplines in its collaboration.

While the design specifications address the load requirements, these load requirements are typically a function of the research and service performance available at the time of construction. With natural hazards, research has been refined over the years to address unknowns that have been experienced through losses. For wind loads, studies are ongoing to determine the influence of variable drag coefficients and associated definitions of projected area and solidity. Extreme variability exists in current design applications. Earthquake loads have been refined over the years for materials that possess certain qualities of ductility and postelastic energy dissipation. This research impacts strength design on process structures and not necessarily serviceability design.

Each natural hazard event poses new failures and lessons that are researched and studied. This research is provided to the industry who must ultimately determine through probability, monetary and political experience if it will be utilized in revised design standards. However, this is a complex process exercised over a long duration and ultimately left to the discretion of the designer for application and rarely applied retroactively.

2.2 Safety Factor

Codes and standards specify minimum failure strengths for structural products. Steel, wood and concrete, even properly manufactured and graded, will vary in strength over a predictable minimum range. The material specification normally accommodates the normal range, with the safety factor providing for variations beyond it. The variability not only goes to the material, but also goes to the installation.

What is designed and specified, may not be what is actually built. Significant encroachment on structural safety factors is not common. On the other hand, unintentional encroachments with the non-structural process or building components are very common. Building waterproofing, process sampling lines, piping insulation, pumps, motors, fireproofing, routing of electrical lines, etc. are normally seen as non-life-safety-related, and may be given less oversight and control in adherence to design specifications. Furthermore, quality of the installation is generally hidden after completion of the assembly. Performance of the completed product depends upon the installation of the product, that is, the close conformance to all of the installation criteria.

As materials age, the effects of load cycling, corrosion, expansion and contraction, leaks, condensation, UV exposure, temperature fluctuations and many similar factors all take a toll. For structural framing components, these effects are normally quite limited and are protected by the cladding. For cladding components, the effects are usually much higher. Fasteners are subject to significant movement from wind induced "flutter," expansion and contraction forces, and deflection of attached components. They can easily be made to "wiggle" within the component, slowly loosening their grip. Corrosion can also degrade the fastener as well as the component, weakening the integrity of the grip. Condensation or moisture within the components can weaken some insulations, making them more prone to breakage, or pulling over the fastener.

Codes are written in anticipation of resisting design loads of a likely nature, knowing full well, that there is a reasonable possibility that heavier loads may occasionally occur. It is expected that when such moderate increases in loading occur, the safety factor will prevent failure. The code may provide guidance for a 100-year natural hazard event, knowing that there is reasonable history to indicate that stronger winds, stronger earth movements or higher floods, might infrequently occur. For example, with all other risk factors (as discussed, above) accounted for, a residual 1.5 safety factor would indicate a reasonable probability that a structure designed for the load from a 115-mph wind, could possibly handle the load from a 140-mph wind, before ultimately failing. However, there are no assurances of the ability of the assembly to resist the additional load, based on the safety factor actually being 1.5. And no designer, who feared, or was asked to design for the 140-mph wind, would depend upon that safety factor to provide any confidence of getting there.

Designing for a moderate natural hazard event is only made reasonable by the knowledge that this design carries a full safety factor. It is reasonable to expect that these higher loads could be resisted, at least with reasonable confidence, only if the process structure, building, components etc. is well built, well maintained, is relatively new and fully incorporates its entire safety factor.

2.3 Natural Hazard Probabilities

Many models exist that predict probabilities that a specific natural hazard event of a particular magnitude will happen within a geographical area over a specific time period. These models may yield very different probability assessments depending upon the time period of record or time period desired. Each model contains certain constraints based on physical or historical data and are usually perceived as random acts. Design specifications attempt to provide a realistic probability that a certain event of a certain magnitude will occur over a geographical area.

Wind design specifications utilize maps provided in ASCE 7. These maps are based upon hurricane analysis predictions of 50- and 100-year return period peak gust wind speeds and weather measurements obtained through the National Weather Service with a minimum 5 year historic period. This analysis utilizes data from past storm events and through statistical probability, estimates the likely wind speeds that a geographical area will meet *or exceed* in a specified period of time. These wind speeds are peak gusts winds and not sustained wind speeds commonly referenced in hurricane tracking and intensity reports. A longer period of record or changes in atmospheric conditions (as commonly seen in current research on global warming and climate changes) will influence probability statistics.

Numerous models also exist which make estimates of earthquake probabilities. Seismic hazard assessments estimate the *mean* earthquake frequencies in order to produce estimates of ground shaking probabilities. These mean frequencies are less specific, that is, they refer to expected values for rate of earthquakes averaged over many time intervals rather than rate associated with the immediate future or any other specific time interval. These events are also perceived as random acts knowing full well that events will occur concurrently or within a short time period. In addition, a longer period between events increases the likelihood of an event occurring due to the dynamics of plate tectonics and stress and strain buildup.

When determining flood occurance intervals and flood zones, model results of the multiple repetitions of storm events are subsequently analyzed to compute frequency-of-occurrence relationships for storm effects. Because multiple life-cycle scenarios are simulated, mean value frequencies are computed along with error estimates of deviation about the mean. Deviations for storm surge may be in excess of 3-ft i.e., storms of 100-yr probability may be exceeded in 50% of 100-yr occurrences by 3-ft or less than published levels. Various deviations exist for various exposures depending upon the data set of record.

Flood Insurance Rate Maps (FIRM) offered by Federal Emergency Management Agency (FEMA) offer *guidance* in determining flood hazards. These maps are commonly utilized by users and designers as hazard maps in depicting the probability that a flood will occur over a set time period. As advocated by FEMA, these maps are insurance rate maps established for determining federal flood insurance premiums within its various zones. FEMA does not advocate its direct use as hazard maps.

As an example, following Hurricane Katrina, Advisory Base Flood Elevation maps were provided to communities to assist in recovery and redevelopment and general avoidance of future flood damages. These maps do not necessarily provide actual flood elevations exhibited during Hurricane Katrina and Rita. As such, areas flooded with floodwaters in excess of 5-ft may be shown within the maps as having flooding levels at 3-ft or less. This is attributed to the perceived level of levee protection for its respective areas. Knowing full well that the levee system is inadequate in height to provide such protection, a decision was made to assume a level of risk until such protection can be provided. However, the level of risk computed by FEMA is constrained by the terms and conditions of the flood polices it offers. It is up to the user to fully understand their specific site risks. Communications of this concept is lacking. In perspective, most federal flood insurance allotted to chemical and industrial sites will cover less that 1% of the total loss that can be seen during such events. The remaining 99% of the loss will be allocated back to the chemical and industrial sites property insurance program or within the retention of the sites owner.

The most common misconception on natural hazard probabilities is the application of probability definitions such as 10-yr, 50-yr, 100-yr, 500-yr etc. This is commonly perceived as an event occurring **only** once in the defined return interval. It is rarely understood that these definitions refer to an **annualized** chance that the event, of a predetermined magnitude, will be met or exceeded. For example, a flood having a recurrence interval of 100 years will occur on an average of once every 100 years, or on an average of 10 times in 1,000 years. It may also be referred to as the 1% flood, as there is a 1% chance of it occurring in any one year. It does not occur at regular 100 year intervals. For instance, it may not occur for 50 years, and then occur several times in successive months. When averaged over the life span of a chemical and industrial site (for example 50 years) it would have a 40% chance of occurring. If the life span was extended to 100-yrs, this event would have a 63% chance of occurring. This is represented by the following formula:

$$= \{1 - [1 - (1/\text{recurrence interval})]^x \} *100$$

where x is the desired time interval.

With such high probability and design vulnerabilities, it is expected that chemical sites with natural hazard exposures will develop a process to manage and mitigate such hazards and resulting risks. PSM programs are specifically designed to accomplish such tasks.

3. Natural Hazards and Process Safety Management Elements

Mitigating natural hazards to minimize or prevent episodic releases or events that can cause significant property damage and business interruption requires a complete understanding of the event and its potential damaging effects. Process Safety Management (PSM) has been developed to guide industries handling highly hazardous chemicals toward safer facilities and contains the commitment levels and evaluation tools needed for identification and mitigation. It can and should be considered the foundation of all loss prevention activities including natural hazard mitigation.

The following sections provide supporting information into incorporating natural hazard events into existing and effective PSM program elements. While the focus of this paper is placed on elements outlined in the Center for Chemical Process Safety PSM Model, guidance can be associated with various other models as deemed fit by the user. Site specific conditions may also warrant the omission or inclusion of some or all of the elements discussed below. The following information provides a basis to consider inclusion of natural hazards into each PSM element as deemed fit by the user.

3.1 Accountability and Responsibility

The corporate policy statement should include recognition of natural hazards as events that can lead to process incidents which should be addressed along with normal process hazards.

3.2 Process Safety Knowledge

Process safety knowledge is the basis for understanding the natural hazards impact on the process. To improve process safety knowledge, accountable personnel should pursue complete understanding of the natural processes that produce hazards in the unit and understand how and when natural processes become hazardous, including the damage caused by wind, inundation, freeze and sudden earth movement. This is achieved by acquiring process, design and natural hazard information and using this knowledge while conducting process hazard analyses. The process reaction under these loads should be well defined. The serviceability of process structures and equipment during these events, as well as their design limitations should be understood.

In acquiring such knowledge, the primary decision to be made is on the continued operation of the unit and/or site during an event. Natural hazards such as windstorms, floods and freezes provide ample warning time before impact and thus allow such decisions to be made. However these decisions are commonly made under misconceptions of the event and its impact on the process unit. For example, the level of wind speed design of a facility may not be equated directly to the event predicted within an area. If a site is designed to a 130 mph peak gust wind, this generally equates to an intense Category 2 or low intensity Category 3 storm. Decisions to cease operations should be made with this understanding.

Earthquakes are sudden and unexpected with relatively no warning time and as such the decision for continued operation is predetermined and/or integrated into the process safety logic. Due to the relative unknowns during earthquakes, emergency shutdowns and subsequent isolation should be initiated. Faults and/or process upsets should have established set points or logic made with the premise of ground shaking. The probability of cascading failures also increases with such events. Subsequent operation should consider these potentials.

The collection of natural hazard information and expected process reaction provides the process safety knowledge needed in subsequent PSM steps (as well as a record of the original review process) so that the process can respond throughout its intended life to such events without an unanticipated incident. The information is documented and made part of the overall process safety management package, which will eventually also include data from the process hazard analysis and process risk management steps. This is then used for employee training, future process changes, etc.

Over the life of the site, new technology in natural hazard design, process operation under such conditions, inherent safety, or mitigation techniques may be developed. While not known or cost effective during initial plant design, they may become so later in the life of the site. It is important for an organization to stay fully abreast of new technology aspects and increase frequencies and apply it as appropriate.

3.3 Facility Siting

Facility Siting guides plant siting decisions defined in the corporate guiding principles or business objectives. The plant siting decisions should also be defined by the potential natural hazard effects. Sites chosen should be selected to avoid or minimize exposures by floods, windstorms, earthquakes, freezing, lightning, snow, volcano, etc. In addition, accessibility to emergency response assistance, accessibility for recovery operations, availability of operational manpower, accessibility to the site under adverse conditions (riot, traffic, damaged roads, etc.), uncertainty in the event probabilities and , accessibility of critical infrastructure and/or lifelines on and off site including power, steam, water, natural gas, nitrogen, etc should all be considered in such siting decisions.

In addition to facility siting, equal consideration should be applied to "process unit layout." Critical decisions should be made in the design phase for the location of process units, control rooms, motor control centers, instrumentation and electronic rooms, pipe racks, storage buildings, storage tanks, material loading and unloading facilities, pipeline or tank farm exposures, utilities, utility lines, reactors, waste water treatment facilities, etc.

Deficiencies presented by natural hazards may be acceptable if sufficient funding is allocated to overcome potential exposures such as raising key equipment above the maximum probable flood, providing more fasteners for wind loads, and improving bracing schemes for earth movements.

3.4 Process Hazard Analysis

Process hazard review is a key and important element of a process safety management program. It is an essential role in identifying natural hazard impacts within process units and response activities to such hazards. For existing facilities, multiple natural hazard factors should be considered in a hazard analysis. HAZOP, What If, Checklist, Failure Modes and Effects Analysis (FMEA), and more quantitative analysis methods might be used. Several factors incorporated into each method may be as follows:

1. Floodwaters present unusual loading conditions on process and storage structures and typically are based upon the amount of material contained within the unit, structure, pipeline, tank, etc... Structures should be designed to properly protect against progressive collapse due to all flood-related loads and conditions. Adequate anchorage should be provided for storage tanks, silos, sealed conduits and pipes, duct banks, lined pits, sumps, and all other structures that are subject to flotation or lateral movement when subjected to flood-related loads. Worst case conditions

should be considered (e.g., fully loaded with material to prevent overturning, empty to prevent sliding, etc.)

2. Water shock on hot surfaces such as reformer tubes, refractory, etc. can lead to significant damage and material breakout. Cooling schemes and processes should fit the projected timeline of advance notification.

3. While design of containment equipment is based on the premise to prevent materials from breaking out (outward pressures), this equipment may not prevent failure due to insufficiently inward pressure design. Flood waters of various levels, velocity, wave impacts, etc. create significant variable inward pressures on such equipment, pump seals, sampling lines, vents, rupture disks, safety relief systems, etc. Other aspects of material and structure stabilization may be impacted.

4. Contaminants such as high salinity in flood waters may adversely impact process units. Contamination impacts operating conditions and it greatly influences damage potential and subsequent recovery. Contamination to fabricated equipment and process machinery is influenced by age, maintenance, and process conditions of the equipment. Potential for under-insulation corrosion will necessitate replacement. Accelerated corrosion may be seen due to high salinity necessitating sandblasting. Asbestos removal will also hamper recovery. Equipment, stainless or not, with moving parts may require significant reconstruction. Salt water contamination requires cleaning with demineralized water since the salt makes it conductive and the deposits create paths for tracking, arcing or corona. Unique decontamination methods such as bleach washing or sandblasting and recovery worker vaccinations may be needed.

5. Flood damage can be significant on instrumentation and control components in direct or indirect (humidity) contact with flood waters. This may be quite limited if control rooms, I/O field interfaces and field sensors are sufficiently elevated. The complexity of control may also dictate salvage potential.

6. Windstorms can create significant impact damage due to the potential of flying debris (landscaping, tree limbs, scaffolding, ladders, signs, insulation, roof gravel, etc.). Covered cable trays and cable groupings with inadequate cable ties can be damage by flying debris, pressurization or wind induced flutter. Flying debris may also impact fire protection equipment setting off deluge systems with subsequent exhausting of fire protection water supplies. Unique ceramics, such as insulation and electric insulators, may experience flying debris impacts requiring replacement.

7. Inadequate building cladding design, commonly due to omission of reinforcements in key pressure zones (such as perimeter and corner zones) or improper installation, will render a building's waterproofing inadequate. This will allow wind driven rain associated with windstorms to enter the building damaging building occupancy. Control rooms, while pressure resistant, are typically not water resistant. Inadequate waterproofing can lead to loss of process control.

8. Cooling towers while structurally designed to the wind loads are highly effective in channeling high winds to cooling tower fill, deck fans and shrouds. Loss of cooling towers and subsequent cooling water is likely during and immediately following a windstorm.

9. Seismic protection should be incorporated into process emergency shutdown systems using shake sensors tied into the safety system. However, systematic shutdown should be incorporated into a hazard review based on the likelihood of failure during shaking (such as unbraced gas piping systems, cooling lines, inhibitor feed lines, etc). Since shaking events can be sudden and without warning followed by repetitive events (aftershocks), thorough hazard reviews and shutdown sequences under such circumstances are needed.

10. Recovery usually requires extensive erection, installation, supervision and engineering costs. Resources needed for such events should be identified in a hazard review.

11. Loss of power during a natural hazard event will typically initiate an emergency shut down. Processes will fail safe. Thermal shock will also occur where refractory material is present. Flaring of process material will occur under an emergency shutdown.

In addition to the above factors, the plant's hazard review technique (HAZOP, What If, etc.) normally applied to individual processes, should also be applied to critical infrastructure and/or lifelines lost during the event and for extended periods following the event. Critical infrastructure and lifelines would include, but not limited to, power, steam, water, demineralized water, natural gas, nitrogen, oxygen, waste water systems, etc. The hazard analysis should identify these at a systems level, using integrated models of interdependent systems. This will identify vulnerabilities which can subsequently be addressed. Protecting critical infrastructure provides a solid foundation from which the site can respond to hazards rapidly and effectively.

All critical infrastructure and critical lifelines should be hazard reviewed as not being available for several days/weeks following a natural hazard incident. Black start needs and capabilities should be closely reviewed by hazard review teams. In a major regional event, neighboring suppliers are forced, through their own vulnerabilities, to shut down. An area-wide pipeline loss of natural gas, water, nitrogen, etc. may eliminate padding on critical product vessels. Emergency connections, emergency access to connections (boats), and inherently-safe plant design should be provided to prevent material breakout.

Site inaccessibility following natural hazard events should be considered on inhibitor and catalyst inventories. Products reactive to air that require nitrogen and/or product dilution with solvents should consider multiple sources or high inventory levels of solvents, nitrogen and inhibitors. Loss of power, cooling water, steam, air, interruption to raw materials from suppliers, loss of cable trays and other critical utilities should be strongly considered in any hazard analysis especially under the following conditions:

- Molten materials (molten salt system for heat exchange, molten material furnaces with refractory)
- High temperature furnaces (olefin furnaces, ammonia furnaces, vinyl chloride cracking furnaces)
- Potential for pluggage and/or solidification (precipitation, loss of agitation, loss of circulation leads to pluggage and/or solidification)
- High value batch processes or extensive process times
- Power generation within plant (exposure to turbine rotor/turning gear, exposure to energized transformers, etc.)

As an example, if circulation pumps are inundated on a storage tank which contains an unstable chemical that is stabilized with another chemical. The circulation pumps are needed to keep stabilizer mixed if high temperature is detected, inhibiting a polymerization reaction. A process normal PHA would find this arrangement adequate with adequate tank monitoring. However, in the post hurricane recovery, the PHA should consider lack of power in combination with other failures resulting in motors to circulation pumps inoperable because they are inundated. Instrument air for the isolation valves is also unavailable; the DCS for the temperature detection is out. A dedicated SOP is needed to handle this unstable chemical. Issues that might need to be addressed could include:

1. The tank may be underwater for several days. Is there sufficient stabilizer?
2. A plan for a back-up pump and motor is needed.
3. A plan for emergency power for the circulation pump is needed.
4. A plan for a manual temperature detection system is needed.
5. A plan to ensure the short stop is in place after the storm.
6. A plan for operators to conduct regular rounds of the tank farm for temperature is needed.
7. A laboratory test may be needed to verify the stabilizer content.
8. A plan to have personnel available to complete these tasks must be in place.
9. The consequences of the event must be quantified and hazard ranked.

In addition to the unstable chemical, the site likely would have other competing exposures to be addressed. Hazard reviews will help set the priorities in your facility. Each facility needs to identify exposures and prioritize actions according to the availability of their post incident man power and post incident emergency inventory. These activities should not be solved during or after the event.

3.5 Process Risk Management

Recognizing the potential hazards, the resulting consequence and the probability of the events (as outlined in Section 3.2), effective mitigation steps can be developed. If the natural hazard information data is available at very early stages of a plant design, inherent safety features can be incorporated. In most cases, a combination of passive and active protection and procedural improvements are usually required.

Process risk management requires screening, ranking, and engineered assessment tools. A high level assessment, such as Quantitative Risk Analysis (QRA) may be needed to make final decisions. Regardless of methods, documentation of the basis for risk decisions is important. Timelines and priorities needed for recovery should be outlined.

Mitigation options such as requirement for flood barriers, need for reroofs, wall cladding enhancements, elimination of potential flying debris (tree limbs, loose landscaping, elevated signs, roof ballast, asbestos materials, etc.), equipment relocation or protection (floodwalls, levees, etc.) , provision of seismic sensors, need for additional seismic bracing, etc. are decided in process risk management. Designers, employees and management in partnership determine the level of protection needed to satisfy each natural hazard risk level and loss exposure goals. Ultimately the exposure is improved through fixed protection and management systems, is transferred through insurance, or is completely avoided by facility siting. Data and information from process knowledge gathering and hazard analysis activities must be evaluated as to economics and potential for risk reduction. Not all risk can be eliminated or reduced through engineering. Process risk management ensures that a balance of inherent or engineered safety and risk transfer is maintained and that all mandatory regulations, corporate standards, and industry and insurance guidelines are met.

3.6 Management of Change

Evaluating changes to facilities prone to natural hazards at the earliest possible stage for its potential impact on property loss prevention is critical. Changes in building and or process components, such as equipment bracing, building cladding, new roof covers, power distribution, piping insulation, equipment elevations, material storage etc. or any other component affecting the natural hazard exposure should follow the management of change process. Any change from original design intent represents a deviation. If the impact of this deviation is not fully understood, the changes even if minor, such as roof flashing over control rooms, can defeat mitigation items outlined in preceding PSM elements and cause a significant peril. Appropriate process hazards management systems ensure that hazards associated with a change or deviation are identified and controlled.

3.7 Process and Equipment Integrity

Key natural hazard mitigation components such as floodwalls, levees, roof systems, equipment bracing, seismic sensors, etc. can be monitored using a similar process as the process and equipment integrity program. This would include documented inspections, tests and corrective action reports.

Planning and quality control are needed on natural hazard mitigation features to ensure the proper installation as specified in the design or the instructions provided by the manufacturer.

Preventative maintenance should be completed on key mitigation components including levees floodwalls, roof covers, equipment bracing, seismic sensors, etc. Uncontrolled vegetation on levees, reduction in elevations through subsidence, floodwall and /or equipment elevations, destabilization on protection structures, roof cover deterioration, corrosion on equipment bracing, uncalibrated seismic sensors, etc. could occur.

3.8 Incident Investigation

All incidents including actual natural hazard events or threats should be promptly and comprehensively investigated. The depth of investigation is commensurate with the level of complexity and size of incident. This will ensure that lessons learned can be quickly applied within the facility. Lessons might manifest themselves as physical changes, process control changes or startup changes as needed at a fully integrated site. Documentation and follow-up to complete recommendations as part of MOC should be provided.

In addition, immediately following an event, a comprehensive pre-startup review should be completed to ensure all process critical systems are serviceable and not weakened.

3.9 Training and Performance

Training is site wide and should incorporate a Standard Operating Procedure (SOP) for natural hazard events. Every conceivable process condition and reaction during a natural hazard event cannot be disseminated through a process hazard analysis. However a defined SOP will help mitigate any hazards and formulate operator response to such events.

3.10 Human Factors

Natural hazard events create very stressful work environments. Personal concerns outweigh business and operational concerns. This stressful environment can defeat mitigation aspects that rely on human intervention such as flood pump operation, flood door closures, boarding up windows, manual isolation of energy systems, etc. Mitigation aspects that rely on human intervention should be avoided where possible. Passive protection schemes should be utilized such as automatic flood pumps, elimination of openings in barriers, impact resistant glass, automatic valve closures, etc.

3.11 Standards, codes and laws

Standards, codes and laws are intended to adequately resist natural hazard-related loads and conditions as deemed practical including appropriate load combinations. Consider the impact of multiple natural hazard events simultaneously (e.g., flood and wind) and whether existing standards, codes and laws adequately address such factors to allow full functionality of the process following such events.

Design considerations should also account for all applicable wind-related, flood-related and earth movement related loads and other loads (e.g., gravity etc.) that will act concurrently on the process unit, building and other critical process components. Use load combinations, load factors, and resistance factors as specified in the governing model codes and standards with the intent of full functionality immediately following an event. Decisions made on this level should be captured to account for changes in science.

3.12 Audits and corrective actions

Preferably, compliance audits should be completed by sites with similar natural hazard exposures. This will facilitate hazard knowledge and mitigation components.

3.13 Emergency Response Planning.

Emergency response planning for natural hazard events is quite common. A well developed emergency response plan will consider all hazards, risk factors, mitigation components and typically rely on recovery and site stabilization. It is not intended for emergency response plans to form the foundation of mitigation. Well conceived passive and active mitigation systems developed through the PSM process will effectively manage such risks with supplemental activities provided by emergency response teams.

4. Conclusion

While incorporating natural hazards into key PSM elements may seem daunting, it is through this process that the true understanding of these hazards and resulting risk can be achieved. Mitigation opportunities will also be identified and as such, risk levels can be lowered to prevent episodic releases or events that can cause significant property damage and business interruption. It is a complete understanding of all the perils related to site specific natural hazards and its process variables, process conditions, safety knowledge, and ultimately mitigation measures that will influence the natural hazard exposure and resulting risk. Consideration of natural hazards in key PSM elements will assist in this understanding.

5. References

[1] U.S. Department of Labor - Occupational Safety and Health Administration Regulations, "*Title 29 of Code of Federal Regulations Part 1910.119 - Process Safety Management of Highly Hazardous Chemicals*", www.osha.gov (accessed Dec 14, 2006)

[2] FM Global Data Sheet 7-43, "*Loss Prevention in Chemical Plants*" May 1999, Revised January 2001.

[3] American Society of Civil Engineers 7-05, "*Minimum Design Loads for Buildings and Other Structures*" American Society of Civil Engineers, 2006.

LOSS
PREVENTION
SYMPOSIUM

Damage to Petrochemical Facilities during Earthquakes

M. Cem Ozbey and Ali Sari
ABS Consulting
14607 San Pedro Ave. Suite 215
San Antonio, TX 78213 U.S.A.
Phone: 210-495-5195
mozbey@absconsulting.com

Prepared for Presentation at
American Institute of Chemical Engineers
2007 Spring National Meeting
41st Annual Loss Prevention Symposium
Houston, TX
April 22–26, 2007

UNPUBLISHED

Damage to Petrochemical Facilities during Earthquakes

M. Cem Ozbey and Ali Sari
ABS Consulting
14607 San Pedro Ave. Suite 215
San Antonio, TX 78213 U.S.A.
Phone: 210-495-5195
mozbey@absconsulting.com

Abstract

Petrochemical facilities suffered great damage during earthquakes, not only due to the strong ground shaking, but also because of the extensive uncontrolled fires caused by the earthquakes. 1999 Kocaeli (Turkey), 1964 Niigata (Japan), and 1964 Alaska earthquakes caused significant damage to petrochemical facilities. Tupras oil refinery, which was designed and constructed in 1961 according to the U.S. standards of that time, suffered significant damage after the Kocaeli earthquake. Prior to the earthquake, the Tupras refinery produced 220,000 barrels of oil-related product per day, which was approximately 1/3 of Turkey's total output. The strong shaking damage and the resulting fire damage to the Tupras refinery included port facilities, storage tanks, cooling towers, stacks, and crude-oil processing units. This paper will give detailed information about the damage observed and the consequences of the damage in the Tupras refinery and will give useful recommendations for the oil refineries located at the west coast of the United States.

1. Damaged to Petrochemical Facilities from Past Earthquakes

Large-magnitude earthquakes close to urban areas have the capability to cause extensive human and property loss in a matter of seconds. Petrochemical facilities around the world have experienced severe damage during earthquakes, not only because of the strong ground shaking, but also because of the fires caused by the earthquakes. Some of the past earthquakes that have caused damage to petrochemical facilities are;

- 1933 Long Beach earthquake (M_s=6.3), California, USA
- 1960 Chile earthquake (M_w=9.5), Chile
- 1964 Alaska earthquake (M_w=9.2), Alaska, USA
- 1964 Niigata earthquake (M_w=7.6), Japan
- 1978 Miyagi-Oki earthquake (M_w=7.5), Japan
- 1999 Kocaeli earthquake (M_w=7.4), Turkey
- 2003 Tokachi-Oki earthquake (M_j=8.0), Japan

Observed damage at the petrochemical facilities during the past earthquakes includes building collapse, failed storage tanks and pipelines, collapsed cooling towers and stacks. Fires caused by the failure of structures or components have contributed to the overall damage significantly in the past. During the 1964 Niigata, 1999 Kocaeli, and 2003 Tokachi-Oki earthquakes, sloshing caused by seismic waves ignited the oil inside the Naphtha floating roof tanks and caused fires that lasted for days. The fires spread to adjacent tanks and resulted in extensive fire damage at these petrochemical facilities. Sadly, the same ignition mechanism caused the fires for all three earthquakes, which could be prevented if lessons were learnt from the 1964 Niigata earthquake. During the 1964 Alaska earthquake, in Seward port an oil tanker was wrenched loose from a pipeline and erupted in flames. The fire spread to the nearby oil tanks causing widespread damage at the port.

2. Seismic Risk at Western U.S.

For a facility located in high and moderate seismic regions, the probability of having a major earthquake might be higher than having a major explosion. For example, according to researchers at the Southern California Earthquake Center, a catastrophic earthquake at the right spot along the San Andreas Fault could significantly reduce energy and water supplies, at least temporarily, and there exists an 80% to 90% chance that an earthquake of 7.0 or greater magnitude will strike Southern California before 2024. According to research conducted after the 1989 Loma Prieta earthquake in the Bay Area, it was found that there is a 62% probability that at least one earthquake of magnitude 6.7 or more would strike the Bay Area before 2032.

The seismic hazard map of U.S. is shown in Figure 1. As can be seen from Figure 1, Western U.S. has high seismic hazard, with the coast area having the highest hazard. Figure 2 shows the probability of having an $M_w > 6.5$ earthquake within 50 km of Southern California in a random 30-year period according to U.S. Geological Survey. A magnitude 6.5 earthquake with a 50 km source-to-site distance can cause significant damage to structures. It should be noted that the probabilities shown in Figure 2 is higher than 50% for most areas in Southern California.

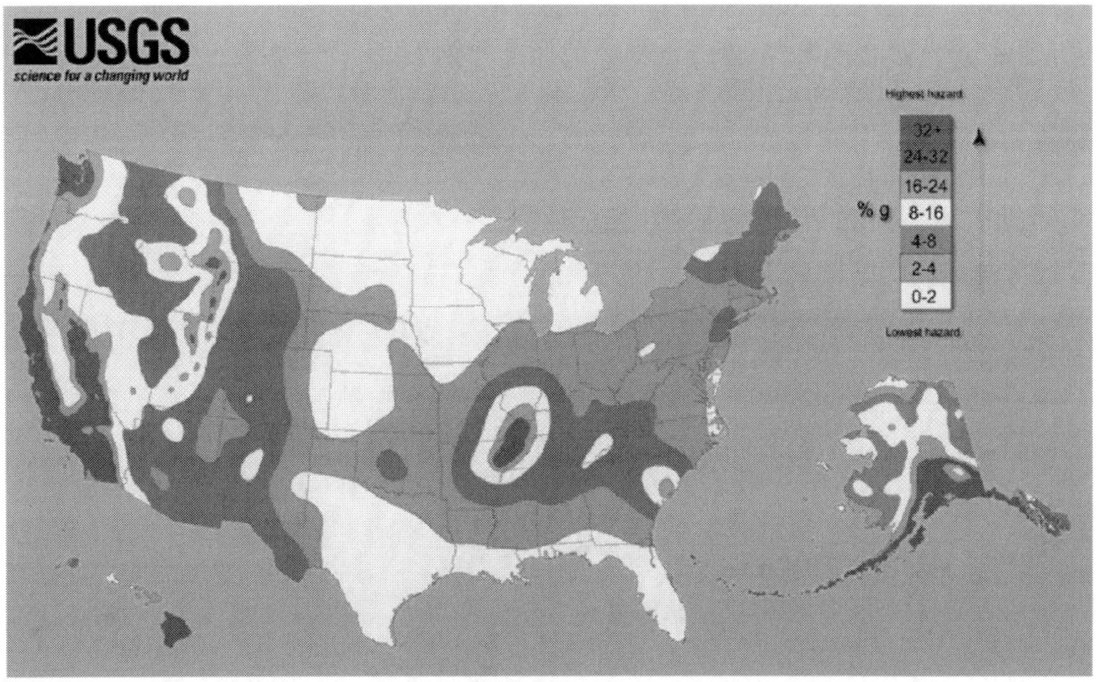

Figure 1. Seismic Hazard Map of United States (2002, USGS)

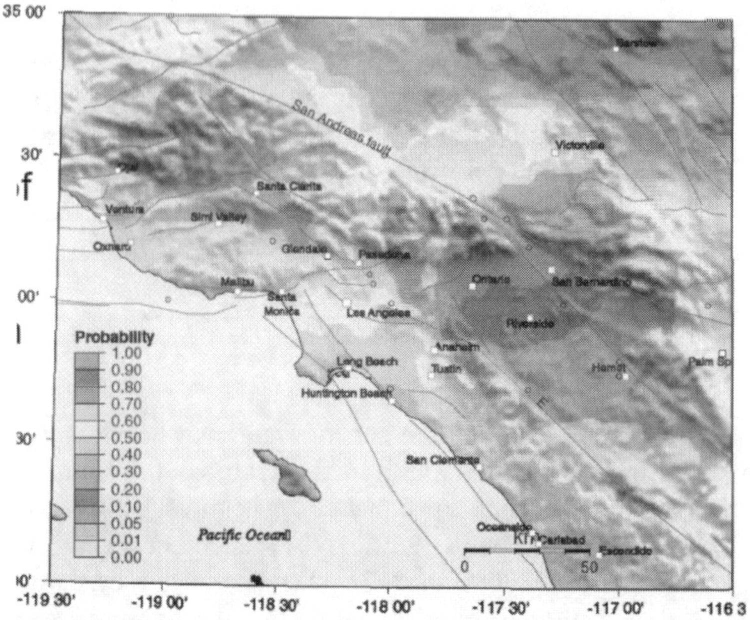

Figure 2. Probability of having a Mw>6.5 Earthquake within 50 km of each location in a random 30-year period (Harmsen & Leyendecker)

Considering the high probability of having a large magnitude earthquake, designing and upgrading petrochemical facilities to withstand large earthquakes is crucial for Western U.S.

Tupras refinery suffered great damage during and after the 1999 Kocaeli earthquake. Kocaeli earthquake occurred on the North Anatolian fault system, which is a major fault system in Turkey. Figure 3 shows a comparison of the North Anatolian and San Andreas Fault systems (USGS 2000). The resemblance between the two fault systems is apparent, concluding that future earthquakes that may occur on the San Andreas Fault system might have similar characteristics as the earthquakes occurred on the North Anatolian Fault system. (Sari and Manuel 2003, Sari 2003)

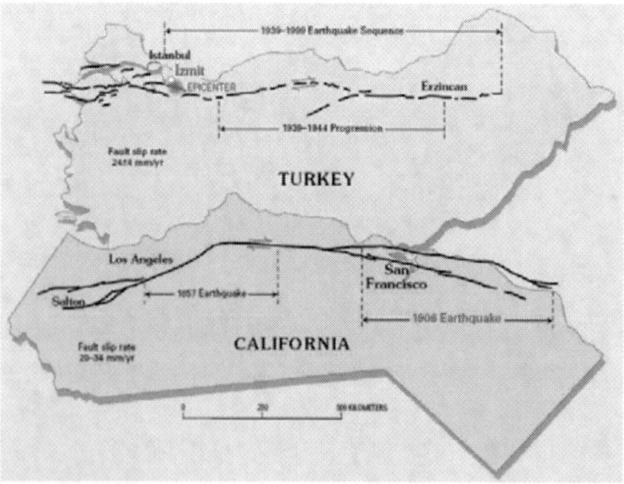

Figure 3. Comparison of San Andreas and North Anatolian fault systems (USGS)

3. Damage to Tupras Refinery during the 1999 Kocaeli Earthquake

3.1 1999 Kocaeli Earthquake

On August 17, 1999, an M_w=7.4 earthquake struck northwestern Turkey causing widespread property loss and claiming over 30,000 lives. The earthquake occurred at the west part of the North Anatolian fault system (Figure 3), with its epicenter about 100 km southeast of Istanbul. The area struck by the Kocaeli earthquake supported approximately 40% of Turkey's heavy industry.

3.2 Tupras Refinery

Tupras (Turkish Petroleum Refineries Corporation) oil refinery was owned by the state in 1999, and was sold to a private company in 2006. It was designed and constructed by a U.S. company in 1961 according to the U.S. standards of that time. Additions were made to the facility in the following years. Tupras refinery is important in the sense that it is similar to Western U.S. refineries built at its time as it was designed according to the same building standards as the U.S. refineries.

Prior to the earthquake, the Tupras refinery produced 220,000 barrels of oil-related product per day, which was approximately 1/3 of Turkey's total output, making it the 7th largest refinery in Europe. It also serves as a major port for LNG and natural gas distribution within and outside Turkey.

3.3 Extend of Damage at Tupras Refinery

Tupras refinery was located approximately 2 km from the fault rupture. Its close location resulted in intense observed seismic motions at the site. One strong ground motion station installed at the refinery recorded very large velocity pulses at the site. Large velocity pulses affect mainly long period structures, and were the cause of the 380-ft-tall heater stack collapse. The strong shaking and the resulting fire damage to the Tupras refinery included port facilities, buildings, storage tanks, cooling towers, heater stacks, and crude-oil processing units. Figure 4 shows an aerial view of Tupras refinery after the Kocaeli earthquake. Fire continued more than 5 days, and was under control after 3 days. All fire water on site (50,000 yard^3) was used on the first day.

Figure 4. Aerial View of Tupras Refinery after the Kocaeli Earthquake

3.4 Tank Farm Fires

There were a total of 45 storage tanks in three tank farm areas at Tupras facility, where each contained naphtha fuel, crude oil, or LPG. Fire started at naphtha fuel tanks due to the sparks created by the hitting of the floating roof to the side walls, caused by sloshing. Figure 5 shows a picture taken after the earthquake at the tank farm area. Sloshing is the vibration on the free surface of a liquid contained in a vessel by the vibration of the vessel itself. Due to the shaking of the floating roof, the naphtha fuel flowed out of the top of the tank and was ignited by the sparks generated by the collision between the floating roof and the side walls. As a result, the most affected area was the naphtha tank farm area. The fire melted six steel storage tanks in the naphtha tank farm area (Figure 6) and burned more than 23,000 yard^3 of the stored naphtha. Overall, 30 of the 45 floating roof tanks were heavily damaged. Most of the damaged tanks were built according to Western U.S. design codes of their time of construction. The tank farm fire also spread to an adjacent cooling tower and destroyed it.

Figure 5. Tank farm fires

Figure 6. Collapsed tanks in the tank farm due to fire

3.5 Collapse of the Heater Stack

The upper two-thirds of a 380-ft-high heater stack (25F-5) collapsed during the Kocaeli earthquake causing both damage to its surroundings and starting fires due to failed piping. The stack was built in 1978 according to U.S. standards (ACI 307-1969). The collapse was mainly due to large near-fault velocity pulse experienced at the site and a large opening between 100-120 ft elevations. Debris from the collapsed stack caused extensive damage to the pipe-rack and the pipes at the crude oil unit (Figure 7), which led to fires at the unit. The debris damaged 63 product and utility pipelines, and a charge heater. Compared to the tank farm fire, the crude unit fire was brought under control relatively quick. Although being under control, the fire could not be extinguished as the supply lines from the Naphtha tanks could not be shut off as there were no block valves before the broken pipelines.

Figure 7. Damage to the pipe-rack and pipes from collapsed stack

3.6 Main Warehouse Fire and Collapsed Cooling Tower

A fire started at the main warehouse, which is believed to be started by spilled chemicals which fell from the storage shelves. The fire was small compared to the crude oil unit and the tank farm fires. Due to limited fire water and manpower, the warehouse was let to burn.

A wooden cooling tower was collapsed and reinforced concrete cooling towers were damaged due to failed supports (columns). Damaged reinforced concrete cooling towers can be seen in Figure 8. Piping connecting to the cooling towers were also damaged and separated from the towers.

Figure 8. Damaged Cooling Towers

3.7 Spillage at the Tupras Port

Two incidences of chemical spillage into the Izmit Gulf due to Kocaeli earthquake were observed. The first spillage was caused by a ship at the loading/unloading naphtha port terminal. The ship tore loose and broke the pipe connections, causing a leakage of naphtha to the Izmit Gulf. The second incident was caused by a broken LPG loading/unloading arm, which released approximately 35 tons of LPG into the bay at the jetty. Due to these two incidents there was a thick layer of oil on the shoreline of the Marmara Sea the days following the earthquake.

3.8 Aftermath

Immediately after the earthquake, operators shut down all the processing units that they could to reduce the spreading of fire. Rupture of the water pipelines that provide all of the refinery's supply severed the refinery's firefighting capabilities. Due to the absence of water on site, foam and chemicals were sprayed from airplanes to keep the fires in control. Portable pumps were transported to the refinery to extinguish ongoing fires. A 3-km region of the refinery was evacuated due to the fires. The fire at the tank farm area continued more than 5 days, and was under control after 3 days.

The process to clean the oil spills started the day following the earthquake. International clean-up teams also helped the Turkish teams. The clean-up lasted one week.

Tupras refinery was completely shut-down for over 3 months. According to officials, the total damage was around 500 million US dollars. Complete recovery took more than a year.

4. Conclusions and Summary

Large earthquakes are rare events in even seismically active regions, just like accidental explosions or strong hurricanes. Although the effects of a large earthquake and a strong hurricane are widespread in a region, earthquakes differ from hurricanes as they cannot be predicted before they happen. Accidental explosions are similar to earthquakes in the sense that they cannot be predicted ahead of occurrence. However, earthquakes have some additional consequences that may contribute to the extent of the damage and loss when compared to accidental explosions.

Considering the fact that probability of having a large magnitude earthquake is higher than having an accidental explosion at a petrochemical facility in Western U.S., being prepared for a large magnitude event is more important than being prepared for an accidental explosion. Just because a large earthquake did not struck a populated region in Western U.S. in the last hundred years, does not mean that seismic hazard is not present for the region. Some of the additional consequences of large earthquakes when compared to accidental explosions are:

- Loss of electrical power
- Damage to water supply pipelines
- Limited outside help due to overall demand for manpower and extensive loss of life and injuries in the area
- Blocked access roads
- Multiple affected sites within a facility
- Lack of communications
- Fleeing personnel in search for their families
- Hazard from adjacent facilities
- Large aftershocks

As Tupras refinery was designed according to U.S. codes and built by a U.S. company, also considering the similarities between the North Anatolian and San Andreas Fault systems; the damage experience in Tupras can take place in many petrochemical facilities in Western U.S. It should be also noted that the financial looses will be higher for a U.S. facility, considering the relatively-high cost of recovery and possible law suits due to environmental pollution. Also, as Tupras was the most important industrial facility in the region and other chemical facilities did experience relatively low damage, all the local and outside fire fighting efforts were concentrated in Tupras. This may not hold true for a petrochemical facility in Western U.S., where limited fire fighting capabilities can exist for a relatively long period of time.

The biggest portion of the fire in Tupras was started by the ignition of Naphtha in some of the floating roof tanks. Sloshing caused by seismic waves ignited the oil inside Naphtha floating roof tanks and caused fires that lasted for days during numerous earthquakes, the earliest being the 1964 Niigata earthquake and the latest being the 2003 Tokachi-Oki earthquake. This shows that, unfortunately, no lessons were learnt from the previous earthquakes in a 40-year period.

During sloshing, the floating roof is shaken vigorously by the earthquake together with the uneven settlement of the tank caused by the ground liquefaction, which resulted in ignition of the oil by sparks generated by the collision between the floating roof and the side wall (Figure 9). The sparks were generated by the metal touch sealing between the floating roof and the side wall. To prevent ignition during sloshing, soft touch sealing composed of synthetic rubbers or urethane foams should be used instead of the metal touch sealing. In addition, basic foundation of tanks should be sufficiently improved especially in the areas of ground liquefaction. The improved basic foundation suffers little damage from the uneven settlement.

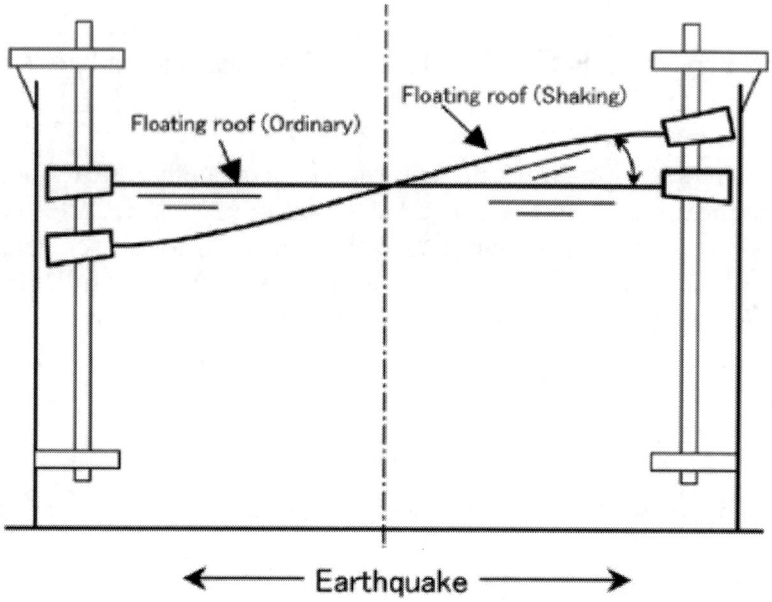

Figure 9. Sloshing of an Above Ground Storage Tank during an Earthquake (Akatsuka and Kobayashi)

5. References

[1] "USGS," http://earthquake.usgs.gov/ (accessed Feb. 1, 2007).

[2] USGS, "Implications for Earthquake Risk Reduction in the United States from the Kocaeli, Turkey, Earthquake of August 17, 1999," *U.S. Geological Survey Circular*, No. 1193, 2000.

[3] Sari, A., Manuel, L. "On the Use of an Energy-Based Ground Motion Parameter in Seismic Hazard Studies for Northwestern Turkey," *Proceedings of the 5th National Conference on Earthquake Engineering*, Istanbul, Turkey, Paper AE-040, 2003.

[4] Sari, A., "Energy Considerations in Ground Motion Attenuation and Probabilistic Seismic Hazard Studies," *PhD Thesis, University of Texas at Austin, Austin, Texas*, 2003

[5] Harmsen, S., Leyendecker E. V. "USGS Seismic-Hazard Web Products: Deaggregation & Ground-Motion Calculator," *Third ATC-35/USGS National Earthquake Ground-Motion Mapping Workshop*, California, 2006.

[6] Akatsuka, H., Kobayashi, H. (http://shippai.jst.go.jp/en/Detail?fn=0&id=CB1012035&) (Accessed Feb. 1, 2007).

Additional references not cited:

[7] Sezen, H., Whittaker, A. S. "Seismic Performance of Industrial Facilities Affected by the 1999 Turkey Earthquake," *Journal of Performance of Constructed Facilities*, Volume 20, No. 1, 2006.

[8] Johnson, G. S. "Refinery Damage and Emergency Response in the 1999 Izmit, Turkey Earthquake," Prevention First 2002, California.

[9] Scawthorn, C., Johnson G. S. "Preliminary Report; Kocaeli (Izmit) Earthquake of 17 August 1999," *Engineering Structures*, Volume 22, No. 7, 2000.

[10] Akinci, O., Kilic, S. "Non-linear responseof the 115 m Tupras refinery stack subjected to the 1999 Marmara Earthquake," *CICIND Report, International Committee on Industrial Chimneys*, Volume 18, No. 2, 2002.

LOSS PREVENTION SYMPOSIUM

The Interaction of Earthquakes with Process Equipment in the Framework of Risk Assessment

Ernesto Salzano
Istituto di Ricerche sulla Combustione
Consiglio Nazionale delle Ricerche
Via Diocleziano 328, 80125 Napoli, Italy
Phone: +39 0817621922 – E-mail: salzano@irc.cnr.it

Antonio Di Carluccio
Dipartimento di Ingegneria Strutturale
Università degli Studi di Napoli, Italy

Anita Garcia Agreda
Istituto di Ricerche sulla Combustione
Consiglio Nazionale delle Ricerche

Giovanni Fabbrocino
Dipartimento S.A.V.A.
Università del Molise, Campobasso, Italy

Gaetano Manfredi
Dipartimento di Ingegneria Strutturale
Università degli Studi di Napoli, Italy

Prepared for Presentation at
American Institute of Chemical Engineers
2006 Spring National Meeting
41st Annual Loss Prevention Symposium
Houston, TX
April 22–26, 2007

UNPUBLISHED

The Interaction of Earthquakes with Process Equipment in the Framework of Risk Assessment

Ernesto Salzano
Istituto di Ricerche sulla Combustione
Consiglio Nazionale delle Ricerche
Via Diocleziano 328, 80125 Napoli, Italy
Phone: +39 0817621922 – E-mail: salzano@irc.cnr.it

Antonio Di Carluccio
Dipartimento di Ingegneria Strutturale
Università degli Studi di Napoli, Italy

Anita Garcia Agreda
Istituto di Ricerche sulla Combustione
Consiglio Nazionale delle Ricerche

Giovanni Fabbrocino
Dipartimento S.A.V.A.
Università del Molise, Campobasso, Italy

Gaetano Manfredi
Dipartimento di Ingegneria Strutturale
Università degli Studi di Napoli, Italy

Abstract

Earthquakes can produce structural damages on industrial equipment. Whenever prevention and mitigation systems are ineffective, catastrophic accidental scenarios as fires, explosion and dispersion of toxic substances may occur, thus increasing risks to people, environment and properties either within the plant borders or in the close surroundings of industrial plants. This paper deals with a simplified procedure for seismic risk analysis of typical industrial installations containing atmospheric storage tanks, pressurised horizontal tanks and pumps, based on historical database and literature data on natural-technological (Na-tech) accidents. The methodology can be applied for any industrial equipment and implemented in any existing code or procedure for risk assessment. Results are presented in terms of fragility curves with respect to a single degree of freedom for earthquake characterisation (Peak Ground Acceleration, PGA), which is assumed as relevant earthquake intensity measure. Preliminary numerical validation of the methodology has been also carried out via structural analysis that takes into account a single failure mode (Elephant Foot Buckling) that affects atmospheric storage tank, anchored or unanchored to ground.

Starting from fragility data, probit analysis has been performed in order to obtain PGA thresholds values for earthquake-related structural damages and loss of containment. These results are certainly of interest for active and passive mitigation actions development, and for the implementation of Early Warning System, i.e., the set of actions that can be taken from the moment when a seismic event is triggered with a significant reliability to the moment the earthquake strikes in a given location.

1. Introduction

The interaction between earthquakes and industrial installations cannot be neglected when industrial risks are of concern (Na-Tech analysis). Any risk evaluation (e.g., Quantitative Risk Analysis or Probabilistic Safety Analysis) should then consider the consequences and probability of occurrence of accidental scenarios triggered by the natural event [1]. Quite obviously, sound Na-Tech procedure needs a multi-disciplinary effort, starting from the definition of probability of occurrence of earthquake intensity (i.e., Probabilistic Seismic Hazard Analysis or PSHA, see [2,3]), to structural analysis for the equipment interaction with seismic actions, and to industrial engineering, which analyses the specific response of the industrial process after structural damages of equipment shell or supports and evaluates risks in global terms. But that is not the only option. An interesting approach to Na-Tech consists of seismic Early Warning Systems (EWS), i.e., a set of actions that can be taken from the moment when a seismic event is detected with a significant reliability, to the moment the quake strikes in a given location [4]. Again, the implementation of EWS needs strong interaction with seismological infrastructures and competences, but can prevent disasters and mitigating effects of earthquakes by means of the activation of preventive effective countermeasures either for the structural damages (e.g. by safety interlock systems which limit or avoid the accidental release of relevant amount of gas or liquid flammable or toxic substances from damaged equipment) or for the mitigation of accidental scenarios.

This study is focused on steel structures for the storage of hazardous substances as atmospheric (anchored or unanchored) storage tanks, as well as pressurized cylindrical horizontal vessels containing liquefied gases and pumps. Evaluations are made with respect to Peak Ground Acceleration (PGA) for the characterisation of earthquake intensity. Finally, based on available data, PGA threshold values for structural damage and accidental scenarios are defined and discussed aiming at EWS implementation and for the identification of cry-wolf issues.

2. Earthquake characterisation and prediction

Measured ground motions refer to seismic waves radiating from the earthquake focus to the site and should be related to the earthquake source, to the path for the seismic wave from the source to site, and to the specific geomorphologic characteristics of the site where the Na-Tech is performed, in order to summarize all the random features of earthquakes, including energy, frequency contents, phases and others which may affect the structural response of structures. Currently, the problem of definition of good predictors for inelastic seismic behaviour of structures is one of the main topics of earthquake engineering. However, empirical vulnerability analyses are often carried out in terms of Peak Ground Acceleration (PGA), mainly because it is relatively easy to infer by earthquake intensity conversion. Furthermore, extensive historical database on damages are usually defined by PGA and transformation from typical earthquake magnitude (e.g., Modified Mercalli Scale or Richter Scale) to this variable are straight.

Given any location, Probabilistic Seismic Hazard Analysis (PSHA), i.e., the exceedance probability of PGA occurrence calculated over 1 year or 50 years basis, which is the reference

curve for structural design purposes (Technical Service Life, TSL) is now simply at reach of any analyst. Indeed, local authorities provide tools for PSHA both in Europe and USA (see for instance, http://www.usgs.gov]. On this subject, it is worth noting that from a structural standpoint, many industrial installations are characterized by specific design requirements and can become obsolete. Thus in many cases, the design reference period should be defined as Functional Service Life (FSL), which is generally lower than TSL (see ISO 15686-1 [5]).
When Early Warning System are needed, the time of arrival of seismic shaking from the focus point is necessary. Hence, more details on waves should be given. In the framework of industrial risk analysis only two types are of primarily interest: P (primary) waves, which travel faster and are the first waves to be recorded, and S (secondary) waves, that together with other type of waves (surface waves) produce most of damages and destruction. P waves can travel through any medium, whereas S waves can only travel through solids. Surface waves are slower than P and S waves and can only travel on the surface of the Earth.

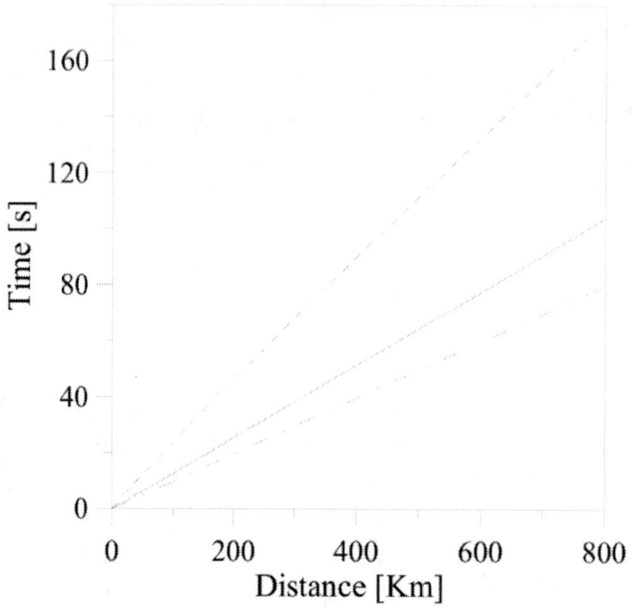

Figure 1 - Time of arrival of seismic waves with respect to distance of earthquake focus point [6]. – – – : S waves; —— : P waves; –·– : (S – P) waves

The speed of a seismic wave is not constant but is dependent upon many factors. Speed changes mostly with depth and rock type. P waves travel between 6 and 13 km/sec. S waves are slower and travel between 3.5 and 7.5 km/sec. The distance of seismic recording station to the earthquake epicentre can be know by using the times of travel of the S and P waves [4]. Figure 1 shows the time of arrival of seismic waves for any location at distance from focus point, which can be usefully adopted for early warning issues.

In the same figure, the difference on arrival time of P and S wave of an earthquake is also showed. In fact, several seconds may elapse between the arrival of the first P wave at the monitoring station and the arrival of the damaging S and surface waves and this time interval increases with the distance from the focus of the earthquake.

3. Fragility analysis for industrial equipment

The effects of seismic actions on any structure are generally related to structural damage and its reparability [7]. However, all typical accidental scenarios in the process industry (explosions, fires, toxic dispersions) depend basically on the total amount of released dangerous substance. Accordingly, seismic vulnerability of equipment should be given either in terms of structural limit states of interest or in terms of content release (or RS, release state). Therefore, in the mainframe of industrial risk analysis, existing data concerning post-earthquake damage observations for industrial equipment have to be reviewed in order to optimize the limit state classifications of equipment response [8]. Quite clearly, because of incomplete descriptions of the actual damage to equipment into empirical database considered, the definition of damage state DS and/or RS is somehow left to judgment.

The structural damage produced by seismic actions on the equipment may be referred as "damage state" (DS) [9]. According to HAZUS [7] damage classification, DS values may rank from DS1 to DS5 where the absence of damage is identified as DS1, slight damages to structures as DS2, moderate damages as DS3, extensive damages as DS4 and the total collapse of structure as DS5. The DS values is an alternative formulation of the classical "limit state" definition, which has been extensively used to evaluate from a structural perspective the economical effort needed to repair and restore the tank structures. On the other hand, all typical accidental scenarios in the process industry (vapour cloud explosions, flash fires, tank and pool fires or toxic dispersions) depend on the total amount of released dangerous substance [8,10]. Accordingly, seismic vulnerability should be expressed in terms of content release when industrial large scale QRA is of concern rather than exceedance of a given structural state of the equipment. For DS or RS, the probability of occurrence of any limit state is generally assessed by means of fragility curves F expressed in terms of lognormal cumulative distribution *f*, characterised by mean μ and standard deviation β:

$$F[\mathrm{RS\ or DS} \mid \mathrm{PGA}] = f(\mu, \beta, \mathrm{PGA}) \qquad (1)$$

In Eq. 1, PGA is the realization of the seismic intensity that is assumed to trigger the failure corresponding to the pre-assigned limit state. For the definition of threshold values, probit analysis for each limit state DS or RS can be usefully defined, as it allows fast recognition of any hazard by means of the two constants k_1 and k_2 as follows:

$$Y = k_1 + k_2 \ln(\mathrm{PGA}) \qquad \forall \mathrm{DS, RS} \qquad (2)$$

The probit variable Y is related to fragility by means of simple integration. Details of the entire statistical procedure are reported elsewhere [1,8,10,11]. Quite clearly, fragility F can be used in risk analysis by evaluating the probability of the system reaching any limit state for all possible values of seismic intensity (IM) combined with the probability of occurring of the specific seismic intensity.

$$\mathrm{Failure} = \bigcup_{i=1}^{\infty} \mathrm{Failure} \cap \mathrm{IM} \qquad (2)$$

Since collapse cannot take place for a given IM if another value already has led the system to failure, events in the previous equation are mutually exclusive and union of events probability is given by the sum of the probabilities:

$$P_f = \sum_{\text{All IM}} P[D > C \mid IM]\, P[IM] \qquad (3)$$

where P[IM] is the outcome of Probabilistic Seismic Hazard Analysis, the ground motion intensity measure IM is Peak Ground Acceleration (PGA) and D and C are respectively the seismic demand and the capacity of structure. In the following, the coefficients for the reproduction of fragility curves and probit variable for atmospheric storage tanks, pressurised equipment of different design and fill level, reactor and pumps for pressurised liquid are showed, on the basis of "experimental" observation and literature data for PGA lower than 2 g.

3.1. Atmospheric storage tanks

Dynamic behaviour of atmospheric storage tanks when subjected to earthquake is characterised by two predominant modes of vibration: the first is related to the mass that rigidly moves together with the tank structure (impulsive mass), the other corresponds to the liquid sloshing (convective mass) [12]. Seismic response of steel tanks depends however on complex fluid/structure interaction that may result in global overturning moments and base shear induced by horizontal inertial forces. Overturning moment causes an increase of the vertical stress in the tank wall and even uplift of the base plate, while the base shear can lead to relative displacements between the base plate and the foundation.

Failure modes reflect these specific aspects of the seismic demand on the structure and depend basically upon the type of interface at the tank base and the presence of mechanical devices are used to ensure an effective connection between the base plate and the foundation (un-anchored or anchored). When unanchored tanks are of concern, the friction at the base is able to ensure the needed stability of the structure under environmental actions, i.e., wind, but can be ineffective when strong ground motions take place, thus generating large relative displacements. Indeed, tank sliding reduces the maximum acceleration suffered by the equipment, however relatively small frictional factor may produce large relative displacements, hence large deformations and even failure of piping and connections can occur. In addition, another large-displacement mechanism is the partial uplift of the base plate. This phenomenon reduces the hydrodynamic forces in the tank, but can increases significantly the axial compressive stress in the tank wall and the possibility that a characteristic buckling of the wall (Elephant Foot Bucking – EFB) occurs. EFB is normally associated with large diameter tanks with height to radius (H/R) ratios in the range 2 to 3, whereas another common buckling mode, known as diamond shape buckling (DSB), is associated with taller tanks, that is H/R ratios about 4. While EFB is associated with an elastic-plastic state of stress, the DSB is a purely elastic buckling. Other structural damage are the collapse of support columns for fixed roof tanks, tank failures due to foundation collapse, splitting and leakage associated only with bolted and riveted tanks [13]. Liquid sloshing during earthquake action produces several damages by fluid–structure interaction phenomena and can result as the main cause of collapse for full or nearly full tanks. Historical analysis and assessment of seismic damages of storage tanks have shown that only full (or near full) tanks experienced catastrophic failures. Low H/R tanks only suffered cracks in conical roof connection, or damage by floating panel sinking. A very common shell damage is the EFB. For unanchored tanks and H/R < 0.8, EFB is typically not experienced but the base plate or the shell connection can fail causing spillage [14]. A full stress analysis is certainly the more accurate way to design and to evaluate the risk of steel tanks under earthquake loads. This approach leads to the direct computation of the interaction between shell deformations and content motion during earthquakes [15]. For base constrained and rigid tanks (anchored), a complete seismic analysis requires solution of Laplace's equation for the motion of the contained liquid, in order to obtain the total pressure history on the tank shell during earthquakes (Eurocode 8 [16]). When flexible tanks are considered, a structural deformation term must be also added to take account of the "impulsive" and "convective" contributions. Actually, the quantitative assessment of risk within a complex industrial installation needs the analysis of a large number of components. Hence, in the light of simplification, statistical and empirical tools derived from post-accident analysis are useful to define easy to manage and general vulnerability functions. As already mentioned above, existing data concerning post-earthquake damage observations for steel tanks have been reviewed in order to optimize the limit state classifications of equipment response [1,8]. Then DS values have been reviewed as three levels of intensity of loss of containment, defined as RS (Risk State): no loss – RS1, moderate loss – RS2, extensive loss of containment – RS3. Results of observational analysis based on large number of water storage tank are reported in Table 1 and Table 2, depending on the design of tank and its fill level. Details are reported elsewhere [1,8]. In particular Table 1 and Table 2 reports the coefficients μ and β of cumulative log-normal distribution for the probability of occurrence of RS limit state for anchored and unanchored storage tank respectively, and the correspondent coefficient k_1 and k_2 for probit function, with respect of PGA expressed in terms of g (acceleration of gravity). Threshold values for the PGA for each limit state (PGA_k) are also reported for the sake of EWS definition.

Table 1. Seismic fragility and probit coefficients for anchored atmospheric steel tanks.

Limit State (RS)	*Fill level*	*μ (g)*	*β (g)*	k_1	k_2	PGA_k *(g)*
≥ 2	Near full	0.30	0.60	7.01	1.67	0.074
3	Near full	1.25	0.65	4.66	1.54	0.275
≥ 2	≥ 50%	0.71	0.80	5.43	1.25	0.110
3	≥ 50%	3.72	0.80	3.36	1.25	0.577

Table 2. Seismic fragility and probit coefficients for unanchored atmospheric steel tanks.

Limit State (RS)	*Fill level*	*μ (g)*	*β (g)*	k_1	k_2	PGA_k *(g)*
≥ 2	Near full	0.15	0.70	7.71	1.43	0.029
3	Near full	0.68	0.75	5.51	1.34	0.118
≥ 2	≥ 50%	0.15	0.12	20.83	8.35	0.113
3	≥ 50%	1.06	0.80	4.93	1.25	0.164

A first comparison of the analysis is given by Talaslidis [17] by numerical analysis for atmospheric storage tank, 50% filling level, with volume of about 37000 m^3. Similarly to the present approach, five limit states for the structural damage were defined and fragility curves were produced, one for each limit state, although no reference is made to the loss of containment. The basic failure modes considered by Talaslidis is material yielding, whereas local buckling (EFB) and other failure modes as sliding, sloshing and uplifting have been neglected. As a consequence, the values obtained as threshold PGA are higher than those obtained by means of historical analysis, which considers all possible failures.

Results reported in Tables 1-2 can be used as the starting point for the prediction of accidental scenarios as fire (pool fire, flash fire, tank fire), explosion (in the case of formation large vapour cloud) or, when toxic vapour are formed, for the dispersion analysis.

3.2. Pressurised horizontal tank and reactors

Although a large amount of data and analyses are reported in literature for atmospheric storage tank, a few information are found for pressurised equipments and reactor vessels, which may be useful for complete risk assessment. Results reported in Tables 3-5 have been derived from the data obtained by Seligson et al. [18]. In this work, Mercalli-modified intensity scale (MMI) to PGA transform is required and obtained according to Trifunac equation [19] as reported in ATC-13 [20]:

$$\log(\mathrm{PGA}) = 0.014 + 0.3\ \mathrm{MMI} \tag{4}$$

Table 3. Seismic fragility and probit coefficients for pressurised horizontal steel storage tanks.

Limit State (DS)	*Limit State (RS)*	μ *(g)*	β *(g)*	k_1	k_2	PGA_k *(g)*
≥ 1	≥ RS1	0.83	0.99	5.36	1.01	0.069
≥ 2	≥ RS2	1.85	0.85	4.50	1.12	0.196
3	= RS3	4.91	0.84	3.39	1.12	0.526

Table 4. Seismic fragility and probit coefficients for pressurised reactor.

Limit State (DS)	*Limit State (RS)*	μ *(g)*	β *(g)*	k_1	k_2	PGA_k *(g)*
≥ 1	≥ RS1	0.79	0.92	5.46	1.10	0.080
≥ 2	≥ RS2	2.02	0.78	4.36	1.22	0.249
3	= RS3	6.35	0.95	3.30	0.99	0.526

Table 5. Seismic fragility and probit coefficients for pumps.

Limit State (DS)	*Limit State (RS)*	μ *(g)*	β *(g)*	*k1*	*k2*	*PGAk (g)*
≥ 2	RS2	0.81	1.29	5.31	0.77	0.032
3	RS3	2.44	1.00	4.30	1.00	0.195

It's worth noting that in the cases reported above, DS1 corresponds to slight movement of tank support or low probability of failure of some connecting pipe, with negligible release of content (RS1), whereas DS2 corresponds to failure of most connection pipes and tank support system, with high likelihood of release of tank content from pipes (RS2). Finally, DS3 and RS3 correspond to total failure of tank (with buckling). It's worth noting that in the case of pressurised flammable or toxic gases, even small crack on tank surface would mean large release of pressurised gas and catastrophic accidental scenario.

4. Fragility analysis by means of structural modelling

Early investigations on the seismic behaviour of atmospheric storage tanks were carried out by Housner [21]. The author, based on experimental and theoretical observations, proposed a simplified model for seismic analysis of anchored tanks with rigid walls. It was assumed that when a tank is subjected to horizontal ground acceleration (with free liquid surface), only a fraction of the total liquid volume participates to the motion as rigid mass, while the remaining fraction starts to oscillate, thus determining dynamic forces on the tank walls. This force is equivalent to the one applied to a lumped mass, known as convective mass, laterally restrained by a horizontal spring. Newmark & Rosenblueth (1971) [22] modified the expression initially suggested by Housner to estimate the convective and rigid masses and gave updated formulations for the evaluation of the seismic design forces of liquid storage tanks. In 1983, Haroun [23] developed a model to evaluate of the seismic response of storage tanks including the deformation of the wall. In this model a part of the liquid moves independently of tank shell, again convective motion, while another part of the liquid oscillates at unison with the tank. If the flexibility of the tank wall is considered, a part of this mass moves independently (impulsive mass) while the remaining accelerates back and forth with the tank (rigid mass). Figure 2 shows the idealised structural model of a liquid storage tank. The lumped masses model can take account of damping, depending on the selected mass. However it is neglected in the following analysis, without any loss of generality and relevance of the results. Total liquid volume can be divided into the following lumped masses: m_c convective, m_i impulsive and rigid m_r masses.

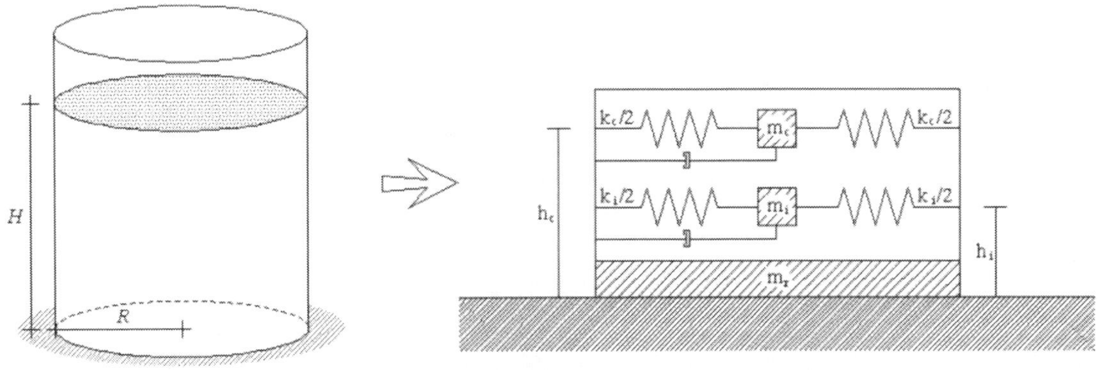

Figure 2 - Lumped masses model for atmospheric storage tank. m_c convective, m_i impulsive and rigid m_r masses.

The convective and impulsive masses are connected to the tank wall by different equivalent spring having stiffness k_c and k_i, respectively. This model for anchored storage tank has been extended to analyse unanchored and even base-isolated liquid storage tanks [24]. The effective masses are defined in terms of liquid mass m in Eqs. (2-5) where Y_c, Y_i, and Y_r are the function of the H/R ratio, which is a filling coefficient and ρ_w is the liquid's specific weight:

$$m_c = mY_c \quad (5)$$

$$m_i = mY_i \quad (6)$$

$$m_r = mY_r \quad (7)$$

$$m = \pi R^2 H\rho_w \quad (8)$$

The natural frequencies of convective mass, ω_c and impulsive mass, ω_i are given by Eqs. (6) and (7):

$$\omega_c = \sqrt{1.84\left(\frac{g}{R}\right)\tanh\left(1.84\frac{H}{R}\right)} \quad (9)$$

$$\omega_i = \frac{P}{H}\sqrt{\frac{E}{\rho_s}} \quad (10)$$

where E and ρ_s are the modulus of elasticity and density of tank wall respectively and P is a dimensionless parameter which is a function of liquid height to radius tank ratio H/R. This is the structural model herein used for the incremental dynamic analysis (IDA) aimed at the estimation of the seismic demand. The investigation was repeated changing the geometry (three different volume capacity were considered) of the tank and then considered filling ratio (Table 6). Besides, for unanchored tank the investigations were performed depending on friction coefficients.

The first step in the IDA analysis (Figure 3) procedure consisted of the acquisition of suitable set records. For the analysis presented in the paper a set of 300 European records was considered. The selected earthquake ground motions records are all stiff soil records with a broad range of magnitude and distances. All accelerogram used in the present study have been retrieved from the European Strong Motion Database (www.isesd.cv.ic.ac.uk). To obtain the seismic demand at selected ground motion intensity levels, seismic records are scaled in terms of Peak Ground Acceleration (PGA).

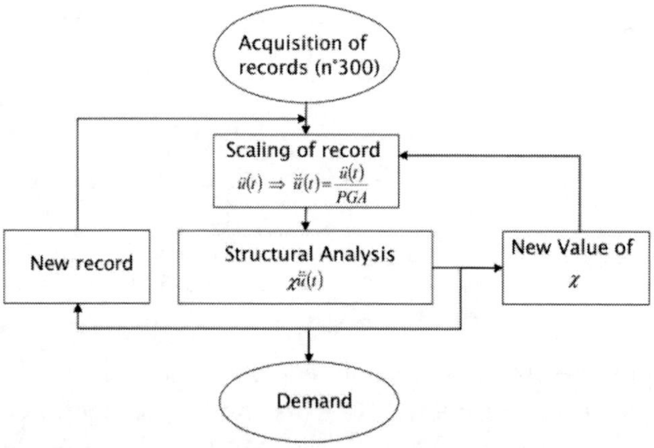

Figure 3 - The incremental dynamic analysis (IDA).

The scaling factor χ varies to get the PGA from 0.05 g to 2 g. A procedure to solve the equation of motion was implemented in a MatLab computer code using the Wilson theta method. Results of present analysis are reported in Table 6, assuming Eurocode 8 [16] guidelines to evaluate the ultimate stress. This approach will be extended to all relevant failure modes of atmospheric storage tanks reported above. Similar results are obtained using different guidelines for the

ultimate stress estimation as that reported by Kim & Kim (2002) [25]. From a structural perspective, it is worth to analyse the results of the structural dynamic analysis carried out to derive vulnerability functions for the anchored storage tanks reported in Figure 3. Probability of failure is plotted on the right side depending on PGA, which in turns affects the demand of the structure as demonstrated by the left plot. The latter shows numerical data and trends (in terms of average and average ± standard deviation) of the demand in terms of vertical shell stress and confirms that a correlation between parameters exists. As a consequence, vulnerability evaluations carried out in terms of PGA can be effectively used in QRA contexts. In the next future, similar analysis will be performed for any failure mode for either anchored or un-anchored storage tanks and for pressurised equipment (either horizontal or spherical).
On the right of the Figure 4, the observational fragility (F_{obs}) and the relative ratio of EFB as calculated in this work (F_{calc}) with respect to the observational fragility are also showed . The specific failure mode here analysed is more effective at high PGA whereas at low PGA. Similar effect can be showed for any tank described in Table 6, where the threshold values reported and the coefficients of probit function show clearly that the greater the tank size, the lower the value of PGA_k and the EFB contribution to the collapse of tank at least at low values of PGA. Finally, the higher the fill level, the higher the effect of EFB and the value of PGA_k.

Table 6. Seismic fragility and probit coefficients for EFB of different atmospheric anchored tanks at different fill level for structural damage level DS2 and loss of containment level RS2. PGA_k is the threshold value for the damage state.

V [m³]	*Tank Height (m)*	*Tank Radius (m)*	*Filling level (%)*	*μ (g)*	*β (g)*	*k₁*	*k₂*	*PGA_k (g)*
250	7.55	3.25	50	DS1	DS1	DS1	DS1	DS1
			80	2.55	0.57	3.05	1.61	0.50
5000	10.8	12.25	25	DS1	DS1	DS1	DS1	DS1
			50	DS1	DS1	DS1	DS1	DS1
			80	1.74	0.76	4.53	1.34	0.25
30000	18.5	22.75	25	DS1	DS1	DS1	DS1	DS1
			50	2.08	0.79	4.33	1.24	0.26
			80	0.78	0.97	5.46	1.05	0.07

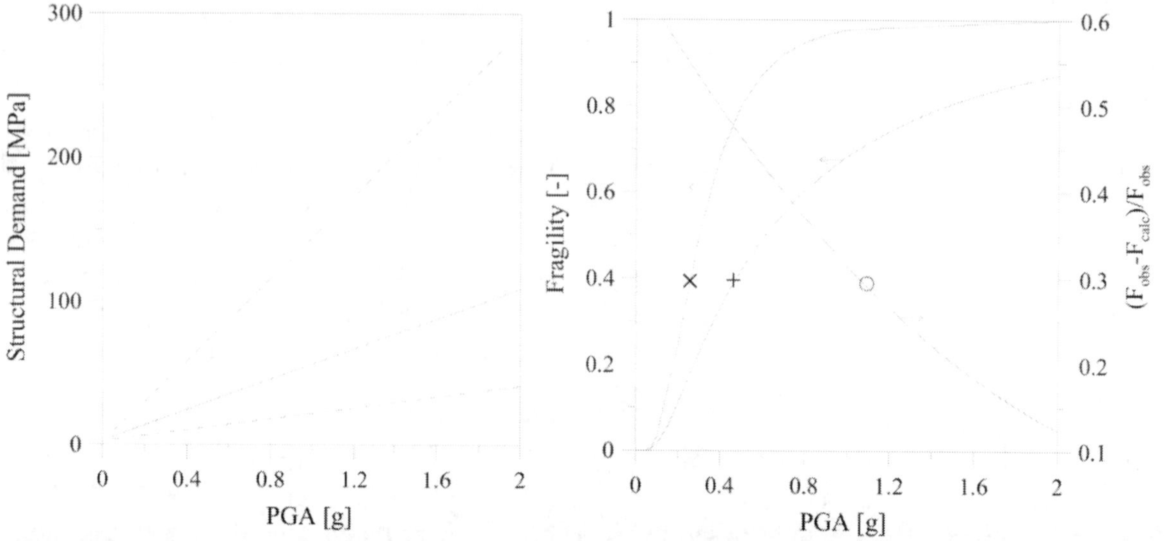

Figure 4 - Left: Structural demand (vertical shell stress) of anchored atmospheric storage tank with V = 30000 m^3, fill level = 80%, with respect to PGA. Dotted line represents standard deviation. Right: For the same tank, probability of failure in terms of calculated (F_{calc}, +) and observational (F_{obs}, ×) fragility and relative contribution of EFB over the total fragility (○) with respect to PGA.

5. Seismic Early Warning Systems for industrial installation

Early warnings of earthquake are recognized as a focal point for the prevention of nuclear accidents, whereas little attention is given to the seismic safety management in the wider case of chemical processes where large inventories of hazardous chemicals are stored.

On the other hand, providing a reliable warning on the order of a few seconds to a minute before ground motion arrives is now possible as showed in Figure 5 [26]. Hence, action addressed to reduce significantly equipment damage, losses of containment for systems, tanks and reactors, and eventually accidental scenarios which may affect the surrounding structures and population, may be specifically designed.

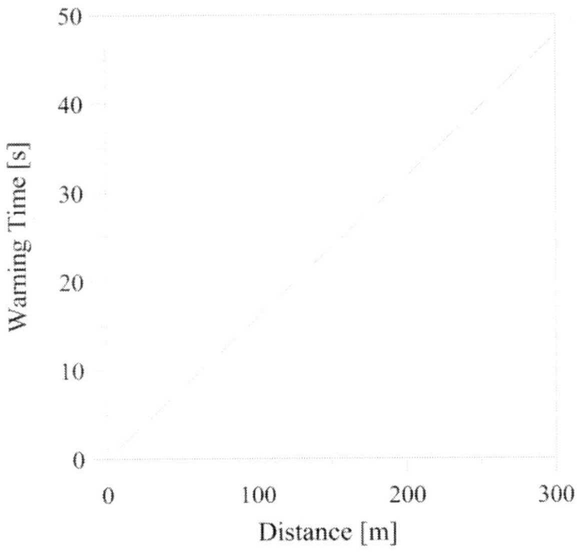

Figure 5 - Warning time varies with respect to epicentre distance [26].

To this regard, active protection system as for instance safety interlock systems (SIS) are normally installed for the shut-off of plants for relatively fast, process-related loss of control, whereas the time allowed for the seismic wave to reach the installation location are rarely considered for mitigation of earthquake effects on plants, processes and storage systems. A safety interlock system, or any emergency shutdown system, is one of the most important protective measurements in process plants, and provides automatic actions to correct an abnormal plant event that has not been controlled by basic control systems and manual interventions [27,28]. The design and maintenance of SIS present a special challenge created by the infrequent need for these systems to act; i.e., a SIS is only needed on those rare occasions when normal process controls are inadequate to keep the process within acceptable bounds [29]. In the normal process-life, SIS failures are likely to be followed by environmental damage or loss of human lives.

The main aspect of EWS to take into account for the development of mitigation actions for earthquake and its relation with process industry is the characteristic time for earthquake scenario to strike in a given location, which is now defined as τ_{WARN}, to be correlated to the characteristic time of EWS action, τ_{ACT}. A rational approach to the problem can be based on the definition of an adimensional number for process industry (EPW) given by the ratio of these two characteristic times and which allows addressing safety management to mitigation actions:

$$EPW = \alpha \frac{\tau_{WARN}}{\tau_{ACT}} \tag{11}$$

where α is a coefficient which depends on the earthquake intensity parameter as discussed in the following. Quite clearly, the greater the EPW the stronger the possibility of preventing damages to plant installations and possibly mitigating the effects of earthquake. Table 7 reports the ranges of value which characterize the possibility and the effectiveness of earthquake mitigation action on the process plant.

The definition of both characteristic times is the key point of this analysis. Indeed, both times are strongly dependent on location, type of process, and often many unpredictable figures. Large variations are then expected. To this regard, a main issue is that only industrial accidents which are followed by the release of large amount of hazardous materials or energy into the surrounding environment will be considered. To be more specific, our purposes is not addressed to the mere limitation of effects of earthquakes (the structural damage to equipment) but to the mitigations of effect on human body due to toxic dispersion of vapour or gas substances, due to fire of liquid releases on the ground or the fire and explosion of flammable clouds formed after the dispersion of gas or vapour from the system of containment. Finally, following the definition of EPW, it's likely that spurious shutdowns may result.

Table 7: Early Warning System correlated by means of the EPW number, the type of action and the earthquake intensity.

EPW	Type of action	Earthquake intensity
<< 1.0	Short arrival time of earthquake and/or slow mitigation action	Earthquake intensity does not affect the mitigation action
≈ 1	Long arrival time of earthquake and/or little time needed for mitigation action	Earthquake intensity affects the mitigation action if multiple failures of containment systems and mitigation safety auxiliary systems (e.g. sprinkler systems) are affected. However, several action prior to the arrival of earthquake are possible to avoid further consequences
>> 1	Long arrival time of earthquake and/or fast and effective mitigation action	Earthquake intensity and mitigation action are independent. Effective actions prior to the arrival of earthquake are possible

The time of arrival of earthquake τ_{WARN} has been previously discussed and it is not dependent on the analysed equipment. On the other hand, the time for action after earthquake alarm depends on many factors and it's clearly dependent on the type of process considered. Indeed, quite obviously, each process has its own peculiarity and general assumptions are often not applicable. In the case of atmospheric fuel storage tanks, unless earthquake magnitude is so high that the rapid catastrophic failure of one or more tanks is observed, the release of liquid after the failure is typically slow and potentially mitigable. The shutdown of operation should be directed essentially to the lock of operation of loading/unloading of fuels from/to tank truck or boat and the SIS consists simply of safety valves which section pipelines. After release, fires related to the presence of fuel in catch basin are the most credible events, whereas explosions are very unlikely because early ignition of vapour cloud is likely in case of earthquakes. Eventually, the typical time of 10 min for safety valve isolation (see for instance [30]) can be considered. Moreover, if major losses of containment are of concern, mitigation interventions (e.g., foam) for the slow release are always possible in the catch basins. The foaming action is of course difficult if the earthquake is so energetic that the entire system has failed structurally. In the case of pressurised equipment or reactor, the time for response is definitely smaller, as automatic safety valve are typically installed. Hence, τ_{ACT} of 3 min can be adopted [30]. Finally, for the definition of EPW, the factor α takes into account earthquake intensity. To this regard, the threshold values defined above for the loss of containment (RS)) can be adopted for the effectiveness of early warning action. The value of α can be then considered as a function of PGA following, in first approximation, the fragility function. Hence:

$$\alpha = \begin{cases} \rightarrow \infty & PGA \ll (PGA_k)_{RS\geq 2} \\ f(PGA, \mu, \beta) & \\ \rightarrow 0 & PGA \gg (PGA_k)_{RS=3} \end{cases} \tag{12}$$

where μ and β are the median and the standard deviation and f is the reciprocal of the log-normal standard distribution for RS2 and RS3 with respect to PGA.

The value of $\alpha = 0$ means clearly that no intervention keeps the plant in a safe state as it exists a probability greater than zero for the total destruction of equipment or even for the entire process plant, including auxiliary system and fire system. Quite clearly, when very high values of α are of concern, the EPW number is close to zero, thus representing no damaging earthquakes in terms of loss of containment.

The following Figure 6 reports the EPW number for atmospheric storage tank taking into account a number of different earthquake arrival times and considering a manual, long operating time safety valve (10 min), which is closed after the earthquake alarm.

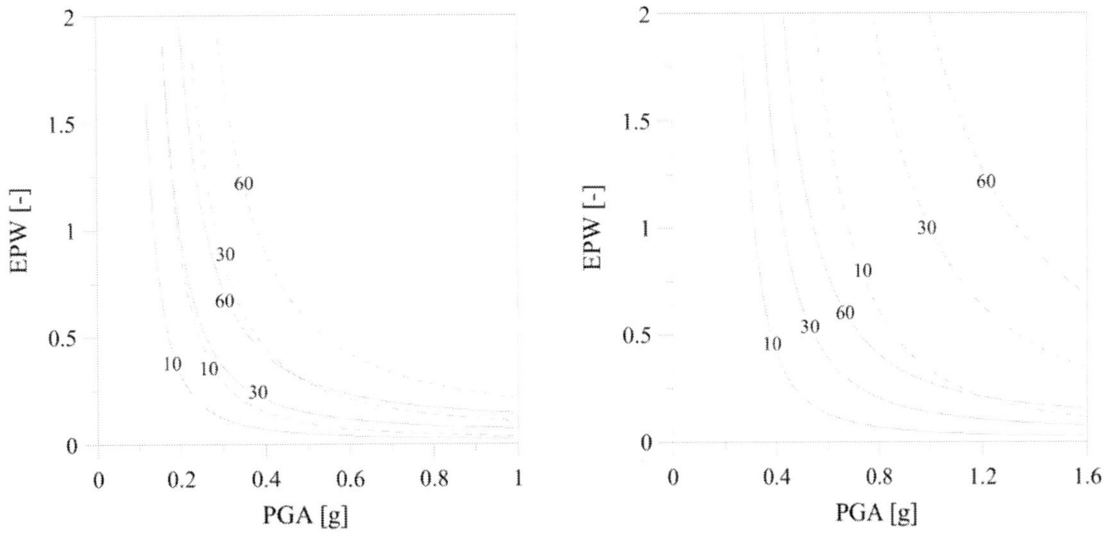

Figure 6 - Calculated EPW for unanchored (left) and anchored (right) atmospheric storage tank by varying earthquake intensity and time of arrival of seismic wave from 10 s to 60 s to the installation location. Time of intervention = 10 min.

Results show that for un-anchored tanks, as expected, process safety and effectiveness of warning systems are strongly dependent on earthquake intensity when long-duration (manual) intervention are present, i.e., damage and loss of containment threshold values for PGA identify the level of effectiveness, regardless the time of arrival of earthquake.

5. Conclusions

Results showed the effectiveness of EWS number to identify priorities when earthquake mitigation systems are needed. For anchored atmospheric storage tank, the use of automatic fast safety systems (sprinklers, foaming) would allow the almost absolute safety when low intensity earthquake are likely. Cry-wolf issues can be overtaken by threshold values for PGA, with respect to any limit state (risk state RS) for the release of containment.

6. Acknowledgments

The authors wish to thank Dipartimento della Protezione Civile della Presidenza del Consiglio dei Ministri for financial support either for Na-Tech Center of Competence or RELUIS Line 9 Research Project "Monitoraggio ed early warning di strutture ed infrastrutture strategiche".

7. References

[1] G. Fabbrocino, I. Iervolino, F. Orlando, E. Salzano, Quantitative risk analysis of oil storage facilities, in seismic areas, *J. Hazard. Mater.*, 12 (2005), 361-69.

[2] C.A. Cornell, Engineering seismic risk analysis, *B. Seismol. Soc. Am.*, 58 (1968), 1583-1606.

[3] J.J. Bommer, Deterministic vs. probabilistic seismic hazard assessment: an exaggerated and obstructive dichotomy, *J. Earthquake Eng.*, 6 (Special Issue No. 1), (2002), 43-73.

[4] I. Iervolino, V. Convertito, M. Giorgio, G. Manfredi, and Zollo, A., Real-time risk analysis for hybrid earthquake early warning systems, *J. Earthquake Eng.*, 2007, in press.

[5] ISO 15686-1, ISO/TC 59/SC 14, *Design life, Buildings and Constructed Assets Service Life Planning – Part 1 General Principles*, 2000.

[6] F.J. Simons, B.D.E. Dando, and R.M. Allen, Automatic detection and rapid determination of earthquake magnitude by wavelet multiscale analysis of the primary arrival, *Earth and Planetary Science Letters*, 250 (2006), 214 – 223.

[7] HAZUS, *Earthquake Loss Estimation Methodology*, National Institute of Building Science, Risk Management Solutions, Menlo Park, CA, 1997.

[8] E. Salzano, I. Iervolino and G. Fabbrocino, Seismic risk of atmospheric storage tanks in the framework of quantitative risk analysis, *J. Loss Prevent. Proc.*, 16 (2003), 403–409.

[9] M.J. O'Rourke, M. Eeri, and P. So, Seismic fragility curves for on-grade steel tanks. *Earthquake spectra*, vol. 16, 4 (2000), 801 – 815.

[10] F.P. Lees, *Loss Prevention in the process industries*, II ed., Butterworth-Heinemann, Oxford, 1996.

[11] D.J. Finney, *Probit analysis*, Cambridge University Press, 1971.

[12] P.K. Malhotra, T. Wenk, and M. Wieland, Simple procedure for earthquakeic analysis of liquid-storage tanks, *Structural Engineering International*, 3 (2000), 197 – 201.

[13] ASCE, American Society Of Civil Engineers, *Guidelines for seismic evaluation and design of petrochemical facilities*, Task Committee on seismic evaluation and design of petrochemical facilities of the Petrochemical Committee, Reston (Virginia), 1997.

[14] D. Ballantyne and C. Crouse, Reliability and restoration of water supply systems for fire suppression and drinking following earthquakes, *National Institute of Standards and Technology*, NIST GCR (1997), 97 – 720 .

[15] M.A. Haroun, Implications of recent nonlinear analyses on earthquake standards of liquid storage tanks, *Proceedings of 5th US conference on lifeline earthquake engineering*, TCLEE 16, ASCE Seattle, USA, (1999).

[16] Eurocode 8, *Design of structures for earthquake resistance - Part 4: silos, tanks and pipelines*, UNI ENV 1998, 1998.

[17] D.G. Talaslidis, G.D. Manolis, E. Paraskevopoulos, C. Panagiotopoulos, N. Pelekasis, J.A: Tsampoulos, Risk analysis of industrial structures under extreme transient loads, *Soil dynamics and earthquake engineering*, 24 (2004), 435 – 448.

[18] H.A. Seligson, R.T. Eguchi, Tierney, K.J. and K. Richmond, *Chemical Hazards, mitigation and preparedness in areas of high seismic risk; a methodology for estimating the risk of post-earthquake hazardous materials release*, Technical Report NCEER-96-0013, National Center for Earthquake Engineering Research, State University of NY at Buffalo, 1996.

[19] M.D. Trifunac, A Note on the Range of Peak Amplitudes of Recorded Acceleration, Velocities and Displacements with respect to the Modified Mercalli Intensity Scale, *Earthquake Notes*, 47 (1976), 9-24.

[20] ATC-13, *Earthquake Damage Evaluation Data for California*, Redwood, CA: Applied Technology Council, 1985.

[21] G.W. Housner, Dynamic behaviour of water tanks, *B. Seismol. Soc. Am.*, 53, (1963), 381 – 387.

[22] N.M. Newmark, and E. Rosenblueth, *Fundamentals of Earthquake Engineering*, Englewood Cliffs, NJ: Prentice-Hall, 1971.

[23] M.A. Haroun, Vibration studies and test of liquid storage tanks, *Earthquake Eng. Struc*, 11 (1983), 179-206.

[24] M.K. Shrimali, and R.S. Jangid, Seismic response of liquid storage tanks isolated by sliding bearings, *Engineering Structures*, 24 (2002), 907-919.

[25] S. Kim and C. Kim, Buckling strength of the cylindrical shell and tank subjected to axially compressive loads, *Thin-Walled Structures*, 40 (2002), 329–353.

[26] R.R. Leach, and F.U. Dowla, Earthquake Early Warning System Using Real-time Signal Processing, Lawrence Livermore National Laboratory, UCRL-JC-123270, 1996.

[27] CCPS, Center for Chemical Process Safety, *Guidelines for Safe Automation of Chemical Processes*, American Institute of Chemical Engineers, New York, 1993.

[28] D.L. Green and A.M. Dowell, How to design, verify and validate emergency shutdown systems, *ISA Transactions*, 34 (1995), 261-272.

[29] P.W. Chung, S.H. Yang and C.H. He, Conceptual design of pressure relief systems, *J. Loss Prevent. Proc.*, 13 (2000), 519 – 526.

[30] C.J.H. Van Den Bosh, and Weterings R.A.P.M., Methods for the calculation of physical effects (*Yellow Book*), Committee for the Prevention of Disasters, The Hague (NL), 1997.

LOSS PREVENTION SYMPOSIUM

Mitigating Hurricane Storm Surge Perils at the DuPont DeLisle Plant

Steven P. Harris
Vice President
ABS Consulting
475 14th St. Suite 550, Oakland, CA 94612 U.S.A.
Phone: 510.817.3105
spharris@absconsulting.com

Duane O. Wilson
Global Project Manager
DuPont Titanium Technologies DeLisle Plant
7685 Kiln DeLisle Road, Pass Christian, MS 39571 U.S.A.
Phone: 228.255.2180
duane.o.wilson@usa.dupont.com

Prepared for Presentation at
American Institute of Chemical Engineers
2007 Spring National Meeting
41st Annual Loss Prevention Symposium
Houston, TX
April 22–26, 2007

UNPUBLISHED

Mitigating Hurricane Storm Surge Perils at the DuPont DeLisle Plant

Steven P. Harris
Vice President
ABS Consulting
475 14th St. Suite 550, Oakland, CA 94612 U.S.A.
Phone: 510.817.3105
spharris@absconsulting.com

Duane O. Wilson
Global Project Manager
DuPont Titanium Technologies DeLisle Plant
7685 Kiln DeLisle Road, Pass Christian, MS 39571 U.S.A.
Phone: 228.255.2180
duane.o.wilson@usa.dupont.com

Abstract

Hurricane Katrina in 2005 caused significant damage to many refineries, chemical plants, and other industrial and harbor facilities along the Gulf Coast causing catastrophic levels of damage. DuPont's DeLisle Plant in Pass Christian, Mississippi, was one of many facilities impacted by this event. DuPont has in place significant operational measures to respond to hurricane events, and post-Katrina has taken measures to mitigate the effects of possible future recurrences through the construction of a higher storm levee. DuPont DeLisle Plant's fast track project to provide a significantly greater level of storm surge protection as well as analyses undertaken to define the extent of the storm surge hazard and design the physical levee improvements are described.

1. Impact and Recovery of Hurricane Katrina on the DeLisle Plant

Over the past half century, development along the Gulf Coast has resulted in substantial concentrations of commercial, residential, and heavy industrial infrastructure within the range of hurricane and storm surge perils. This coastal development has included construction of power and chemical plants, refineries, and ports and terminals in direct proximity to these perils. Hurricane winds and storm surge from Katrina in 2005 resulted in significant damage to many of these industrial and harbor facilities along the Gulf Coast causing many hundreds of millions of dollars in damage. DuPont's DeLisle Plant in Pass Christian, Mississippi, was one of many facilities impacted by Hurricane Katrina's winds and storm surge inundation.

Located adjacent to the Bay of St. Louis, the DeLisle Plant produces titanium dioxide ($TiO2$). The DeLisle Plant is a major employer and source of economic benefit to the Mississippi Gulf Coast. In Mississippi, the DeLisle Plant generates over 2800 jobs, with $191 million of annual income, and provides almost $13 million in annual tax revenues. Storm damage and limitations to the operation of the DeLisle Plant has significant impacts to both DuPont and to the surrounding Gulf Coast communities.

The DeLisle Plant was built in 1978 and was protected by an earthen levee to 20 feet above sea level. The 20 foot levee design was based on benchmark data from Hurricane Camille in 1969. The bulk of the operating Plant is built to elevations 10-15 feet above sea level. The levee lies between the Plant and the Bay of St. Louis to the South placing the site inside a bowl that will retain water that overtops the protective levee or falls as a result of storm precipitation. The Federal Emergency Management Agency (FEMA) provided a recording of Katrina storm surge near the DeLisle site of 25.1 feet. Estimates of the storm surge levels at the Plant site indicated surge of 26-28 feet above sea level. As a result, a significant portion of the Plant and facilities were inundated with water levels as high as 16-18 feet.

Damage to the Plant from Katrina can be summarized in three major categories:

- Damage caused by high velocity water created when storm surge overtopped the levee
- Damage caused by gentler, rising still water after water levels equalized
- Hurricane force wind damage

While detailed analysis has not been done to differentiate the three types of damage, there are ample examples of each causative agent of damage.

1.1 Damage from moving water

Primarily isolated to the southernmost part of the Plant near the edge of the levee, this type of damage is indicated by significant structural damage to certain buildings, washout of railroads, and movement of railcars. In addition, some damage was caused by floating debris washing into the Plant and impacting Plant facilities. Figures 1 and 2 show damage caused by high velocity water and wave action.

1.2 Damage from rising still water

Rising still water was responsible for the most widespread damage at the site. Rising water entered buildings and other structures and caused anything not well anchored to float. Complete destruction of the internal assets of office buildings resulted from still water inundation. Offices, control rooms, computer facilities, and labs all suffered similar effects. While in general walls survived, damage from rising water required complete replacement and/or refurbishment of facility interiors.

Still water was also the primary cause of damage to the Plant's electrical infrastructure. Because the site topography results in somewhat of a bowl, storm surge water levels rose and overtopped the levee. The water receded as the storm surge subsided, but only to the top of the levee at the 20 feet elevation. Water in portions of the Plant at lower elevation of the bowl, below 20 feet, remained until the water on the site could be tested and safely drained. Water stayed in offices, electrical control rooms, instrument control rooms, etc. for many hours after the hurricane force winds dissipated. Refurbishment of electrical equipment after this exposure was not practical. Figures 3 through 4 show damage caused by rising still water inundation at the Plant.

1.3 Wind Damage

Due to factors of safety built into design codes and DuPont Design and Construction Standards, wind damage was limited at the DeLisle site. The primary damage caused by wind in process areas was to the cooling tower structures. Similar cooling tower damage was also reported at other industrial sites along the Gulf Coast. In addition to the cooling towers, some wind damage did occur on building roofs, particularly on older roofs. Figure 5 shows damage caused by direct hurricane force winds to the Plant cooling tower and shroud.

2. DuPont's Hurricane Crew and the Recovery Process

The DeLisle Plant has established facilities and procedures to allow a limited number of Plant staff to remain on site through severe weather events. The Severe Weather Crew enables the most rapid and reliable post-hurricane response and recovery of the facility. The Severe Weather Crew is housed in a specially designed hurricane shelter shown in Figure 6. The shelter is designed to house 24 people and is used as a conference room/training center under normal conditions.

As soon as the Katrina winds subsided enough to allow the Severe Weather Crew to exit their shelter, they began to perform damage assessments. DuPont already had an emergency response team and trailer on the way to assist. Communications with the Severe Weather Crew was the major challenge. The Plant was without power for the first time in its history, communication lines to the outside were badly damaged, and cell phone signals were spotty at best. Initial inspections indicated no environmental impact from the storm surge, and this was communicated to local agencies. The communication was difficult and initially relied on DuPont and local authorities making personal contact.

After the initial crisis management, attention turned to getting as many employees accounted for and back to work as soon as possible. A hot line had been set up to allow employees to report in. Additionally, daily morning rallies were held to go through the list of employees unaccounted for and update that list with their status. Fortunately, no employee or contractor fatalities as a result of Hurricane Katrina. A full accounting of all employees took roughly one week.

Several actions were taken initially to encourage employees to return to work. The more significant DuPont actions included:

- Procurement of travel trailers and the conversion of an employee recreation area into a trailer park to house displaced employees
- Development of a plan to provide rationed gasoline to employees, allowing employees to get to and from work. This was critical in the first two weeks with limited gasoline supplies
- Contracting with a caterer to provide meals for contractors and employees involved in the cleanup
- Procurement of numerous generators to allow some office buildings to be brought back in to service

In parallel, employee teams were formed to bring critical areas back to varying levels of temporary service. These teams were supported by a response team located in Wilmington, Delware staffed 24 hours a day to procure required items. These teams focused on the following areas:

- Utilities – A prioritized list of utility needs was developed and a plan for making needed repairs to allow operations using portable generators. Water and sewer facilities were highest on this priority list.
- Safety equipment – A team was formed to determine safety equipment needs.
- Critical process equipment – It was critical to prevent long-term product interruption and a plan was developed to keep two main reactor vessels active by introducing enough feedstock and utilities to maintain the reactions at a standby state. Resources from other TiO2 sites were brought in to assist in the implementation of this plan.
- Communications/data recovery – While much of the Plant process data, process control, and communications were backed up on data discs, many of the Plant servers were destroyed. An Information Technology team addressed these issues and developed methods to get these systems back up and available. This effort led to routing of communications through servers at other sites and back to DeLisle. Some systems had only been backed up locally, so a process for recovering information on hard drives also had to be developed.
- Damage assessment – DuPont has specialists in almost all aspects of Plant operation. A team was formed to address selection and coordination of these specialists to travel to the Plant to complete their assessments. Logistics of travel, transportation, and housing created challenges for most external visitors. In addition to the trailers set up for employees, several house trailers were brought in to accommodate external visitors in support of Plant activities.
- Community involvement – While the Plant was certainly challenged by issues inside the fence line, the responsibilities of being a good corporate citizen and helping the neighboring community were recognized. A team was formed to organize the outreach program to support local needs, including emergency supply distribution, school facilities, and other local needs.
- Permitting/environmental – Due to the inoperability of many systems on the site, special procedures had to be developed to handle the water brought in to the Plant by Katrina and to deal with other special emissions needs required to stabilize the Plant. A team was formed to stay in close contact with regulatory agencies to assure that everything DuPont did complied with regulatory requirements.
- Engineering – To rebuild the site, a Design Engineering team was rapidly formed and activated. The importance of maintaining the Plant's engineering records in an "As-Built" condition and stored in protected environments for easy retrieval is magnified when something like a major hurricane or fire occurs. The Engineering team immediately began to assemble the information necessary to begin construction work for the rebuild. This required the estimation of the rebuild work tasks and funding authorizations to allow purchasing to begin.

Due to the major damage to almost all of the electrical equipment on the site, the critical path to rebuilding was getting equipment specified, on order, and delivered to the site. Thanks to an up-to-date database of electrical distribution facilities and equipment, the DeLisle site was able to work with national electrical suppliers to have orders promptly placed in their production queues. Switch centers, motor control centers, motor starters, and almost all other types of electrical equipment were needed. Daily review calls were established with each of the major suppliers to assure delivery information was up to date. Due to the timing of this work, several suppliers had to arrange second and third shift work over Thanksgiving and Christmas of 2005. DuPont sourcing provided delivery assistance, and most of the equipment was shipped on dedicated team carriers to prevent field delays.

In parallel to the equipment delivery challenges, labor for implementation was almost as challenging. DuPont utilized several different contract organizations to provide manpower for the rebuild. Entire electrical control rooms had to be gutted except for overhead cabling and new equipment had to be reinstalled. Labor was procured from as far away as Puerto Rico to complete the work. Per Diem and other incentives were required to attract skilled labor, particularly electricians, to the site. Figure 7 shows temporary tent barracks constructed to house workers. These camps, in addition to camper hookups adjacent to the site, provided shelter for the labor force.

The operating personnel at the Plant laid out a prioritized sequence of equipment required to restore the Plant's operation. Much of the Plant laboratory was included in the facilities required for startup, as in-process samples had to be processed, and finished product had to be tested prior to release to customers. In the end, the initial production line resumed operations in late January 2006 and the second operating line in mid-February 2006. Given the amount of devastation, resumption of Plant operation in roughly five months after Hurricane Katrina was a monumental effort.

3. Post-Katrina Mitigation – The Challenge

Very soon after the recovery efforts began to restore the DeLisle Plant to operations, discussions and planning were underway on options to prevent future hurricane damage. Since it was impractical to rebuild the DeLisle Plant to a higher elevation, there was considerable concern that the Plant was still vulnerable, especially given the opinion of the National Oceanographic and Atmospheric Administration (NOAA) and other meteorological experts that the Gulf and Atlantic Coasts have entered a period of heightened hurricane formation that could last for decades.

3.1 Flood Wall Design Selection

DuPont DeLisle partnered with the Baton Rouge office of Arcadis Engineering to assist in development of solutions for increasing the level of protection against storm surge. As the development of the Project unfolded, Dupont leadership, recognizing the importance of the viable operation of DeLisle, introduced to the Project team the challenge of having the Plant protected for the 2006 hurricane season. This gave the Project team about six months to plan and implement a solution.

An effective levee solution really comprises several elements:

- Sufficient height to prevent overtopping of storm surge
- Sufficient strength to withstand the hydraulic pressure created by the storm surge
- Inboard protection to prevent "scouring" of the protection and ultimate failure

Several options existed for flood protection and each was considered as a possible solution:

- Earthen berm – most widely used and practical when adequate footprints exist
- Sheet piling – a common solution in dense or congested areas
- Concrete foundation and wall – typically very expensive and time consuming, but ultimately low maintenance

3.2 Sheet Pile Flood Wall Design and Analysis

After evaluating each of the options, the project team selected the sheet piling option and the Dupont DeLisle Plant embarked on an accelerated mitigation effort. This effort to design and construct a sheet pile wall was enhanced by the availability of limited soils data to allow preliminary design of a floodwall. Arcadis used the limited soils data that existed to size the length, strength, and quantity of sheet piling needed, and DuPont Sourcing went to work to find sheet piling manufacturers that could deliver in the timing needed.

It was necessary to establish an initial wall design height to move forward. The best information available at the time allowed an estimate of possible storm surge at the DeLisle Plant of up to 26–27 feet above sea level. The Project team, therefore, set the design criteria for the top of flood wall at elevation 30 feet. This allowed for some added margin for knows, such as wave action to prevent scouring on the inside of the flood wall, as well as unknowns. The length of sheet piling wall required for this project was about 9,200 feet, or 340 truckloads of sheet piling. As sheet piling was being manufactured, Arcadis completed current soil samples and found that the existing soils had compacted over the years and were actually stronger than the limited data available at the time of sheet piling order. Because of that, Arcadis, partnering with a company called Geocomp, recommended that the piling ordered be driven in a way that actually provided protection to 32 feet above sea level – 2 feet of *Lagniappe*, to use a South Louisiana term. Figure 8 shows a summary of the model that Geocomp developed to model hydraulic loading and design the flood wall.

Coupled with the general analysis of the hydraulic pressures and soil conditions shown above, an analysis of the impact of conditions on the sheet piling was undertaken. To do that, bending moment and shear forces were calculated across the length of the piling to determine if the piling was of sufficient strength to withstand the forces. Figure 9 shows the results of that analysis.

Soil conditions were also analyzed to determine if the hydraulic pressure created by storm surge would tend to cause the entire levee structure to slide laterally due to weak sub-soil conditions with the compacted levee wall above. Figure 10 shows the result of that analysis that calculated the potential for failure of the floodwall. Safety factors indicate this is not a significant risk.

Finally, from an analysis standpoint, sub-soil conditions needed to take into account the porosity of the soil and if the situation exists where the hydraulic pressure from the storm surge could actually set up a sub-surface flow field that results in water flowing under the levee wall and impermeable soil and creates a "blowout" on the interior of the levee wall, resulting in levee failure. Figure 11 shows the analysis for that condition.

Because of the presence of the flow field and relatively low permeability on the inboard side of the levee, relief protection in the form of many shallow relief wells were installed to allow a controlled amount of flow to occur at the height of the storm surge, preventing "blowout" and failure on the inboard side.

Lastly, a concrete scour pad was specified on the inboard side of the flood wall. This serves as a means to prevent washout and undermining in the event of some wave overtopping the floodwall. Since the top of levee elevation is 32 feet above sea level, this is highly unlikely, but this mechanism has been widely reported by researchers as a cause of failures of levees in New Orleans in Hurricane Katrina.

3.3 Railroad Entrance Flood Gate

The other significant challenge in this Project was the design and construction of a flood gate at the only wall penetration required on the Project – the railroad entrance to the facility. Patterned after numerous gates that Arcadis has designed, the gate design was completed quickly and construction commenced shortly thereafter. This portion of the scope was not completed prior to the 2006 hurricane season, but contingency plans were in place to seal the railroad opening in the event of an approaching hurricane. The Figure 12 shows the nearly finished gate with one side of the gate in the closed position and the other side open.

4. Storm Surge Hazard Modeling – Confirmation of the Design

In parallel to the design and construction of the heightened levee wall and entrance gate, DuPont initiated an effort to provide additional confirmatory data for the adequacy of the new fast tracked storm surge levee Project. Storm surge phenomena have been observed and reported around the world for centuries, however, due the rare nature of hurricanes and even rarer nature of extreme storm surges, very limited high quality historical data are available to statistically establish the likelihood and severity of these phenomena at specific locations. In the case of the Gulf Coast, historical records can provide some information on storm surges going as far back as a century with higher quality scientific data for only a half-century.

This paucity of scientific measurements and data requires the application of new computer modeling techniques that rely on a scientific understanding of the regional meteorology, and the physical processes that create storm surge, coupled with modern computer computation capabilities. These computer simulation models utilize the limited historical event data for corroborate hindcasts of past events.

4.1 Storm Surge Phenomena

In the northern hemisphere, hurricane winds have a counterclockwise motion induced by the Coriolis effect. Storm surge is primarily the result of these hurricane winds pushing the shallow coastal waters ahead of the cyclonic winds on the eastern face of a storm resulting in a water surface rise higher than normal sea level. Past hurricanes events have caused storm surges as high as 30 feet in the United States with the highest storm surge ever measured of 42 feet in 1899 in Australia.

Although wind is the major cause of storm surges, other parameters including surface bathymetry and the tropical cyclone minimum central pressure also play an important role. Astronomical tides at the time of peak surge also contribute to higher maximum surge heights. During an intense event prolonged exposure to high sustained winds produces persistent forcing along the coastline, causing the surge to stay inland for hours. The longer the surge duration, the greater the likelihood that peak astronomical tides will occur during the maximum storm surge. Each of the various storm parameters is incorporated into a complex computer simulation model to develop storm surge results from many thousands of simulated synthetic hurricanes.

4.2 Storm Surge Modeling

The capability for computation of and prediction of hurricane surge effects is based primarily on the use of analytic and mathematical models. These types of models allow the estimation of the interactions between cyclonic winds, ocean bathymetry, and other factors that contribute to the phenomena over a wide range of parameters. EQECAT, an ABS Group Company, has developed a storm surge model, that is integrated into the WORLDCATenterprise® (WCe) suite of catastrophe software. The WCe software has the capability to model hurricane winds, earthquakes, flooding, ice storms, and other natural phenomena. The WCe hurricane wind and storm surge models are capable of determining the combined effects of these phenomena on coastal as well as inland structures and facilities. WCe is widely utilized within the insurance industry to simulate the hundreds of thousands of possible storm events that could occur but have not been observed over the limited duration of human habituation of the Coastal United States. This capability to simulate storms and their effects on the manmade built-up environment provides a powerful new tool for risk management.

The WCe model analysis described herein was used to develop estimates of the height of storm surges that are possible at the DeLisle Plant in Pass Christian, Mississippi. In addition to the possible surge heights, the frequencies with which these surges might be expected to occur were also determined. This is achieved by running a suite of tens-of-thousands of simulated possible hurricane events with storm intensities covering the full range of possible events. Storm surge heights at the DeLisle site were computed for each of the thousands of event producing a large statistical database that allowed the determination of the frequency associated with varying heights of storm surge. This information along with estimates of storm rainfall were provided as inputs to validate the design of the storm protective features at the DeLisle Plant.

4.3 Methodology

The WCe storm surge model uses a finite element method to solve the complex governing equations that determine storm surge heights. The model computes sustained wind speeds and other parameters in the open waters using the WCe wind model. For a given tropical cyclone, minimum central pressure, profile factor, translational speed, and radii of maximum wind are all parameters that contribute to the event storm surge. The WCe surge model then computes flooding based upon the surge height, wave period, and the ground elevation.

High-resolution finite element grids perpendicular to the coast along with bathymetry data obtained from National Geophysical Data Center (NGDC) are incorporated into the WCe model. Figure 13 shows the grids in the region around the DeLisle Plant site and Figure 14 show the digitized bathymetry of the region. A fine grid in the model is utilized to represent the near shore bathymetry since surges are sensitive to these shallow waters.

A dominant parameter affecting storm surge is the wind stress coefficient of a storm since tangential wind stress is the one which provides the main driving force. The wind field at the grid points over water are derived using the WCe wind model. This model is a parametric wind field model that uses the central pressure, radius to maximum wind, profile factor, and translational wind speed to compute the sustained wind field at a point.

The model simulates the flow by solving vertically integrated form of shallow water gravity equations. The integration over each element is performed by using Simpson's Rule. The model is fully nonlinear and is forced by wind stress and quadratic bottom stress. Other specific boundary conditions are selected to allow computation over the grid.

4.4 Simulations and Analysis

The WCe model was run using a complete stochastic set of tens-of-thousands of synthetic storms and can also be run using historical storms. The peak storm surge height is modified to account for astronomical tides by specifying the tidal range and period at each coastal location and for each event, determining the combined effect of the storm surge and astronomical effect. Figure 15 shows the results of a hindcast of the storm surge heights for Hurricane Katrina.

For each hurricane event, the probabilistic distribution of storm surge inundation depth is calculated for the site location, using the probabilistic distributions of all important storm and location parameters. Inundation is modeled in two zones: the high-velocity zone (typically the first few hundred meters from the coast), where wave action and debris can severely damage structures, and farther inland, where the main problem is flooding as opposed to structural damage.

5. Summary

5.1 A Successful Mitigation Project

The DuPont DeLisle project team and its contract partners successfully completed this Project in remarkable time. From Arcadis and their design assistance firm Geocomp, to the sheet piling supplier L.B. Foster, installation contractor Cajun Constructors, to the Plant construction forces and Zachry Construction, this Project truly embodied how working as a team with a common, challenging goal can produce dramatic and successful results. ABS Consulting provided state-of-the-art hazard analytics and input information supporting the adequacy of the fast-tracked levee design criteria.

The application of these powerful new computation tools allow facility and risk managers access to high quality risk information and can provide a clearer understandings of benefit cost trade offs in the design of coastal protective measures. The advent of fast computational hardware and user friendly software has allowed the cost effective application of these powerful techniques to site specific problems not possible only a decade ago.

Figure 1: Railcars overturned from siding by moving water

Figure 2: Concrete masonry unit wall failed from storm surge pressures

Figure 3: Site inundation from storm surge

Figure 4: Electrical switch centers flooded by storm surge

Figure 5: Wind damaged cooling tower siding and shroud

Figure 6: DeLisle severe weather crew shelter

Figure 7: Temporary housing camp for site repair and restoration

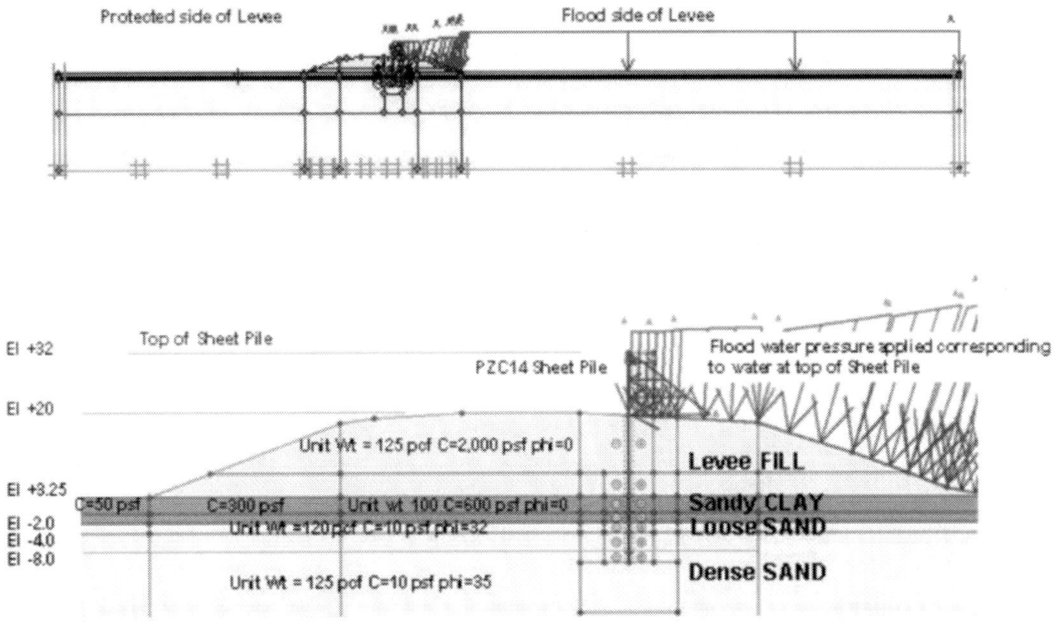

Figure 8: Finite element analysis model – Station + 140 Geometry

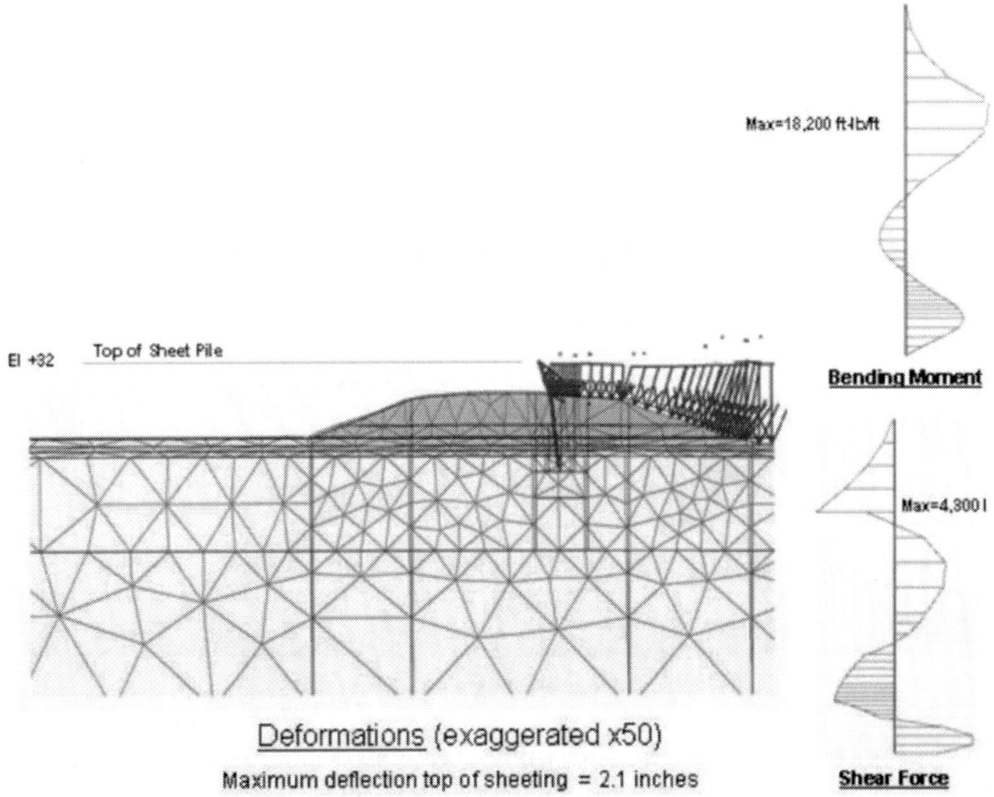

Figure 9: Finite element analysis model – Flood Conditions (No Flow)

Figure 10: Finite element analysis model– Flood Conditions (No Flow)

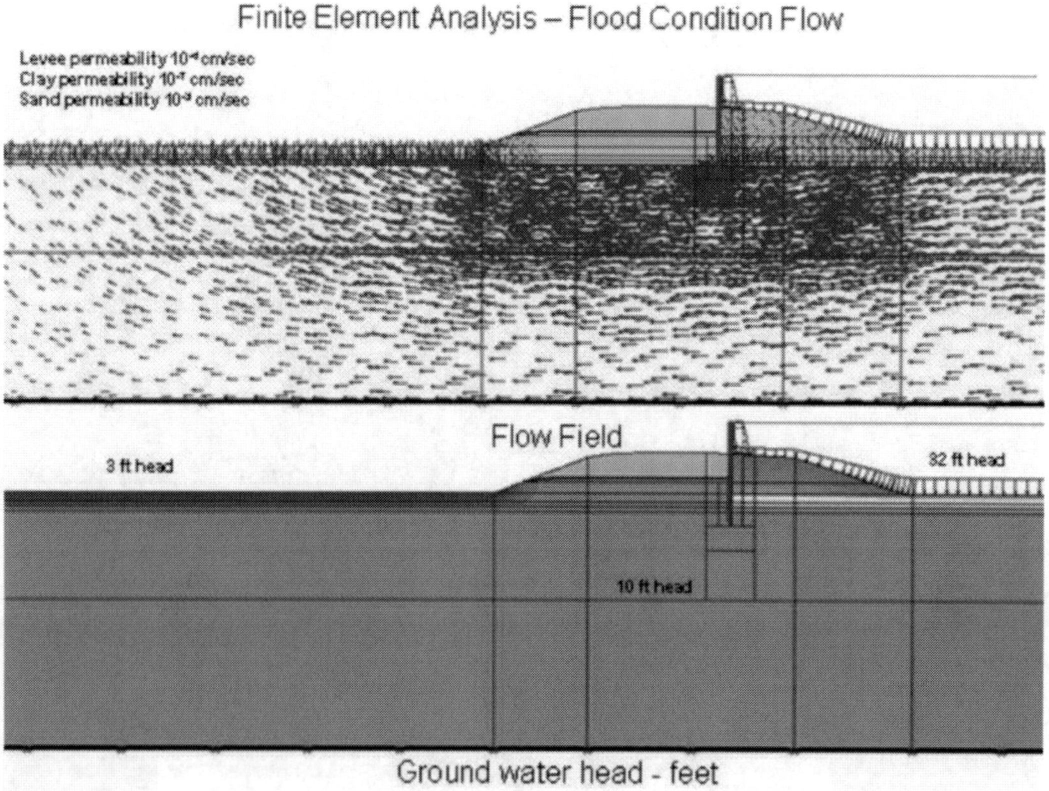

Figure 11: Finite element analysis model– Flood Conditions Flow

Figure 12: Railroad entrance flood gate under construction

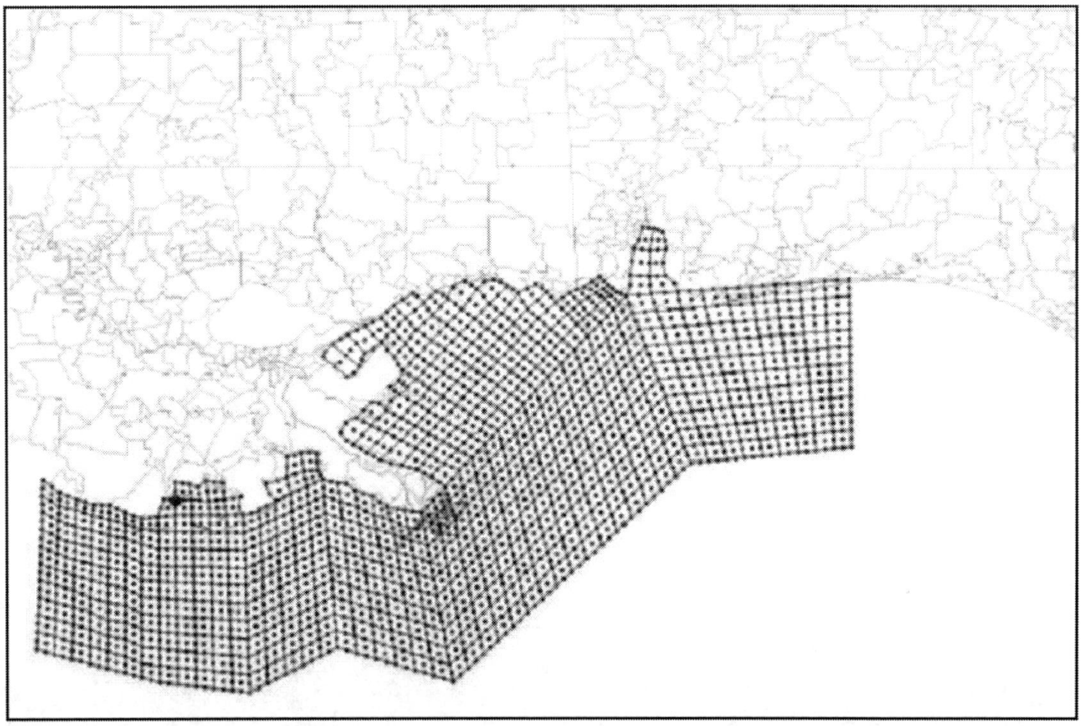

Figure 13: Finite element grids along the Gulf coast used to model storm surges

Figure 14: Digitized bathymetry in the Gulf of Mexico for WCe storm surge model

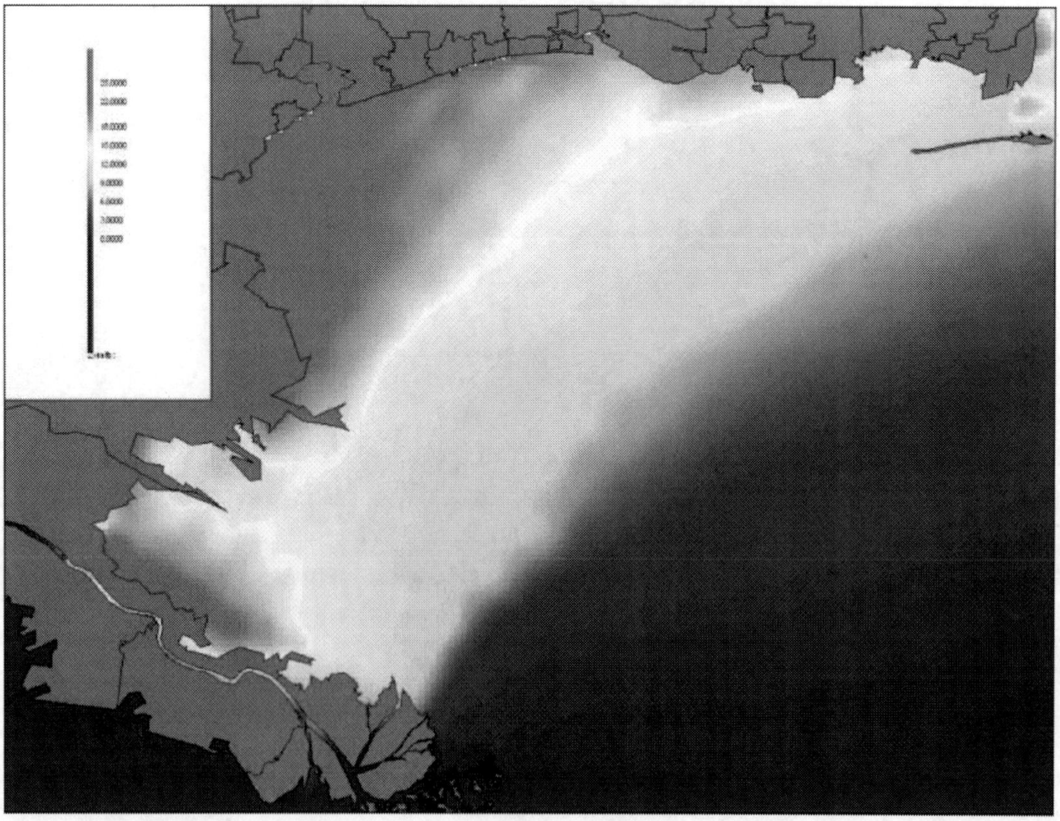

Figure 15: Hindcast of Hurricane Katrina storm surge

6. References

1) "Hurricane Storm Surge Inundation and Advisory Base Flood Elevation Maps," Federal Emergency Management Agency, January 6, 2006.

2) "Performance of Physical Structures in Hurricane Katrina and Hurricane Rita: A Reconnaissance Report," National Institute of Standards and Technology Technical Note 1476, June 2006.

3) "Florida Commission on Hurricane Loss Projection Methodology," EQECAT, an ABS Group Company, February 2004.

LOSS PREVENTION SYMPOSIUM

Preparing for Self-Sufficiency

Christina Francis
The Procter & Gamble Co.
2 P&G Plaza, TE9
Cincinnati, OH 45202

Prepared for Presentation at
American Institute of Chemical Engineers
2006 Spring National Meeting
41st Annual Loss Prevention Symposium
Houston, TX
April 22–26, 2007

UNPUBLISHED

Preparing for Self-Sufficiency

Christina Francis
The Procter & Gamble Co.
2 P&G Plaza, TE9
Cincinnati, OH 45202

Abstract

The City of New Orleans' infrastructure was paralyzed due to the disasters of Hurricanes Katrina (August 29th) and Rita (September 24th) in 2005. New Orleans Fire Department (NOFD) maintained a presence post-disaster despite massive losses to their departments and equipment.

The city infrastructure which Procter & Gamble (P&G) relied on for support was eliminated in these losses. Instead of a normal fire department response time of 3 to 5 minutes, adequate city water pressure, phone lines for automatic alarm dialing, and 911 services, P&G was faced with critical risk control issues. Additionally, lack of housing in the area, necessitated on site housing for employees which included approximately 135 trailers and support infrastructure.

P&G minimized risk to its people, property and profits by becoming self-sufficient for services. This was accomplished by providing on site trained personnel to supplement services lost following the hurricanes.

Paper

Paper not available at the time of publication of the proceedings.

LOSS PREVENTION SYMPOSIUM

Lessons Learned and Mitigation Options for Process Facilities Following the 2005 Hurricane Season

James R. (Bob) Bailey, Ph.D., P.E., F. ASCE
Managing Engineer, Exponent, Inc.
10850 Richmond Avenue, Suite 175
Houston, Texas 77042
832-325-5712 (office)
832-325-5799 (fax)
bbailey@exponent.com

Dr. Marc L. Levitan
Director, LSU Hurricane Center
Associate Professor
Department of Civil & Environmental Engineering
Louisiana State University
Baton Rouge, LA 70803
225-578-4445 (office)
225-578-7646 (fax)
levitan@hurricane.lsu.edu

Prepared for Presentation at
American Institute of Chemical Engineers
2006 Spring National Meeting
41st Annual Loss Prevention Symposium
Houston, TX
April 22–26, 2007

UNPUBLISHED

Lessons Learned and Mitigation Options for Process Facilities Following the 2005 Hurricane Season

James R. (Bob) Bailey, Ph.D., P.E., F. ASCE
Managing Engineer
Exponent, Inc.
Houston, Texas
832-325-5712 (office)
832-325-5712 (fax)
bbailey@exponent.com

Dr. Marc L. Levitan
Director, LSU Hurricane Center
Associate Professor
Department of Civil & Environmental Engineering
Louisiana State University
Baton Rouge, LA
225-578-4445 (office)
225-578-7646 (fax)
levitan@hurricane.lsu.edu

Abstract

The 2005 hurricane season caused unprecedented levels of damage to coastal communities throughout the Gulf of Mexico. Hurricane Katrina and Hurricane Rita were of particular interest to petrochemical and refining companies, given the impacts they had to their onshore operations. Floodwaters and extreme winds from these two hurricanes damaged numerous types of assets, including tanks, cooling towers, generator buildings, process towers, pumps, controllers, utility poles, and transformers. They also caused prolonged power outages, supply disruptions, and widespread employee dislocations, all major contributors to both property and business interruption losses.

Interestingly, the impacts of Hurricanes Katrina and Rita could have been far more devastating had either storm passed further west of their landfall points. This paper will focus on the nature of wind damages sustained by these two events; the history of the design of these facilities in the context of current practices; the likelihood of such events occurring during the operating life of a plant; and options for mitigating future damage. Also discussed are ways to quantify risks in a manner that will enable operators and owners to make informed, cost-effective decisions regarding mitigation planning and associated reductions in loss exposures. Such efforts are important not only in light of the tightening market for commercial insurance, specifically in the form of lower coverage limits, higher deductibles, and reduced underwriting capacities, but to increase the chances that an operation can recover quickly in order to restore services to their customers while also minimizing business interruption losses.

1. Summary of Katrina and Rita Impacts

Process facilities as a whole faired relatively well in terms of wind damage to structures following Hurricane Katrina (damage due to rising floodwaters is not addressed in this paper). Major losses were incurred, however, due to business interruption costs resulting from prolonged power outages and from absent employees displaced by the storm. By contrast, more wind damage was sustained by process facilities following Hurricane Rita, largely in the Sabine Pass Region along the Texas-Louisiana border. Most of the damage was to cooling towers, utility poles, supply/storage (metal) buildings, and insulation on process vessels. As with Hurricane Katrina, major business interruption losses were incurred due to prolonged power outages and to displaced employees. Indeed, restoration of power proved to be the most challenging issue in the aftermath of both storms. Figure 1 shows the wind field maps for hurricanes Katrina and Rita. Interestingly, a majority of process facilities are located about 60 miles on either side of the two storm tracks. Had either storm made landfall further west, the resulting property damage would have been greater and more widespread.

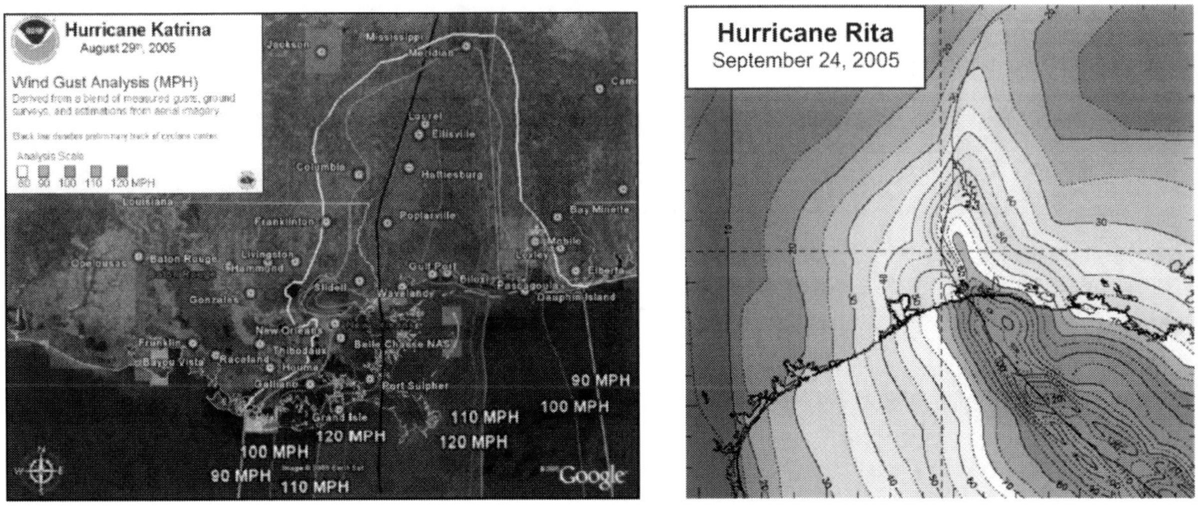

Figure 1. Wind Field Maps of Hurricane Katrina and Hurricane Rita (NOAA).

2. Wind Damage to Process Facilities Components

A process facility typically has some of the following basic components: process structures, process towers, tanks of various sizes, compressors, drums, exchangers, furnaces, pumps, reactors, pipe racks, cooling towers, storage warehouses, control rooms, maintenance buildings, and an administration building. In general, process structures are low-rise, fully engineered, heavy steel buildings, designed for heavy equipment and operating loads. A large percentage of these buildings are open frame structures, with damage thus comprised mainly of roof cladding damage and debris impact damage. Storage tanks vary significantly in size. For example, a refinery will typically have a tank farm comprised of numerous large-diameter tanks (> 100 m) that are open or have floating or fixed roofs, while process tanks are usually smaller in diameter (< 20 m) with fixed roofs.

Most control rooms are either constructed of reinforced cast-in-place concrete or reinforced concrete block, or are fully engineered steel buildings with metal cladding. By contrast, administration buildings are usually either pre-engineered metal buildings, or are constructed of un-reinforced masonry. Maintenance buildings are typically large span, pre-engineered, metal buildings. Process buildings are often fully engineered, heavy steel buildings with average to weak cladding strengths, while storage warehouse buildings are usually large span, pre-engineered metal buildings with average to weaker cladding strengths. Most process facilities were built during a period (pre-1982) that preceded more modern standards for minimum design loads for buildings and other structures with regard to extreme wind events.

2.1 Tanks

Larger tanks tend to fail due to inward collapse of the tank wall caused by lateral wind pressures, while smaller tanks tend to fail due to local buckling of the tank wall cause by wind-borne debris impacts, or in some instances simply toppling over. Tank damage following hurricanes Katrina and Rita was apparently limited to large empty tanks either without a roof, or with floating roofs that rested at the bottom of an empty tank. Such tanks are usually more vulnerable to extreme wind effects due to the absence of sufficient lateral support. By contrast, tanks that have wall stiffeners, fixed roofs, product inside, or a combination of these features, faired better given the added lateral stiffness provided in each case.

Figure 2 shows two cases of a tank failure following Hurricanes Katrina and Rita. In the first case, the fixed roof blew off following Hurricane Katrina. To be sure, in this case it is unclear whether storm surge may have contributed to the roof failure by damaging the tank wall first, which then may have initiated or even contributed to the catastrophic roof failure. The second case shows a tank wall collapse following Hurricane Rita. In this case, the tank was apparently either under construction or undergoing refurbishment at the time of the storm. Hence, it is unclear whether the structure at that time represented its final operating condition. Nonetheless, collapse of the tank wall due to lateral wind loads underscores the effect of the absence of a fixed roof or of wall stiffeners in terms of providing additional lateral stiffness to the tank body.

Figure 2. Tank Damage after Hurricane Rita

2.2 Cooling Towers

Of all the components of a process facility, cooling towers sustained the greatest amount of damage. According to information provided by the Louisiana Chemical Association (LCA), 29 cooling towers sustained some form of damage, including one complete collapse, 12 heavily damaged, and 12 moderately damaged following hurricanes Katrina and Rita. Often constructed of wood with thin-wall shrouds on top, these structures exhibited various forms of damage, from broken slats to collapsed frames, as illustrated in Figure 3. To be sure, designers of cooling towers have begun to comply with the performance-based criteria of ASCE 7, which establishes a minimum design load for buildings and other structures. However, in many instances cooling towers are often older structures (30+ years) that have experience notable material degradation of their wooden members over time, particularly when exposed to coastal environments. Although considered a less costly part of a process plant, successful operation of these towers is crucial to the ongoing operation of a facility.

Figure 3. Cooling Tower Damage

2.3 Process Towers

Damage to insulation was identified in several instances following Hurricane Rita. According to LCA, 30 vessels required some form of insulation repairs. Failures of straps and fasteners resulted in the loss of metal coverings, allowing exposure of insulating materials to the environment. As a result, high winds were able to pull insulating materials, some made of asbestos, away from vessel surfaces. However, no case of a collapsed tower or a tilted tower due to high winds was reported. Given their weight and supporting foundations, this outcome was expected for the wind speeds experienced by these facilities. Figure 4 shows wind damage to the insulation covering of a process tower following Hurricane Rita.

Figure 4. Damage to Insulation on a Process Tower

2.4 Power Lines

Wooden pole failures were prominent throughout the hurricane-stricken areas. Lines on hundreds of these damage poles had to be restrung and pulled. Prolonged outages were attributed, in large part, to the extent of such damage, and the subsequent time it took to repair the downed poles. Experience shows that metal or reinforced concrete poles well anchored into the ground are more likely to withstand extreme wind loadings than their wooden counterparts. The time required to replace these structures, including lines and transformers, depends on the extent of utility crew mobilization and on the priority given to a particular area or site. Emphasis is often placed on repairing downed transmission lines first (large metal frames or multi-pole wooden structures), followed by distribution lines serving areas with critical functions (hospitals, law enforcement, fire stations), then to other areas of a community. Emphasis may also be placed on restoration of a facility if its function is deemed as critical to recovery efforts (e.g., a gasoline refinery). Figure 5 shows an example of common forms of utility pole failures experienced following hurricane Katrina and Rita.

Figure 5. Typical Damage to Utility Poles

2.5 Buildings

Figure 6 shows examples of damage to facilities sustained by hurricanes Katrina and Rita, and also by Hurricane Ivan, which impacted industrial facilities along the Gulf Coast. Corrugated metal and asbestos panels (some panels 50+ years old) pulled away from their fasteners due to outward acting wind pressure. Failures resulted from either shearing of the panel around the fastener, or the fastener itself failing. Material degradation played a major role in the performance of these cladding panels. In many cases, the estimated wind speeds were on the order of a moderate Category 1 storm (85 mph sustained). Damage to built-up roofing systems and to corrugated metal roofing panels was also reported.

Figure 6. Wind Damage to Buildings at Industrial Sites along the Gulf Coast

2.6 Other

Other damage reported following hurricanes Katrina and Rita included a flare tower collapse, a telecommunication tower collapse, and a construction crane failure. The damage photos do not provide sufficient information to determine the cause of failure. Figure 7 shows the damage sustained by these structures. Among the greater concerns regarding tower structures is the threat they pose to nearby structures as lay down hazard in the event they topple over.

Flare Telecommunications Crane

Figure 7. Tower and Crane Failures

In summary, most components of process plants impacted by hurricanes Katrina and Rita sustained nominal, if any, wind damage. Thick-walled steel process towers, pipe racks, smaller process vessels, larger tanks, control rooms, and fully engineered buildings designed to support administrative and operations functions performed relatively well. However, as mentioned earlier, the majority of these facilities along the Texas and Louisiana coastlines did *not* experience a design level wind event. A direct hit by a stronger storm could have resulted in major damage to a substantially larger portion of the components within these plants.

3. Design Practice

Many of the unique types of structures found in process facilities are not addressed by building codes and standards. Until recently there has been little guidance for procedures to estimate wind loads on these aerodynamically complex structures. This lack of guidance has forced operators and engineering firms to develop in-house procedures for such analysis. As documented by the ASCE Wind Loads on Petrochemical Energy Task Committee Report (ASCE 1997), the variation in wind load estimation techniques was extremely large. For example, a comparison study of different company design guides for several configurations of pipe rack structures showed a variation by a factor of five in the wind forces on the structures, when computed using 13 different company design guides for a specified wind speed.

Subsequent research has shown that the correct loads are likely somewhere near or slightly below the middle of that range. This finding implies that while many facilities would have had adequate designs for the specified wind speed, a significant percentage would also be under designed. As a practical matter, the operator of a facility may have specified that it wanted a certain structure to be designed for a maximum wind speed of 120 mph. Depending on which engineering contractor was selected, or which company design guide was used, the facility may actually have been constructed to withstand a wind speed as low as 80 mph or as high as 160 mph. The implications of this issue are discussed further in a paper by Levitan (1999).

The 1997 ASCE report included recommended guidelines for determination of wind loads open frame structures, pipe racks, and vessels. These recommendations have begun to be accepted by the industry, providing at least some level of uniformity and certainty about wind load design moving for new construction. In fact, ASCE has recently formed a new task committee (on which the authors both serve) to create an update to the 1997 report. The updated report will include revised guidelines for the structure types originally addressed, and expand the scope of the guidelines to include other structures as well. New structure types to be addressed in the forthcoming document include cooling towers, pipe bridges, LNG facilities, and tanks. It is anticipated that the updated report will be published by ASCE in 2008.

4. Probability of Occurrence

Knowing the chances that a certain type of event (e.g., a 100-yr event) will occur during the expected or planned operating life of a plant can be beneficial toward understanding the risk exposure of a facility to an extreme event like a hurricane. Table 1 shows the probability that a particular type of event, for example a hurricane, will occur at a site during a certain time period. The calculation is based on a Poison distribution where the probability P_n that an event associated with an annual probability P_a will be equaled or exceeded at least once during an exposure period of n years is given by:

$$P_n = 1 - (1 - P_a)^n$$

Assume that the annual probability of a moderate Category 3 storm striking an existing plant is one percent (1%, or 0.01, i.e., the 100-yr event), and that the plant is scheduled to operate for 30 more years. Therefore, the probability is 26 percent, or about 1 in 4, that the plant will experience a moderate Category 3 (or greater) storm during the next 30 years. When considering applicable design standards and associated factors of safety at the time of construction, such an event represents a reasonable threshold of major damage for many process facilities.

Table 1. Occurrence Probability, P_n (%)

Annual Probability, P_a (%)	Return Period (years)	Exposure Period, n (years)						
		10	15	20	25	30	50	100
10	10	65	79	88	93	96	99	99.9
4	25	34	46	56	64	71	87	98
2	50	18	26	33	40	45	64	87
1.3	75	13	18	24	29	33	49	74
1	100	10	14	18	22	26	39	63
0.4	250	4	6	8	10	11	18	33
0.2	500	2	3	4	5	6	10	18
0.1	1000	1	1	2	2	3	5	10

In recognition of these levels of natural hazard exposures, companies of all types are more readily adopting the practice of risk-based design in recognition that some facilities are more critical than other facilities in terms of their overall contribution to company or agency operations. Examples include structures that contain control systems, unique equipment, or a major hazard to life safety. Such critical facilities require designs that are significantly stronger than conventional facilities. Table 2 shows typical levels of design over a range of environmental conditions for various commercial facilities as required by applicable industry standards. Unless specified otherwise by the operator of a process facility, the designer will typical use the *minimum* require value, which often reflects an un-factored load associated with a 50-year return period event (2% annual probability of exceedance).

Table 2. Risk-based Standards for Critical Facilities

Design Return Period (years : basis)	Environmental Condition	Standard
50-90 : Design	Wind design of ordinary facilities	ASCE 7-05
50 : Design	Minimum cost and unmanned offshore platforms	API RP 2A
100 : Design	Design of offshore platforms for waves and earthquake	API RP 2A
500 : Ultimate Strength	Hurricane Risks for the Cayman Islands, British West Indies	APEC, 1999
500 : Ultimate Strength	Onshore building design for earthquake	UBC
500 : Ultimate Strength	Onshore building design for earthquake	CBC
200 : Design 1,000 : Ultimate Strength	Design of offshore platforms for waves and earthquake	NPD (Norway)
200 : Design 1,000 : Ultimate Strength	Design of offshore platforms for waves and earthquake	API RP 2A
100 : Design 1,000 : Ultimate Strength	Tension Leg Platform (TLP) design for wind and waves	High cost platforms ($1B+)
250 : Design 2,400 : Ultimate Strength	LNG tank design for earthquakes	EEMUA-147
475 : Design 10,000 : Ultimate Strength	LNG tank design for earthquakes	USDOE/NRC
1,000 : Design 10,000 : Ultimate Strength	New design of nuclear power plants for earthquake	NFPA-59A
10,000 : Ultimate Strength	Natural Phenomena Hazards Design and Evaluation Criteria for Department of Energy Facilities	DOE-STD-1020-94

5. Mitigation Options

According to a report by AON, in 2006 Gulf Coast operators with catastrophic exposure that experienced no losses in 2005 could expect a 20 to 35 percent increase in their premiums, while operators with catastrophic exposure and losses could expect 50 percent increases. Moreover, deductibles would increase from two to five percent, with wind sub-limits of $150M or less. Underwriting capacity reportedly diminished, while business interruption coverage was being withdrawn. Such hardening of the insurance market seemed somewhat unusual, given that most petrochemical facilities *did not* experience a design level wind event. However, the damage sustained and associated costs, particularly with respect to business interruption, underscored the apparent uncertainty associated with exposures of these facilities to an extreme wind event. It is believed that had either hurricanes Katrina or Rita struck further west as strong Category 3 hurricanes, the resulting damage would have been an order of magnitude higher.

With a tightened insurance market and ongoing concerns about business continuity, Gulf Coast operators are exploring options to reduce their risk exposures to hurricane. This heightened awareness is not new, just more widespread. Indeed, an operator undertook a series of steps to decrease their vulnerability to storm surge flooding following the hundreds of millions of dollars of losses (property and business interruption) sustained by their refinery located in the central Gulf Coast in the aftermath of Hurricane Georges in 1998. As a result of these measures, the facility withstood the storm surge resulting from Hurricane Katrina and sustained lower flood losses.

The following steps may be taken to determine the best approach for implementing a mitigation plan. First, determine the occurrence probability for the event. In the case of extreme wind, the gust wind is determined as a function of exceedance probability (or return period). Next, determine the level of damage associated with a given wind speed. This effort may require separating the components of a facility in categories of structures and equipment, then determining the associated vulnerabilities based on historical experience, published data, or engineering analyses.

Third, create a loss hazard curve showing the level of damage as a function of exceedance probability for an associated wind speed. Fourth, identify mitigation options for the categories of structures and equipment, where deemed reasonable, the associated cost for each option, and the resulting reduction in estimated losses. Fifth, create another loss hazard curve showing the reduced level of damage given the implementation of mitigation measures as a function of exceedance probability. Finally, integrate the two loss curves (i.e., calculate the area underneath each curve) and subtract the two values to get an estimate of the expected annual loss reduction. Once this value is known, a determination can be made about the economic feasibility of the proposed mitigation measures.

5.1 Example

An example is provided as follows to illustrate this approach. Assume a probabilistic analysis has been performed to determine the wind speed as a function of exceedance probability (or return period). Losses to each component of a plant (control room, process towers, cooling towers, warehouses, etc.) are estimated then summed together as a function of wind speed. Finally the total loss at a given wind speed is assigned to its associated exceedance probability, resulting in the creation of a loss exceedance curve as shown in Figure 8 (the "Without Mitigation" curve).

For the purposes of this example, assume the total plant value, including contents, is $1B. For a 1000-yr return period event (roughly corresponding to a direct hit from a moderate intensity Category 4 hurricane at the location of this facility), the maximum amount of direct wind damage is estimated to equal approximately $180M (or 18 percent of the plant value). From Table 1, the likelihood of the plant experiencing such a storm at least one time during the next 50 years is approximately 5%. (In this example only direct wind damage is being addressed, which represents a percentage of the total loss costs. Other possible losses include, but are not limited to, property damage due to flooding, business interruption, environmental damage, cleanup, and litigation.)

A series of plant-wide wind mitigation measures totaling $5M are then considered as illustrated in Figure 9. The reduced loss hazard curve, also illustrated in Figure 8, is then calculated assuming these proposed mitigation measures are in place. Given these two plots, the area between the two curves is calculated, resulting in a value for the reduction in the expected annual loss of $780,000. Assuming a benefit/cost ratio of two, a discount rate of five percent, and a 25-year operating period, the resulting present value amount that could be justified under these assumptions for the proposed wind mitigation work is about $5.5M. In this case, the mitigation measures as proposed would be viewed as economically prudent to pursue. To be sure, every case is unique, and estimated loss reductions will vary depending on the nature of the vulnerabilities and the proposed mitigation measures.

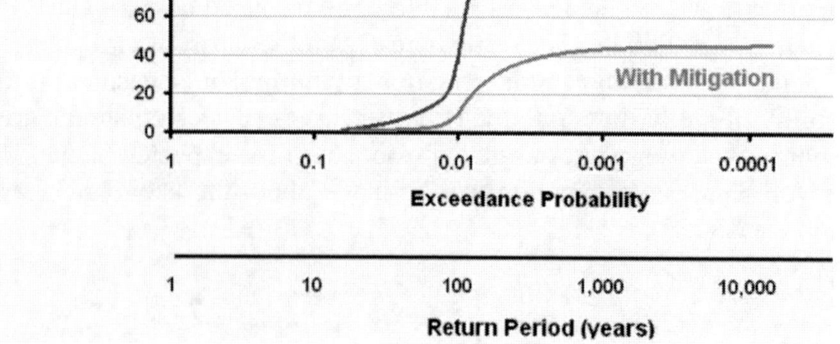

Figure 8. Loss Exceedance Curves

Hurricane-rated Overhead Door

New Hurricane-rated Roofing System

Hardened Control Room

Concrete Cooling Tower

Figure 9. Examples of Proposed Wind Mitigation Measures

6. Summary

Process facilities as a whole faired relatively well in terms of wind damage to structures during Hurricane Katrina, while relatively more wind damage was sustained by these facilities during Hurricane Rita. Most of the direct damage was to cooling towers, utility poles, supply/storage (metal) buildings, and insulation on process vessels. Major business interruption losses were incurred primarily due to prolonged power outages, to displaced employees, and also to the effects of flood damage. Loss of cooling water capacity due to damage sustained by cooling towers also contributed to downtime at some facilities. The majority of process facilities did not experience a design level wind event. Hence, a direct hit by a stronger storm could have resulted in major damage to a substantially larger portion of the components within these plants.

The chances that a specific mean recurrence interval wind event (e.g., a 100-year storm) will occur during the expected or planned operating life of a plant can be determined, and is beneficial toward understanding the risk exposure of a facility to an extreme event like a hurricane. Heightened awareness of natural hazard exposures has led companies of all types to more readily adopt risk-based design practices in recognition that some facilities are more critical than other facilities in terms of their overall contribution to company operations. An example of a method to determine whether a proposed mitigation plan would be economically feasible was provided. This method is one way to enable process plant owners and operators to make more informed decisions regarding their risk exposures, which is of particular interest given recent changes in the commercial insurance market.

7. References

[1] *Minimum Design Loads for Buildings and Other Structures*, ASCE Standard ASCE/SEI 7-05, published by the American Society of Civil Engineers, 2006.

[2] Levitan, Marc L., "Are Chemical Plants at Risk from Hurricane Winds?," Hurricanes and Chemical Hazards Internet Conference, sponsored and archived by PERI, the Public Entity Risk Institute (available at http://hurricane.lsu.edu), 1999.

[3] *Wind Loads and Anchor Bolt Design for Petrochemical Facilities*, Task Committee on Wind Induced Forces and Task Committee on Anchor Bolt Design, published by the American Society of Civil Engineers, 1997.

[4] Simiu, E., and Scanlan, Robert H., Wind Effects on Structures, 3rd Edition, John Wiley & Sons, New York, 1996.

[5] Liu, Henry, Wind Engineering, Prentice Hall, Englewood Cliffs, New Jersey, 1991.

[6] National Hurricane Center, U.S. Department of Commerce, Miami, Fl.

[7] National Climactic Data Center, U.S. Department of Commerce, Asheville, NC.

[8] National Oceanic and Atmospheric Administration, U.S. Department of Commerce, Washington, D. C.

[9] Louisiana Chemical Association, Baton Rouge, LA.

[10] Hurricane Consulting, Inc., Friendswood, TX.

LOSS PREVENTION SYMPOSIUM

Why use Computational Fluid Dynamics for Explosion Studies?

David D. Herrmann, P.E.

Prepared for Presentation at
American Institute of Chemical Engineers
2007 Spring National Meeting
3rd Global Congress on Process Safety
41st Annual Loss Prevention Symposium
Houston, TX
April 22–26, 2007

UNPUBLISHED

Why use CFD for Explosion Studies?

David D. Herrmann, P.E.
DUPONT Engineering Technology
1007 Market Street
Wilmington, DE 19898
(302) 774-1766
david.d.herrmann@usa.dupont.com

Abstract

It takes too long! It is too expensive! This is what the safety analyst hears over and over again when discussing the use of Computational Fluid Dynamics (CFD) for explosion studies. This paper provides an overview of three examples showing the value of CFD in different situations. These examples were relatively simple applications of a CFD that gave convincing evidence to alter the perception of risk from hazardous events. The intent of this paper is to educate the industry as to the possibilities that exist for improving safety when the physics of explosions are better understood. The paper discusses the use of CFD in several examples: (1) vapor overpressure of a large, low pressure, vertical storage tank; (2) a blast wall that (if constructed) would have deflected the blast towards an occupied structure, and (3) an explanation for the damage seen from a distillation column overpressure.

1 Background

There is a perception in industry that computational fluid dynamics (CFD) is expensive and very time consuming. That perception is supported through experience in using CFD for applications where the business stakes are very high, the problem is very complex, and the cost is justified on that basis. Common use of CFD includes such applications as multiphase mixing in reactor systems and polymer flow through dyes. While these applications are extremely valuable, the reputation of high cost and long analysis times from these applications has sometimes limited the use of CFD. In order to get the full potential from CFD as a tool, it is also important to recognize the applications where knowledge can be gained that not only solves a specific problem, but also demonstrates fundamental principles that be leveraged.

The three examples of explosion consequence analyses discussed in this paper took between 4 hours and 3 days to perform. The major amount of time involved is in obtaining the information and setting up the model.

There are a number of commercially available CFD tools on the market. Some are general tools that are applied to a wide variety of applications while others are special use tools. What typically differentiates special use tools from general CFD software is that special use tools have some degree of validation that is specific for the intended end use.

CEBAM (Computational Explosion and Blast Assessment Model) was used for these simulations. There are a number of papers that discuss the validation of the software, two of which are referenced. [1],[2] Rather than discussing the validation of the software, this paper seeks to focus on how the tools can be applied to extract value that otherwise could not be readily obtained.

2 Subtle advantages of using CFD

2.1 Understanding physical differences in consequence scenarios

People are often trained to use simplified approaches and empirical models to perform consequence analysis. They typically have a time and cost advantage over more sophisticated modeling approaches. The main disadvantage is that the users may forget (or never be exposed to) the underlying physics behind the modeling. Several approaches are described in the literature for determining the energy of explosion when a vessel ruptures due to a vapor phase deflagration. Assuming a vessel full of vapor, the most common equations are shown in Equation 1and Equation 2.

Equation 1 $$W = \frac{P_a V}{(\gamma - 1)}\left[1 - \left(\frac{P_{atm}}{P_a}\right)^{\left(\frac{\gamma-1}{\gamma}\right)}\right]$$

Equation 2 $$W = \frac{V(P_g)}{(\gamma - 1)}$$

Where:

P_a = vessel burst pressure (absolute)

P_{atm} = atmospheric pressure (absolute)

P_g = vessel burst pressure (gauge)

V = vapor volume in the vessel

γ = ratio of Cp/Cv for the gasses causing the overpressure (i.e., Cp/Cv for the compressed vapor or gas at the temperature when the vessel bursts)

W = work of expansion

While these work of expansion values are typically multiplied by 2 to account for ground level reflection, the point of looking at the equations is to note that both are dominated by the "PV" term. This implies that a damaging blast wave may occur from a relatively small vessel with a very high burst pressure, or from a vessel with a very large volume and a relatively low burst pressure.

Cone roof storage tanks are examples of vessels with very large volume and when designed with a weak shell to roof seam, a very low burst pressure. While Baker [1] provides a method for estimating the blast pressure in the near field, it is based on the blast wave achieving the speed of sound, which is unlikely in this case.

A simple model was set up to simulate the vessel rupture as the evolution of compressed gas from a cylinder that is 18 feet in diameter and 17.5 feet in height. Ground level probes are set up in the model to track blast pressure and impulse at specific locations. The basic model setup is shown in Figure 1.

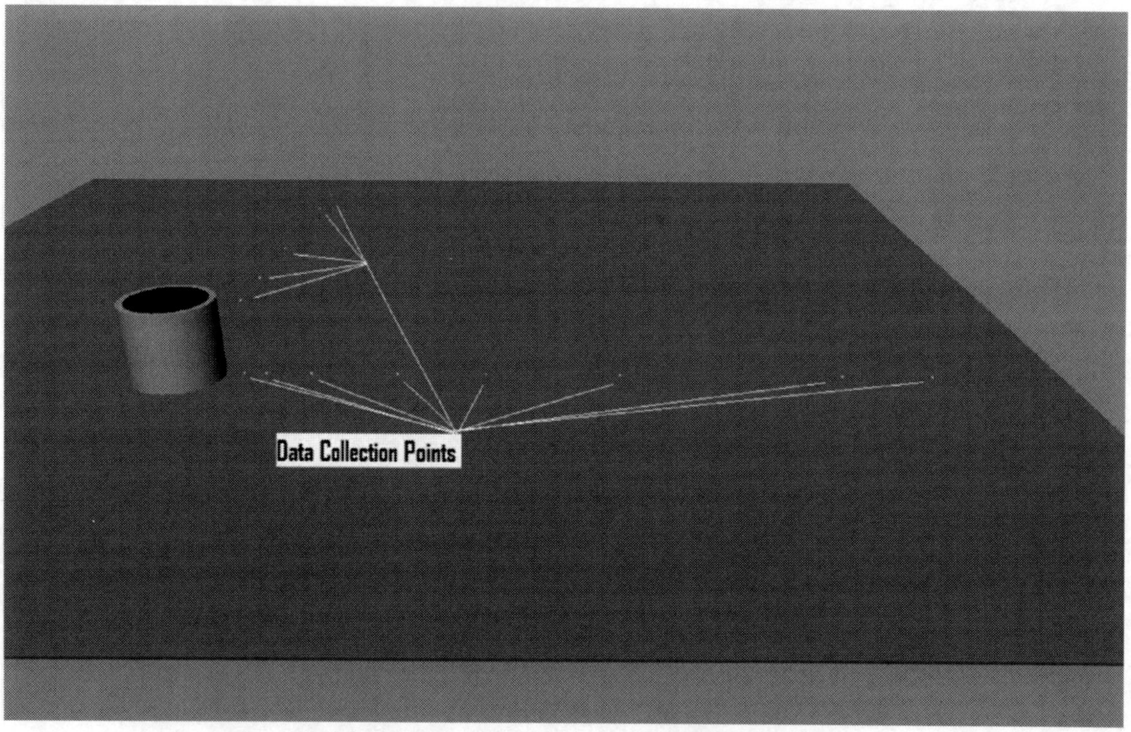

Figure 1 – CFD model setup for vertical storage tank rupture

The initial pressure of the compressed gas in the vessel is set to 3.7 psig. The simulation results are shown in Figure 2, Figure 3, Figure 4 and Figure 5.

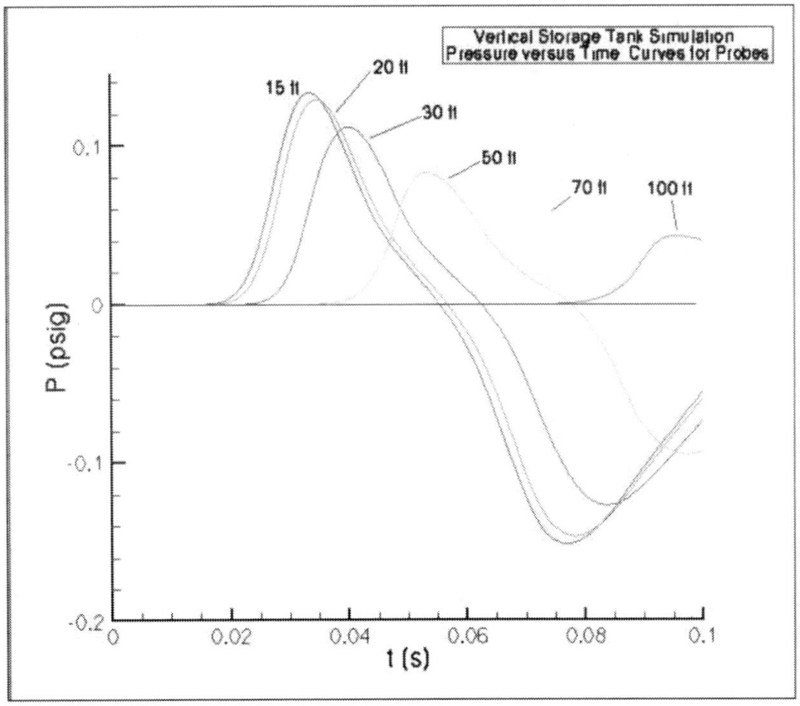

Figure 2 – Pressure versus time with distance from the explosion center.

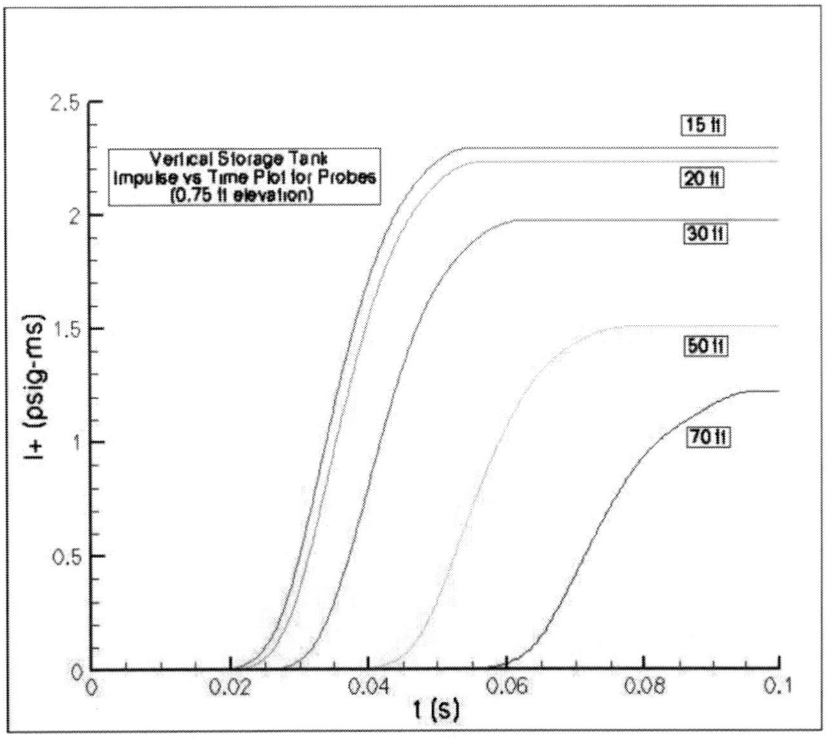

Figure 3 – Development of impulse with distance from the explosion center.

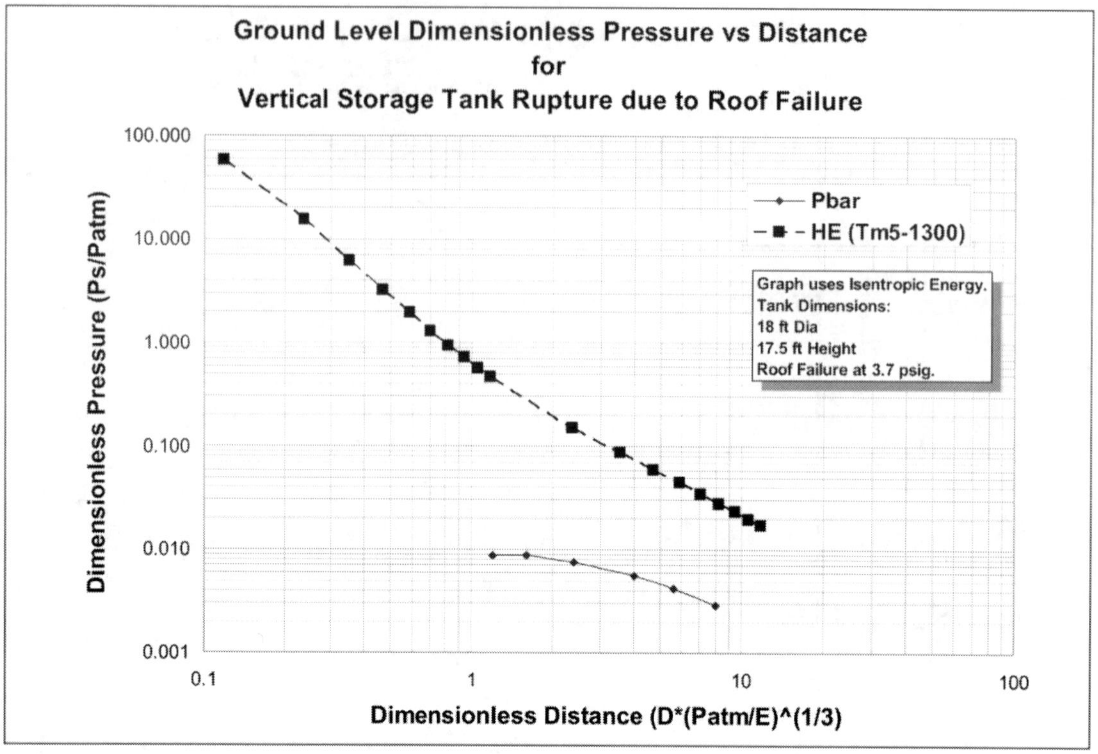

Figure 4 – Scaled pressure versus distance for storage tank simulation.

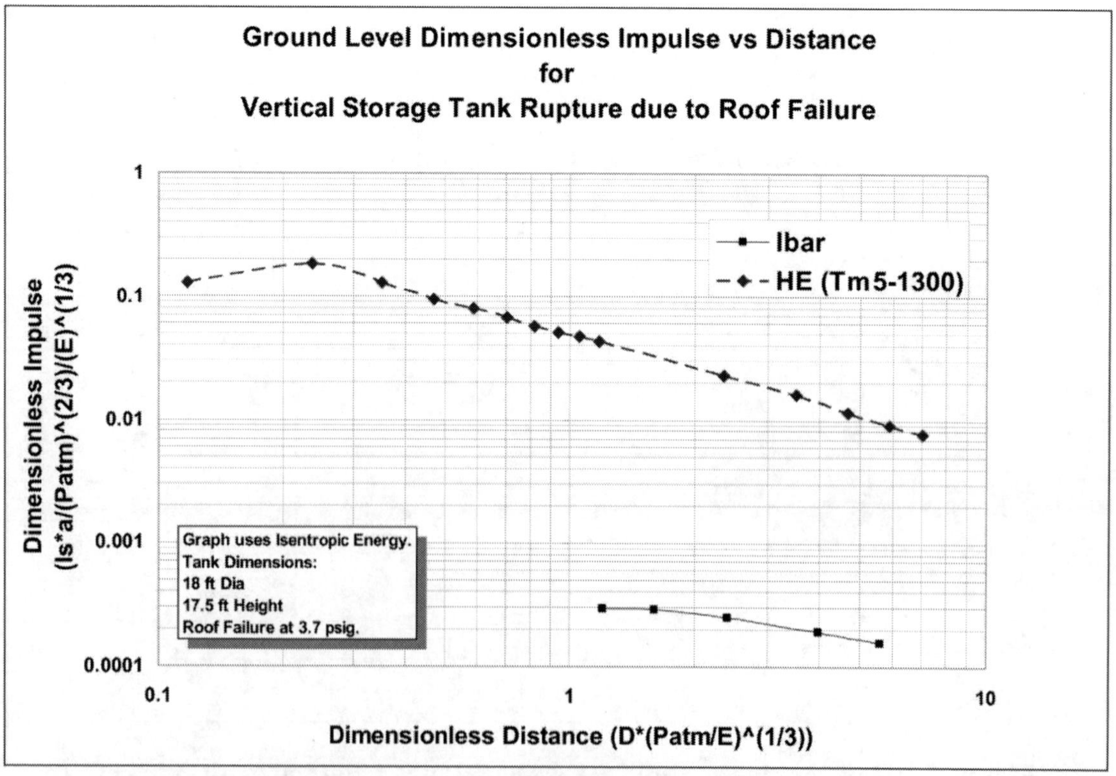

Figure 5 - Scaled impulse versus distance for storage tank simulation.

Figure 4 and Figure 5 compare energy scaled pressure and impulse versus that of high explosives (HE). The curves are similar to predictions of blast curves for vapor cloud explosions where the flame speed is less than mach 1. For a typical blast pressure prediction, a straight line would be drawn from the dimensionless burst pressure (0.27) to the high explosives curve and the blast would be scaled from there. The CFD estimates are at least one order of magnitude lower primarily due to the fact that the blast wave does not reach the speed of sound. Having vessels with such low burst pressure would seem to impart a degree of inherent safety from the standpoint of potential blast effects.

It is important to remember that this analysis does not consider all the potential effects of the vessel rupture that include pool fire radiation and missiles.

2.2 Is a "Blast Wall" beneficial?

In this example, a facility siting team recommended installation of a wall to prevent the blast wave from an explosion in one facility from damaging the neighboring facility. An overview of the site is shown in Figure 6. The source of the explosion is a flammable vapor/air explosion in Area B on the ground floor manufacturing area nearest the proposed wall. The blast wall was being erected to be protect Area A. A third building (Area C) exists. The wall is long with a slight angled bend in the middle. Upon review of the proposal, it was decided to determine if the wall would provide the estimated benefit and if this "subtle change" would have any other effects to the area in the event of an explosion.

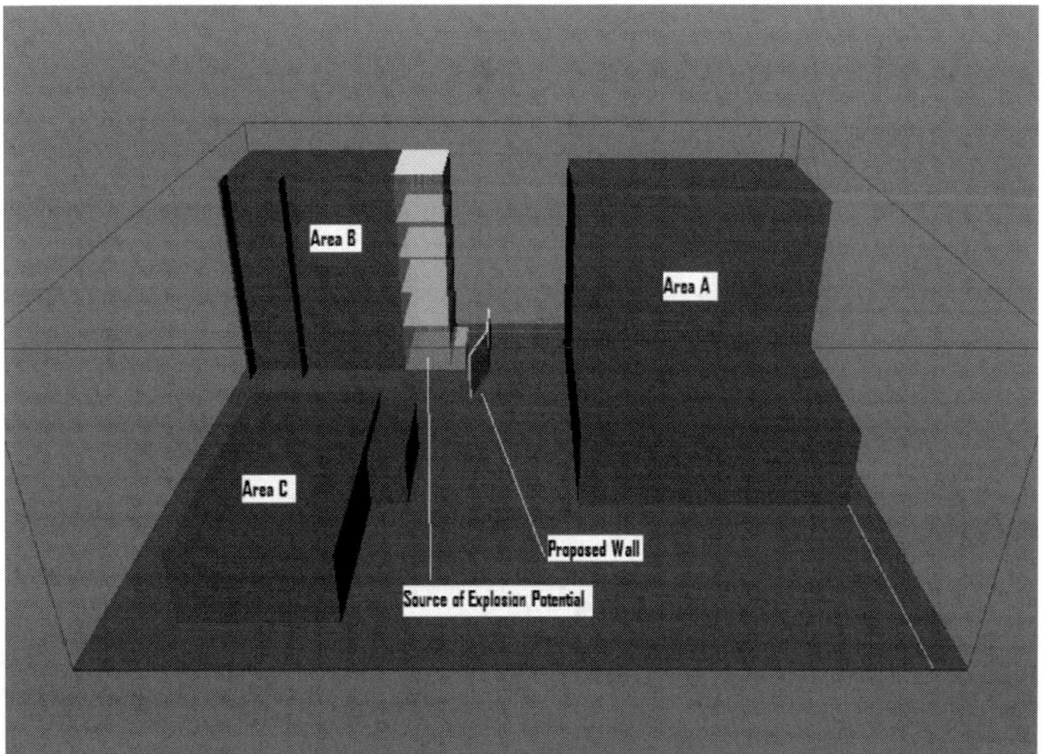

Figure 6 – Plant overview showing proposed blast wall.

An initial consequence analysis had been conducted that estimated the free field blast load on Area A at ground level nearest to Area B to be 2.2 psi and 40 psi-ms without the wall. The CFD analysis required that the flame speed be estimated as an input to the software. In order to produce a comparable analysis, the flame speed input to the CFD software was adjusted in order to create a reflected blast load on the same section of Area A of approximately twice that of the free field study. While the two flame speed values (from the empirical model and CFD tool) were not the same, they were comparable. The (CFD) estimated blast load values closest to the target were chosen to be 3.75 psi and 106 psi-ms. Figure 7 and Figure 8 show the simulation results as the area existed with no blast wall in place. The ignition location was conservatively chosen to be in the lower right corner of the congested volume, where the confinement afforded by adjacent walls would give worst case results.

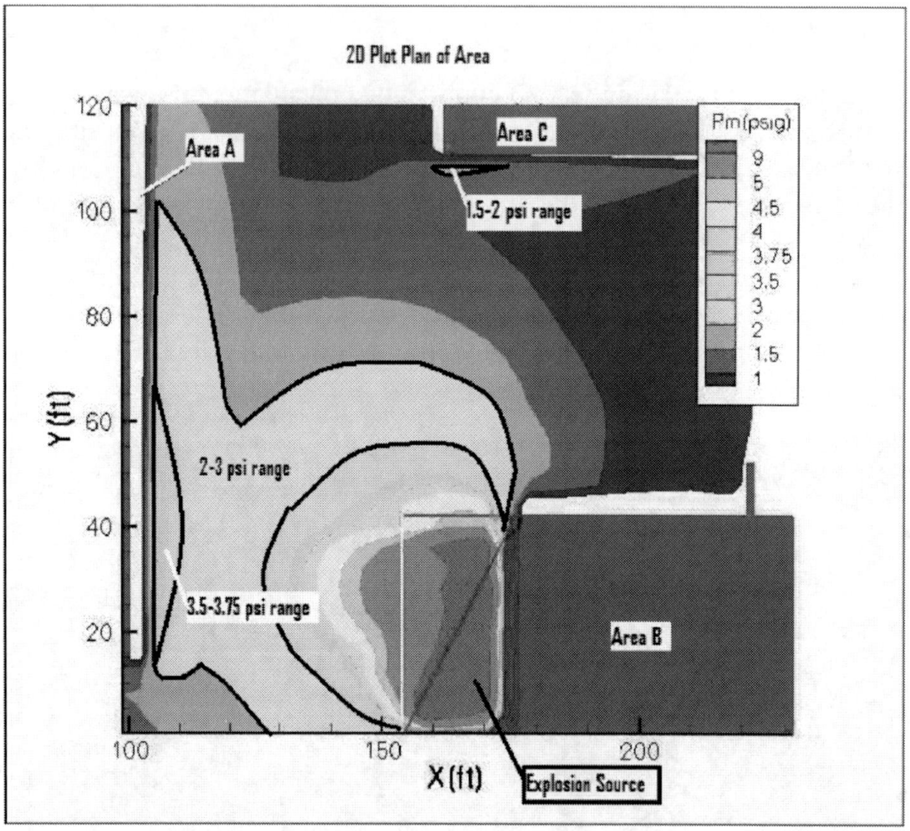

Figure 7 – Explosion pressure contours from “no wall” scenario.

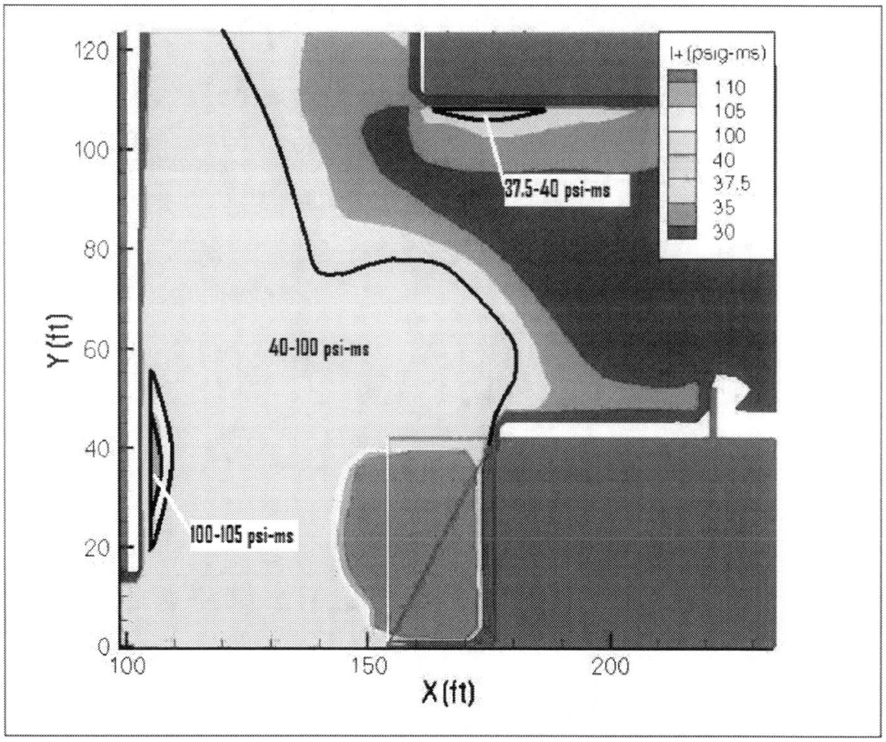

Figure 8 - Explosion impulse contours from "no wall" scenario.

The blast wall was inserted into the model, and the same simulation was run again.

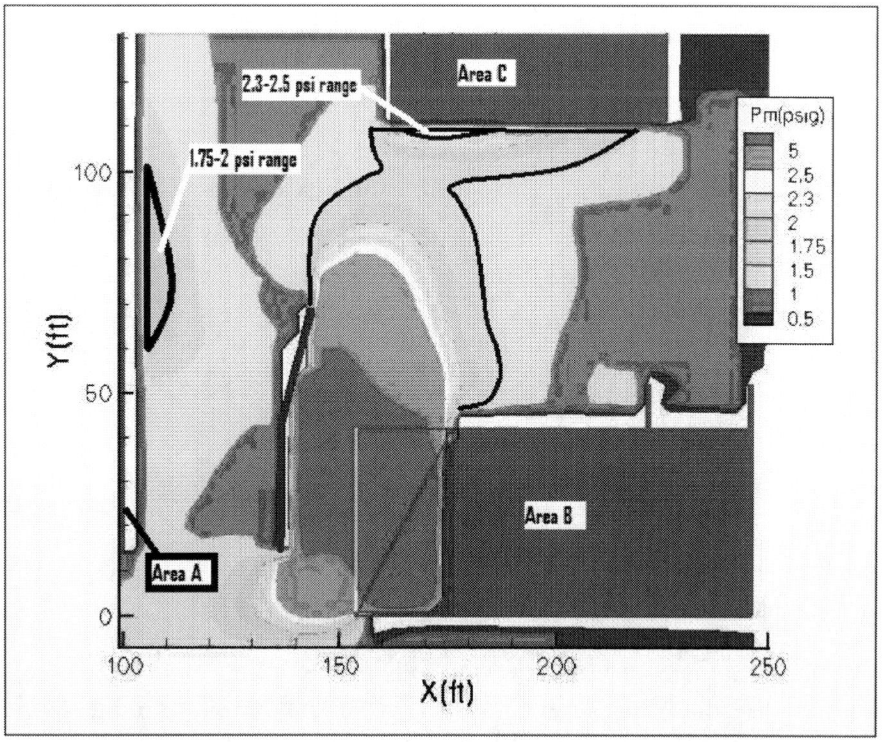

Figure 9 –Explosion pressure contours with blast wall.

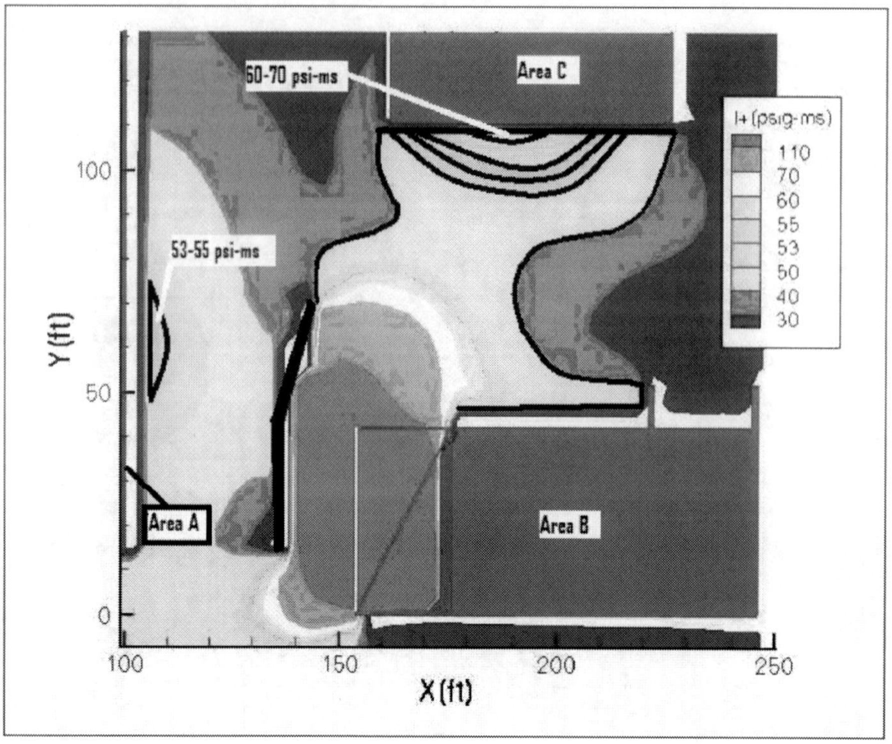

Figure 10 –Explosion impulse contours with blast wall.

Table 1 summarizes the results of these simulations. While the blast wall was somewhat successful in mitigating the ground level blast loads on the building, it refocused the blast towards the Area C building, increasing the potential damage to that building.

Table 1 –Summary of Simulations for Blast Wall Analysis

	Area A		Area C	
Scenario	Pressure	Impulse	Pressure	Impulse
No Blast Wall	3.75	106	1.75	38
With Blast Wall	2.2	54	2.4	65

As a result of this study, it was determined that the proposed blast wall would not be successful at improving the risk at the site. The cost to design and install the wall was in the order of $100,000.

Note: Be careful not to extrapolate this example to "all blast walls are bad". Blast walls need to be carefully applied, and their benefits need be understood. When properly used, they can be an important tool in the mitigation of blast effects.

2.3 Modeling the consequences of column rupture.

In this example, it is necessary to understand the impact of column rupture on a building. The difficulty in this analysis is a general lack of understanding of the gas volume in the vessel that actually contributes to the blast. Past advice for the case of a bubble cap column, has been advised to take the volume of several trays (not specific as to how many). In the case of a packed column, the volume above the packing has been used. Of course the volume used may also depend on the chemicals and initial conditions at the time of the explosion as well and the column design. There have been cases where auto decompositions in packed columns have actually accelerated by virtue of the presence of packing. While this is outside the scope of typical vessel burst blast analysis, it is necessary to recognize possible exceptions to the rules.

In this example, the challenge was to reproduce the effects actually seen in an explosion. This event involved a column filled with packing that ruptured at the top flanged section. During the gas release from the column, the packing was thrown upward.

The column is approximately 200 feet tall with a 7 ft internal diameter. Approximately 30 ft of the top of the column was removed by the rupture.

Since the column packing was ejected by the force of the expanding gas, one assumption is that the entire vapor volume of the column contributes to the blast. Using those assumptions, an estimate of blast load to the building was calculated. The resulting free field blast load was 1.1 psi and 32 psi-ms. Based on this blast load, we would have expected the reflected walls (loading of 2.2 psi and 62 psi-ms) of the building to collapse. This did not happen. In fact the damage to the building from the blast wave was minor.

We decided to model the blast using CFD. A simple 3D model was built showing the building and the column minus the top section that had been removed by the explosion. A cylinder of compressed gas was placed inside the column equal to the entire initial column volume. The model is shown in Figure 11.

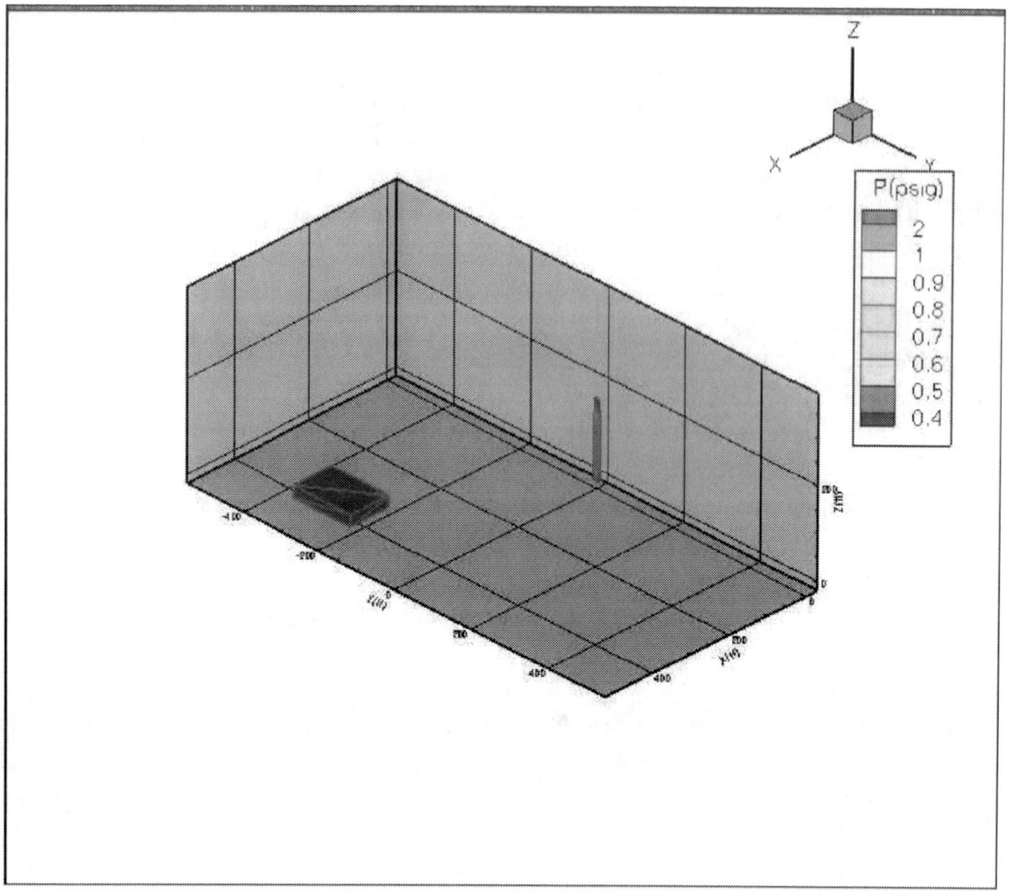

Figure 11 – Column Rupture Model

Movement of the blast wave across the computational domain is shown in the following snapshots (Figure 12). Figure 13 and Figure 14 show the maximum blast loads (reflected) on the building during the simulation. This shows that by modeling the blast in its actual configuration, the estimated blast loads on the building are significantly reduced and in fact, the damage estimated from these blast loads is consistent with the minimal level of blast damage actually seen. Like the earlier example of the storage tank, the primary force of the explosion is upward. In addition, it is not possible for all of the compressed gas to participate in the initial blast wave due to the long length to diameter of the vessel, and the time needed for the entire contents of the gas to escape.

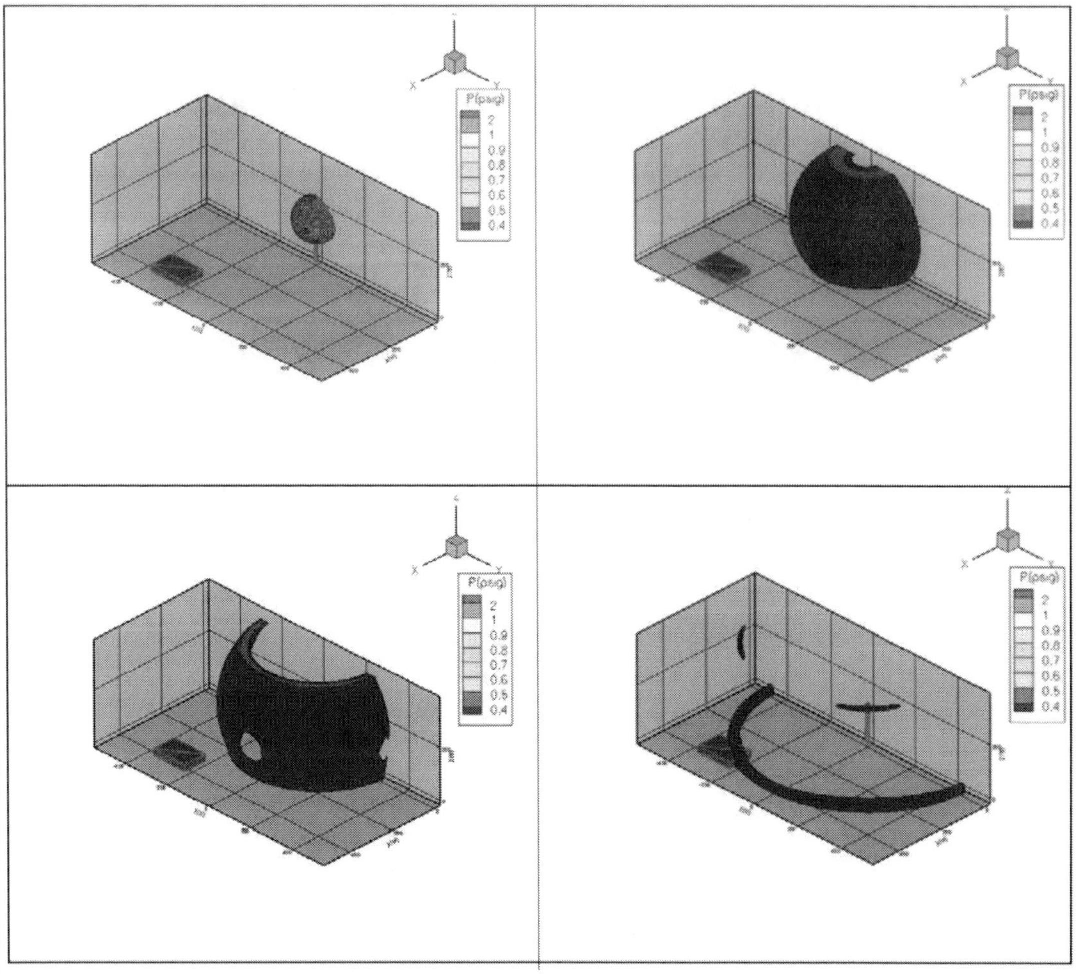

Figure 12 – Column Explosion – Snapshots During Simulation

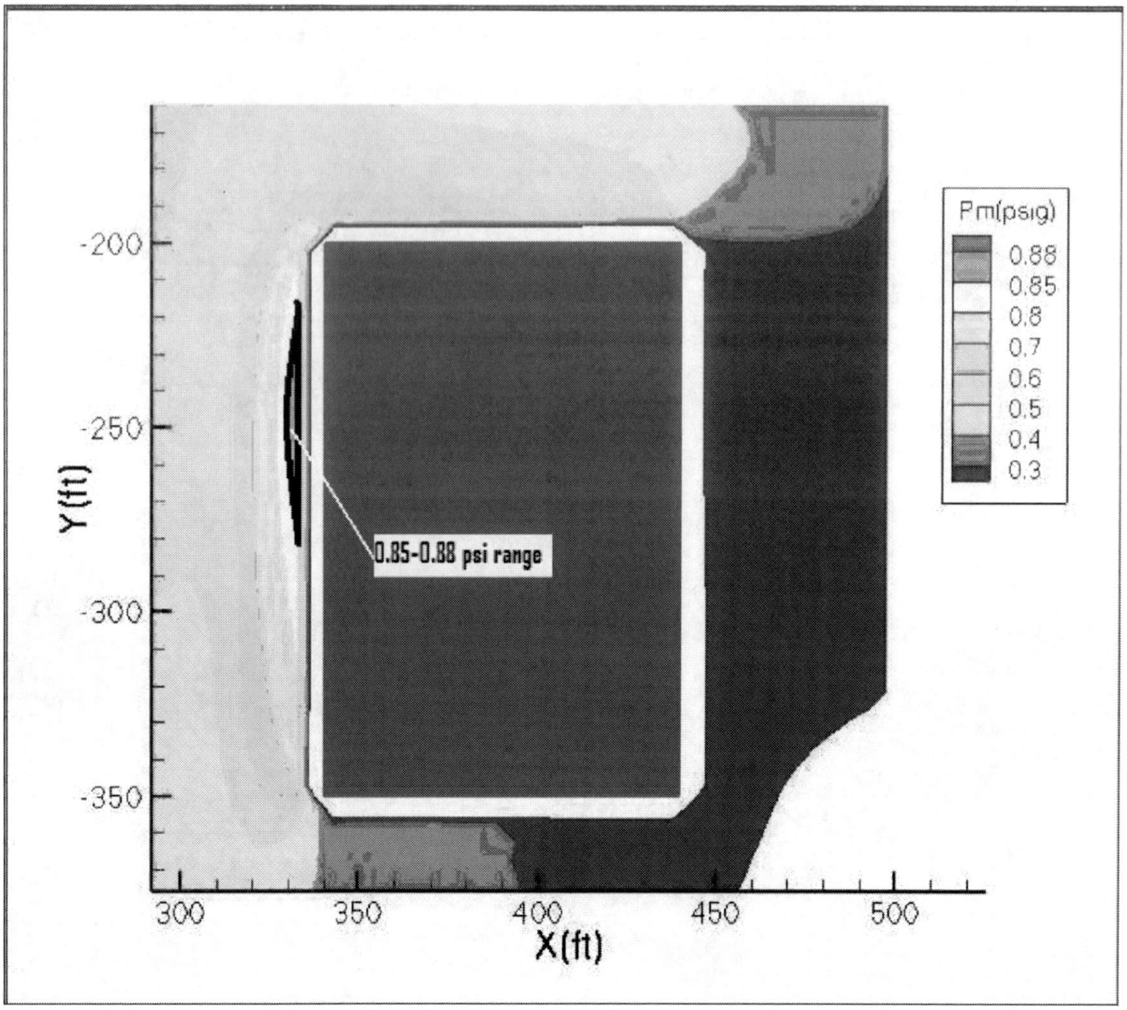

Figure 13 – Column Rupture – Maximum Pressure Contours

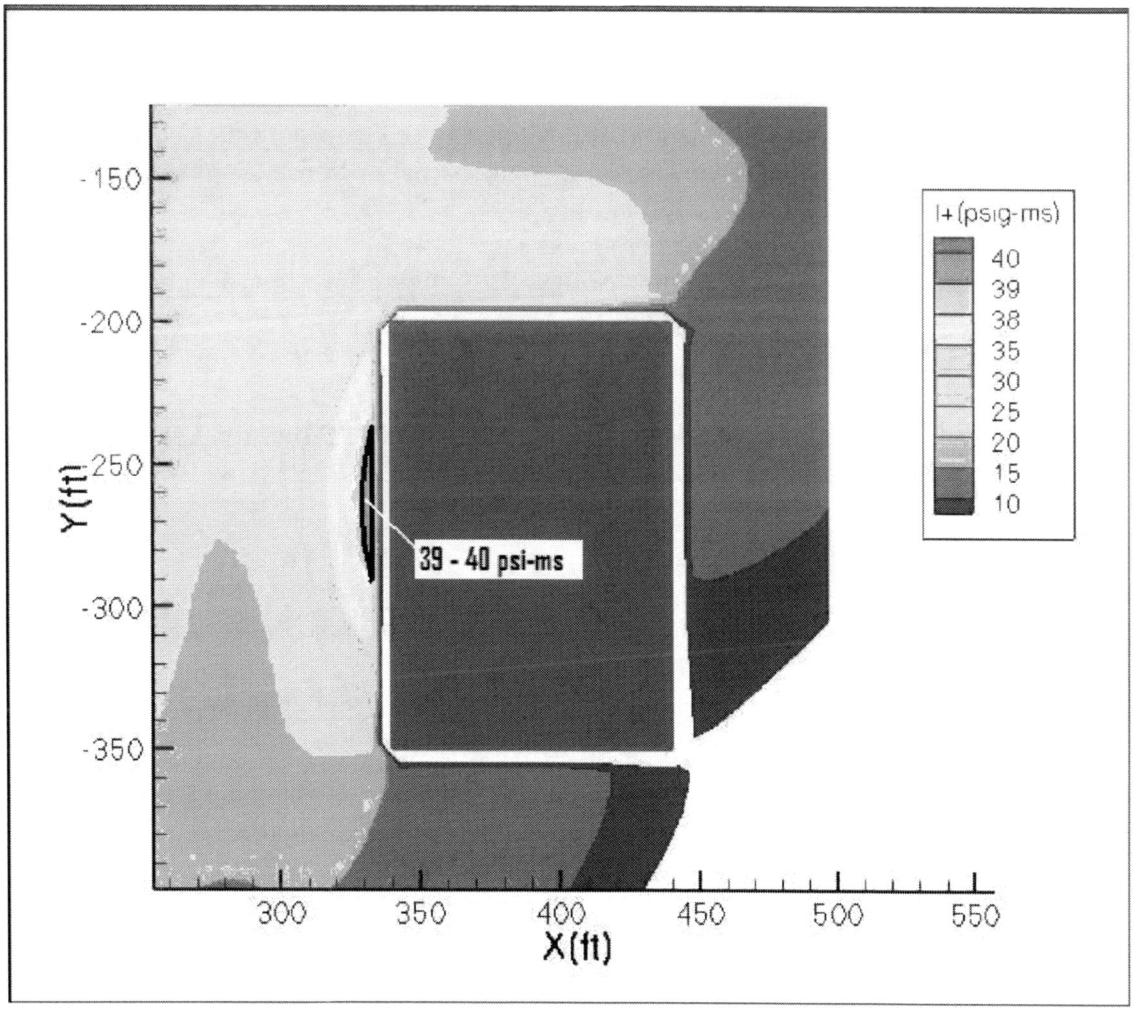

Figure 14 – Column Rupture – Impulse Loads

3 Summary and Conclusions

While computational fluid dynamics finds a great deal of use in solving very complex problems, its value is not limited to complex problems. Many relatively simple problems can be solved quickly and economically with the use of CFD. The 3D representation of the results, as well as the depth of the analysis helps the novice and experienced engineer understand basic principles that can improve their ability to make critical decisions regarding the risk to their operations.

4 References:

1. Clutter, J.K., Mathis, J.T., Stahl, M.W., **Modeling environmental effects in the simulation of explosion events**, International Journal of Impact Engineering, ©Elsvier Ltd, 2006.

2. Herrmann, David D., **Developing a Sound Basis for the Design of Vented Explosion Barricades in Chemical Processes**, Process Safety Progress, Volume 24, Issue 1, ©American Institute of Chemical Engineers, March 2005, pp: 52-58.

3. Center for Chemical Process Safety, **Guidelines for Evaluating the Characteristics of Vapor Cloud Explosions, Flash Fires, and BLEVES**, American Institute of Chemical Engineers, 1994.

The Usefulness of Phenomenological Tools to Simulate the Consequences of Dust Explosions

Ch. Proust, E. Leprette
Institut National de l'Environnement Industriel et des Risques
BP 2, F-60550
Verneuil-en-Halatte
Phone: +39 0817682947
Christophe.proust@ineris.fr

Daubech, I. Sochet
Laboratoire Energétique Explosion Structures, ENSIB
10 Boulevard Lahitolle, 18020 Bourges
Cedex, Franc

Prepared for Presentation at
American Institute of Chemical Engineers
2006 Spring National Meeting
41st Annual Loss Prevention Symposium
Houston, TX
April 22–26, 2007

UNPUBLISHED

The Usefulness of Phenomenological Tools to Simulate the Consequences of Dust Explosions

Ch. Proust, E. Leprette, J. Daubech, I. Sochet

ABSTRACT

Using numerical modelling is a "natural" step in today's engineering work, even in safety. Up to a certain extent, solving more or less accurately the basic Navier-stokes equations has replaced the traditional analytical approximate solutions of the same equations. Doing this, we surely have gained in flexibility, sometimes in accuracy, but we may have lost in expertise and ergonomics. In this paper, different modelling ways are set in perspective for the specific case of explosions, trying to link them to a specific area. It is the opinion of the present authors that simple physical "modelling" is justified in areas where a consensus is required on "basic " approaches as for standards, that complex numerical modelling is particularly fruitful in research and that some intermediate "phenomenological" modelling is possible and proves profitable for process safety where some technical expertise is required. Example of such tools are given and compared to existing data.

Introduction

Dust explosions still represent a major threat. Some accidents, such as the explosion of the silo in Blaye in France (August 1997, 12 victims: Masson, 1998), remind us that explosions of combustible dusts can really have serious consequences. There is an obvious need to be capable of "engineering" the safety to limit the consequences of dust explosions. Modelling approaches have been developed in which rather simple and crude "rules of thumb" neighbour sophisticated full numerical calculation of some fluid dynamic equations (CFD codes). Are the former so primitive that they have to be definitely forgotten? Answering positively would mean that the evaluation of dust explosion is not understandable for normally educated people and would raise difficult questions about the training of personnel in charge of safety. There is thus certainly a need for the coexistence of various modelling strategies and there are reasons for which so different approaches have been continuously developed. The ambition of this paper is to debate on the indications and limitations of some very different modelling approaches. The discussion is centred onto the "simple" and very practical situation of vented vessels.

Engineering laws

This question has received considerable attention since the beginning of the seventies with the pioneering work of Donat (1971). A number of very valuable contributions followed including the experimental works of Bartknecht (1993) and Eckhoff (1991). Analytical models have been used which implement the theory of propagating flames in closed vessels extended to the situation of open vessels. The basic equations and justifications may be found in the work of Bradley and coworkers for instance (Bradley et al., 1978). Experiments were not only planned to produce data in such a controlled way that a direct confrontation with the equations could be made possible but also in real equipments operated in a realistic way. A very large debate in the scientific community resulted. Some consensus arose about the main issues including vent sizing. In Europe the standard EN 14491 has been prepared and in North America NFPA 68 guidelines are regularly updated accordingly. It is not the purpose of the present article to discuss vent sizing methods but to recall some major achievements about the understanding of the development of explosions.

If A is the size of the opening of a vessel of internal volume V, the maximum flowrate Q^- of gases of specific mass ρ through A is classically linked to the maximum internal overpressure ΔPred ("reduced explosion pressure") by :

$$Q^- = C_d \cdot A \cdot \sqrt{\frac{2 \cdot \Delta P_{red}}{\rho}}$$

at least for small overpressures where C_d is the discharge coefficient of the orifice. The maximum rate of production of volume by the flame Q^+ due to the expansion of the burned product may be written as:

$$Q^+ = A_f \cdot S_t \cdot E_{exp}$$

where A_f is the maximum flame area, S_t the burning velocity of the flame and E_{exp} the expansion ratio of the products of combustion. A_f should be linked to some geometrical area of the vessel like its internal area as suggested long ago by Wheeler (1922) on the basis of his excellent experiments. The internal area turns out to be roughly proportional to $V^{2/3}$. The maximum internal overpressure is expected to be reached at the point where $Q^- = Q^+$ so that finally:

$$\frac{A}{V^{2/3}} \propto \frac{S_t \cdot E_{exp}}{C_d \cdot \sqrt{\frac{2 \cdot \Delta P_{red}}{\rho}}}$$

In this expression, the expansion velocity S_t. E_{exp} should mostly depend on the cloud itself (nature of the particles, mass concentration, turbulence level, …) so that for a "given cloud" the explosion overpressures should correlate to the geometrical parameters. Experiments in which the cloud has been produced by "similar means" (Bartknecht, 1986; Donat, 1971) seem to substantiate reasonably this correlation (Figure 1). Such a compact, "self explaining" and experimentally consistent formula is appealing to safety people and this fact may have played a significant role in the development of venting technologies for instance.

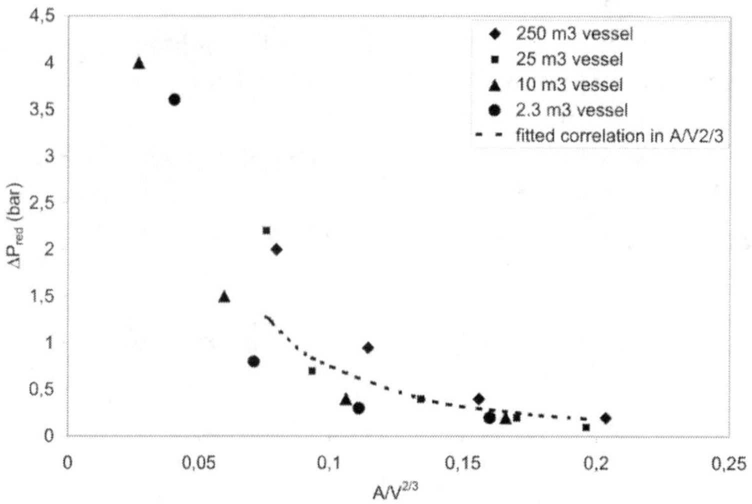

Figure 1. Experimental points of vented explosions from vessel volumes ranging from 2 to 250 m^3. Cellulose dust (K_{st}=200 bar.m/s, P_{max} = 9 bar) dispersed pneumatically by gas pressurised containers

Refined but resembling expressions have been popularised by standards. It is even sometimes said that "most practical situations are covered". However, direct comparison with experiments obtained in real or realistic configurations (Eckhoff et al., 1984; Ekhoff et al., 1986; Ekhoff et al., 1988; Tonkin et al., 1972; Pineau et al., 1985) as depicted in Table 1 and shown in Figure 2 indicate that, if the trends are correctly featured, the predicted values may be significantly above or lower than the measurements.

Table 1. "Real or realistic" vented dust explosion experiments

ref	Author	Equipment	Dispersion of the powder	Injection velocity-pressure (m/s-barg)	Pipe diameter (m)	Volume of the vessel (m^3)	Aspect ratio	Area of the opening (m^2)	Dust
A	Eckhoff, 1988	filter	tangential flow of dust from a pneumatic transport line	35 - 0	0,155	5,8	2	0,11	Maize starch
B	Eckhoff, 1988	filter	tangential flow of dust from a pneumatic transport line	35 - 0	0,155	5,8	2	0,2	Maize starch
C	Eckhoff, 1988	filter	tangential flow of dust from a pneumatic transport line	35 - 0	0,155	5,8	2	0,4	Maize starch
D	Eckhoff, 1988	filter	tangential flow of dust from a pneumatic transport line	35 - 0	0,155	5,8	2	0,55	Maize starch
E	Tonkin, 1972	cyclone	tangential flow of dust from a pneumatic transport line	13 - 0	0,23	1,2	1	0,06	Wheat flour
F	Tonkin, 1972	cyclone	tangential flow of dust from a pneumatic transport line	13 - 0	0,23	1,2	1	0,08	Wheat flour
G	Tonkin, 1972	cyclone	tangential flow of dust from a pneumatic transport line	13 - 0	0,23	1,2	1	0,12	Wheat flour
H	Eckhoff, 1986	silo cell	coaxial flow of dust from a pneumatic transport line	38 - 0	0,155	236	6	3,4	Maize starch
I	Eckhoff, 1986	silo cell	coaxial flow of dust from a pneumatic transport line	38 - 0	0,155	236	6	5,7	Maize starch
J	Eckhoff, 1984	silo cell	coaxial flow of dust from a pneumatic transport line	12 - 0	0,2	500	3,5	2	Wheat dust
K	Eckhoff, 1984	silo cell	coaxial flow of dust from a pneumatic transport line	12 - 0	0,2	500	3,5	3	Wheat dust

ref	Author	Equipment	Dispersion of the powder	Injection velocity-pressure (m/s-barg)	Pipe diameter (m)	Volume of the vessel (m^3)	Aspect ratio	Area of the opening (m^2)	Dust
L	Eckhoff, 1984	silo cell	coaxial flow of dust from a pneumatic transport line	12 - 0	0,2	500	3,5	5	Wheat dust
M	Eckhoff, 1984	silo cell	coaxial flow of dust from a pneumatic transport line	12 - 0	0,2	500	3,5	8,8	Wheat dust
N	Eckhoff, 1984	silo cell	coaxial flow of dust from a pneumatic transport line	12 - 0	0,2	500	3,5	14	Wheat dust
O	Pineau, 1985	featured air mill*	coaxial flow from a pressurised line	250 - 4	0,025	1	3,5	0,03	Wheat flour
P	Pineau, 1985	featured air mill*	coaxial flow from a pressurised line	250 - 4	0,025	1	3,5	0,05	Wheat flour
Q	Pineau, 1985	featured air mill*	coaxial flow from a pressurised line	250 - 4	0,025	1	3,5	0,07	Wheat flour
R	Pineau, 1985	featured air mill*	coaxial flow from a pressurised line	250 - 4	0,025	1	3,5	0,1	Wheat flour

* 1 m^3 vessel with pressurised injection

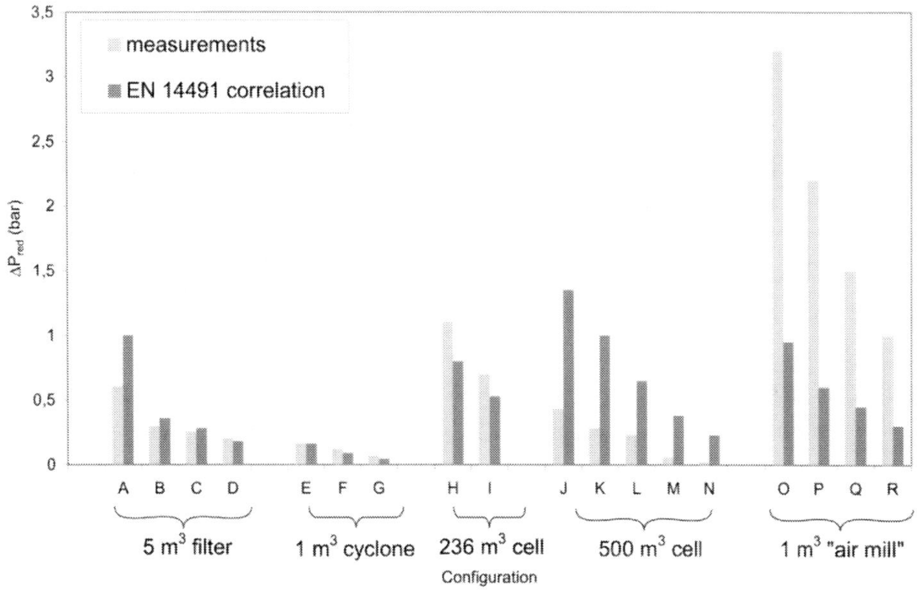

Figure 2. Predicted and measured overpressures for vented dust explosions developing in realistic equipments

One very likely reason for these discrepancies is the significant incidence of the initial turbulence of the cloud (Amyotte, 1985) upon the combustion (S_t). Since the turbulence level is strongly dependent on the flowfield inside the vessel before ignition, it is natural to think that a more accurate estimation of the explosion characteristics would be achievable only if enough details about the flowfield could be fed into the combustion.

CFD codes

Numerical methods which proved successful in predicting flowfields and turbulence in industrial processes appear as a logical start point and some have been modified to add combustion inside. The softwares used today at industrial scale date back from the early 1980s like REACFLOW

(Wilkening et al., 1999), AUTOREAGAS (century Dynamics/TNO, 1999) and FLACS (Bjerketvedt et al., 1997) to name but a few. To our knowledge, only FLACS has been adapted to dust explosions under the name DESC (Skjold et al., 2006). Significant effort has been devoted to their development and a lot of impressive data can be obtained. Our purpose is to discuss the way combustion in propagating flames is being handled.

Basically these codes solve in a "mesh" (finite volumes in space and finite difference in time) the Reynolds averaged Navier-Stokes equations which may read in a very general way:

$$\frac{\partial}{\partial t}(\rho \cdot \phi)+\frac{\partial}{\partial x_j}(\rho \cdot u_j \cdot \phi)=\frac{\partial}{\partial x_j}\left(\rho \cdot \Gamma_\phi \cdot \frac{\partial}{\partial x_j}(\phi)\right)+S_\phi$$

using the standard Einstein formalism where ϕ is a scalar like a velocity, a mass fraction, the enthalpy… $\Gamma\phi$ is the related diffusion coefficient and $S\phi$ the relevant source (sink) term. As far as the consumption of the reactants is concerned, $S\phi$ is roughly calculated as (Arntzen, 1993):

$$S_\phi \propto \frac{S_b}{n \cdot \Delta}$$

where n is a preset number of cells of size Δ typically 3 to 5. The burning velocity S_b is calculated from a library of data like Bray or Gülder models (Bray, 1990; Gülder, 1990) for the turbulent regime but the initial laminar flame propagation in the vicinity of the ignition source is also described. To be physically meaningful, the scale on the denominator should have been a flame thickness and the cell chosen size even smaller. But the choice of the cell size depends also on the dimensions of the process to be modelled and on the computer resources bearing in mind that a 1 million point calculation is significant. Δ is often of the order of a few tens of cms and so that the "numerical flame thickness $n\ \Delta$" may be strongly disconnected from the burning scales (in the laminar regime the flame thickness is of the order of a fraction of mm). To allow the calculations anyway, the burning is being artificially spread over an optimised number of cells and properly tuned so that the calculated flame velocity is in agreement with the data from the library.

In this situation, combustion is considered mainly as a source of heat and the interaction of the flame with its environment appears mostly through turbulence. To some extent the flame is rather passive. Today, it is recognised that a flame is able to generate its own flow and turbulence (Bradley, 2001; Bychkov et al., 2000) acting as an amplifier of disturbances through the Landau-Darrieu instability mechanism. A very simple situation is represented by the free upward laminar flame propagation in a vertical tube open at the ignition end (bottom) and closed at the other (top). Van Wingerden (1994) calculated the propagation of a flame with a burning velocity of 25 cm/s in a 16 cm tube. Comparable experimental data obtained with a very lean methane-air flame (Proust, 2004) are presented on the same figure (Figure 3) with the same scales. Clearly the curvature of the simulated flame is much smaller and the flame area accordingly (by a factor 2 at least). In the calculations, the typical flame thickness is about 5 cms. At least for this reason, the flame cannot behave in that tube as a "thin" interface able to develop the classical parabolic shape due to the hydrodynamic instability.

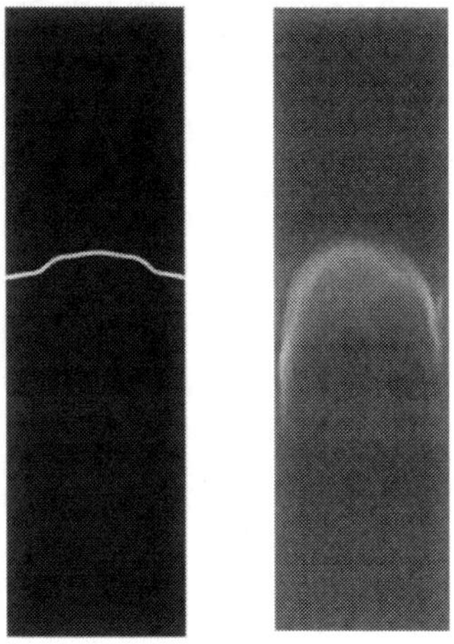

Figure 3. Simulated (van Wingerden, 1994) and observed (Proust, 2004) shape of a lean premixed flame (6.3 % methane in air) propagating in a vertical tube closed at the ignition end (bottom) and closed at the top

Some difficulties may then arise when trying to simulate the propagation of the flame in an initially quiescent mixture without any obvious reasons for a fully turbulent flow to appear. Consider for instance, the explosions of a homogeneous stoechiometric quiescent methane-air mixture in an empty room of 1 m^3 with a 0.15 m^2 open vent (Figure 4). Experiments and numerical simulations (with AUTOREAGAS) have been performed.

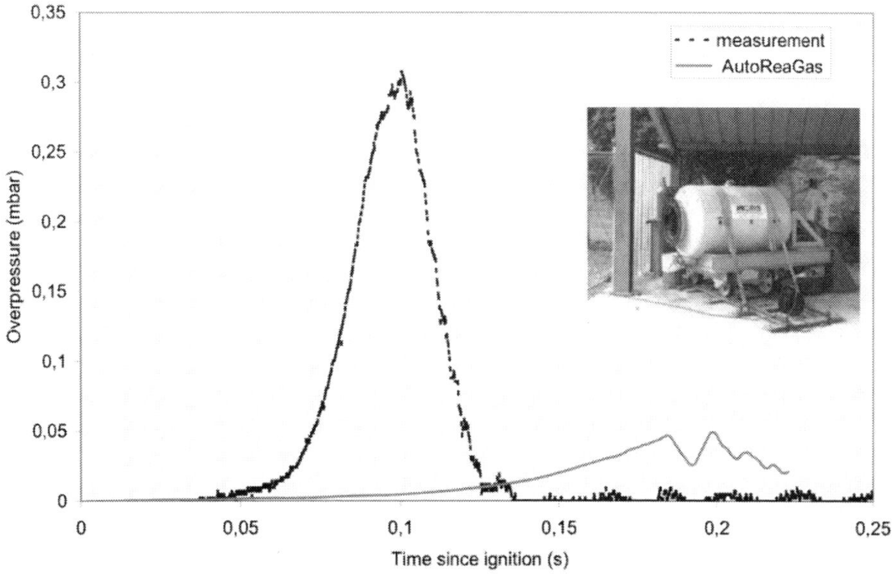

Figure 4. Measured and simulated internal overpressure in the 1 m^3 ISO vessel (0.95 m inner diameter, 1.4 m long) with rear ignition and an open vent of 0.15 m^2 mélange (stœchiometric methane-air)

The calculation underestimate significantly the overpressure inside the chamber. Similar observations have been made previously (Janovsky et al., 2004). It is believed that the reason for this comes from the modelling point of view discussed above. One conclusion is that the present CFD codes may not be used outside their domain of modelling for which it is intrinsically assumed that the combustion is largely dominated by the turbulence of the flow.

Other tricky points exist like the so-called "subgridding" process according to which the effect of any object smaller than the cell size is supposed to generate additional turbulence in the flow. This assumption and the way it is implemented may appear speculative but "subgridding" surely simplifies greatly the input of the geometrical data since the small objects need not be described accurately.

Phenomenological tools

These intrinsic physical limitations may a reason for continuing the development of advanced physical/analytical modelling but may not be the main justification for this. Consider first that in most cases there is not enough geometrical details available to implement a CFD code in a reasonable way (and at an affordable cost) and second that the flame propagation question is not, by far, the only concern. Stakeholders are mainly concerned by mechanical effects linked to the explosion such as disruption of containments, debris projection, blast… to implement any steps of decision making like the choice of proper mitigation techniques. Questions pertaining to flame propagation are seen as a necessary step but regarded as a background.

The necessity of being able to answer these questions with sufficient accuracy became urgent at the beginning of the 1990s. Owing to the complexity of the subject which involves the mechanical resistance of structures, flame propagation and fluid mechanics, etc., a simulation method clearly had to be developed. Neither the classical scaling laws such as those appearing in the venting methods, too coarse, nor the CFD codes, too much targeted on a single aspect of the problem (ex: turbulent driven flame propagation) appeared as reasonable solution. EFFEX originates from this need and was built brick by brick within a few years, beginning by a flame propagation module (the core of the software), completed later by peripheral tools aiming at predicting turbulence characteristics, combustion rates, mechanical resistance, projection of debris and external pressure effects (Proust et al., 2000). EFFEX is not a CFD code but rather precise phenomenological software. To some extent, this tool obeys the same logics as ISOVEX developed by FMRC (Tamanini et al., 1996). As it is difficult to envision carrying out full-scale tests, models have been qualified separately and the overall consistency has been checked by comparison to actual accidents. Most of the physics is derived from fundamental research. The operational version of EFFEX (version 3) contains eight modules which solve at least 40 non-linear time-differential equations. Hereafter are given some details only about the flame propagation modules (TURBULENCE, COMBUSTION and EXPBAT as illustrated in Figure 5). More information is given elsewhere (Proust, 2005).

Turbulence		
Injection mode	*pneumatic and blast*	*gravity*
Injection pressure (mb)	306,00	*************
Port diameter (m)	2,60	0,30
Number of ports	1,00	1,00
Volume injected (m^3)	1000,00	*************
Cross section of the vessel (m^2)	125,00	64,00
Height of the vessel (m)	8,00	39,00
Injection velocity (m/s)	209,08	19,56
u'jet (m/s)	13,23	1,24
Ljet (m)	2,93	0,34
u' (m/s)	13,92	0,73
L (m)	1,12	0,80

Combustion		
Injection mode	*pneumatic and blast*	*gravity*
Preexponent factor	0,70	0,70
Exponent for u'	0,75	0,75
Exponent for L	0,25	0,25
Laminar burning velocity (m/s)	0,23	0,23
Expansion ratio	5,00	5,00
Turbulent burning velocity in jet (m/s)	46,99	5,05
Mean turbulent burning velocity (m/s)	38,45	4,30

```
###### CALCUL DE L'EXPLOSION DANS UN BATIMENT   (XABCDE1E) ######
                    DATE : 12/8/2005 - 15H22

   Legende de calcul : demo

         DONNEES RELATIVES AUX BATIMENTS

   Section droite du batiment AB (m2) :  49.0
   Longueur du batiment Lst (m) :  7.0
   Surface ouverte en permanence Souv (m2) : 18.0

         DONNEES RELATIVES AUX BRECHES QUI SE FORMENT
   Pression d'eclatement de la paroi DeltPrup (hPa) : 15.00
   Masse par unite de surface de la paroi SIG (Kg/m2) : 15.0
   Perimetre des breches (m) : 18.0
   Surface des breches (m2) : 18.0

         DONNEES RELATIVES AU FRAGMENT CONSIDERE
   Dimension caracteristique du fragment Lbrech (m) : 0.50
   Altitude initiale moyenne du depart H (m) : 6.0
   Angle de projection par rapport
     a l'horizontal Alfa0 (degres) : 20

         DONNEES RELATIVES A LA COMBUSTION DU MELANGE
   Vitesse de Combustion St (m/s) : 2.0
   Rapport d'expansion volumique des produits brules Ratio : 7.00


Etes-vous satisfait de ces données ? (Oui=0/Non=1)
(par défaut Oui) :
```

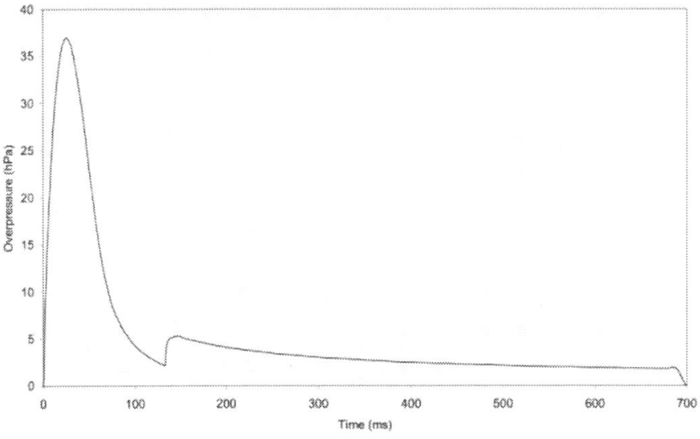


```
                 Signification de l'indice * :
                  1 : Altitude Maxi
                  2 : DeltPmax atteint
                  3 : DeltP=0
                  4 : DerivP<0
                  5 : Changement de Deb (initial
```

Iter	Temps (ms)	Surpres-sion (hPa)	Fraction (1) (%)	Vitesse projecti-le (m/s)
0	0.0	0.00	0.0	0.00
0	0.0	0.00	0.0	0.00
1	0.2	1.00	0.1	0.00
10	2.1	7.83	0.9	0.00
20	4.2	13.83	1.8	0.00
30	6.3	18.36	2.8	0.18
40	8.4	22.60	3.7	0.47
50	10.5	26.15	4.6	0.82
50	10.5	26.15	4.6	0.82
60	12.6	29.10	5.6	1.21
70	14.7	31.52	6.5	1.63
80	16.8	33.44	7.4	2.09
90	18.9	34.91	8.2	2.57
100	21.0	35.95	9.1	3.06
110	23.1	36.61	9.9	3.57
120	25.2	36.91	10.8	4.08
125	26.2	36.93	11.2	4.33
150	31.5	35.90	13.1	5.58
200	42.0	29.64	16.7	7.78
250	52.5	21.10	20.0	9.37
300	63.0	13.80	22.9	10.36
350	73.4	9.17	25.7	10.94
400	83.9	6.54	28.2	11.28
450	94.3	4.93	30.7	11.49
500	104.7	3.84	33.0	11.62
550	115.1	3.06	35.3	11.69
600	125.6	2.49	37.4	11.72
635	132.9	2.17	38.9	11.73
638	133.5	2.15	39.0	11.72
650	136.0	4.61	39.6	11.72
700	146.4	5.28	41.9	11.70
750	156.9	4.99	44.0	11.68
800	167.3	4.74	46.1	11.64
850	177.7	4.51	48.1	11.59
900	188.2	4.31	50.0	11.55
950	198.6	4.13	51.8	11.51
1000	209.0	3.97	53.6	11.47
1050	219.4	3.83	55.3	11.43
1100	229.9	3.69	56.9	11.39
1150	240.3	3.57	58.5	11.35

— Suite —

Figure 5. TURBULENCE, COMBUSTION and EXPBAT modules of EFFEX

The first step is to estimate the relevant parameters of the turbulence from the process characteristics (TURBULENCE module). Only the integral properties like the turbulence intensity u' and turbulence scale L (Hinze, 1975) are required for turbulent flame propagation. Pertinent experimental data are available (Hauert et al., 1994; Roux and Proust, 2003).The integrated "k-epsilon" model proposed by Tamanini is tempting (Tamanini, 1998) but has been found to be insufficiently accurate. Another model has been devised based on the observation that, in most situations, the turbulent movements are caused by jets, sometimes due to explosion and pressure development in the neighbouring vessels. The classical jet theory has been implemented to calculate the recirculation patterns and deduce u' and L. The model has been adapted to various flows like: pneumatic multiport pressurised injection, flame jets coming from a neighbouring vessel, pneumatic feeding... Excellent agreement with existing is usually obtained (Figure 6).

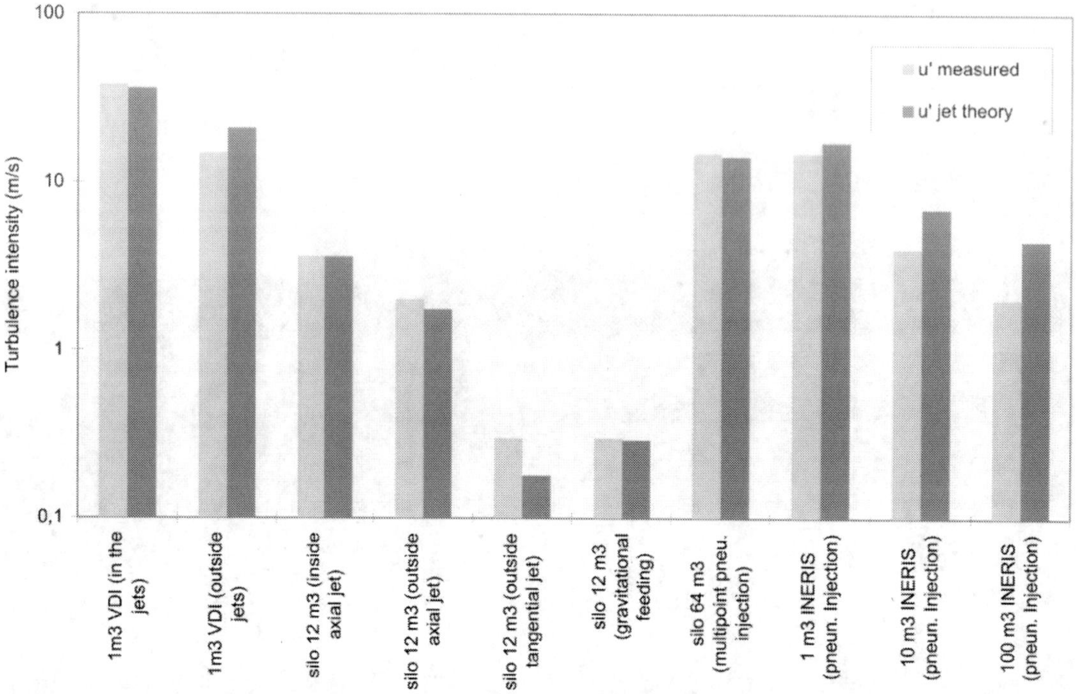

Figure 6. Comparison between turbulence intensity measurements in realistic situations and prediction by means of the present "jet theory")

The second step is to evaluate the burning velocity (COMBUSTION module). Fundamental works (Proust, 2006) suggest strong similarities between premixed gas flames and dust flames. The turbulent combustion rate should depend u' and L and the specific reactivity of the mixture. Correlations similar to those used for premixed gaseous flames (Schneider and Proust, 2006) have been selected. The predictions seem to compare favourably (Figure 7) to experimental data (Proust et al., 2000), showing in particular that the scale effect, through L, seems correctly captured. Nevertheless, more direct measurements are desired at full scale and for other dusts to check this important correlation.

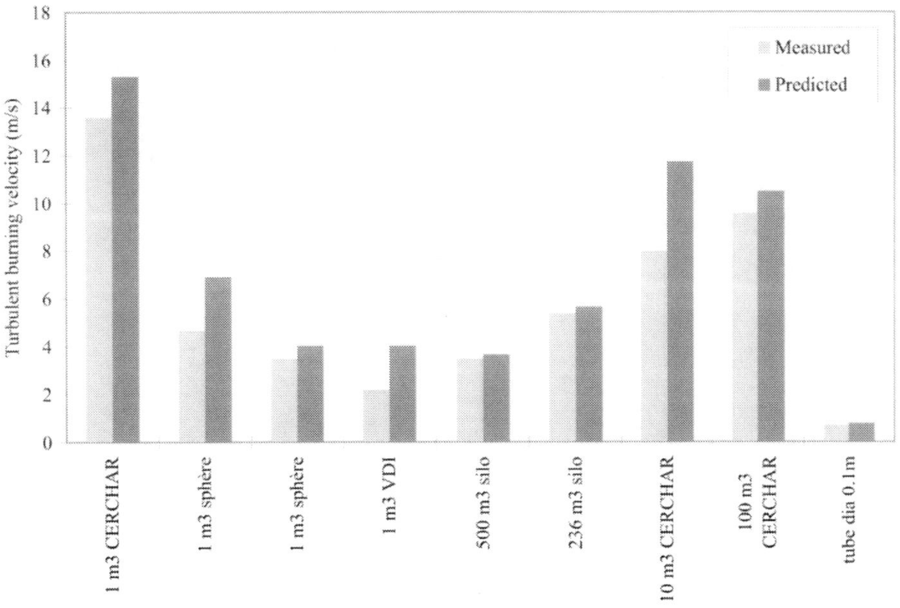

Figure 7. Comparison between predicted and measured turbulent burning velocities (starch-air clouds)

The third step is to calculate the pressure evolution. In a real plant, the flame propagates in vessels and ducts. For gaseous mixtures at least, recent work (Proust, 2004) has clearly shown that the flame propagation and explosion development do not obey the same physics in both configurations. In vessels, the pressure is more or less uniform and the pressure is directly linked to the expansion of the flame ball. The simulation module used for this purpose (EXPBAT) includes conventional time dependant equations which can be used to describe the propagation of flames in a vessel (Bradley and Mitcheson, 1976; Kansa and Perlee, 1978, Lewis and von Elbe, 1987). In practical situations, the initial flame ball development is skipped to concentrate on the subsequent development of the flame when the flame is fully deployed. The flame area is constant, directly linked to the cross section of the vessel via a constant experimental flame folding parameter (Proust and al., 2000). A number of comparisons have been made in extremely varied configurations including those of Table 1 showing a good agreement (Figure 8).

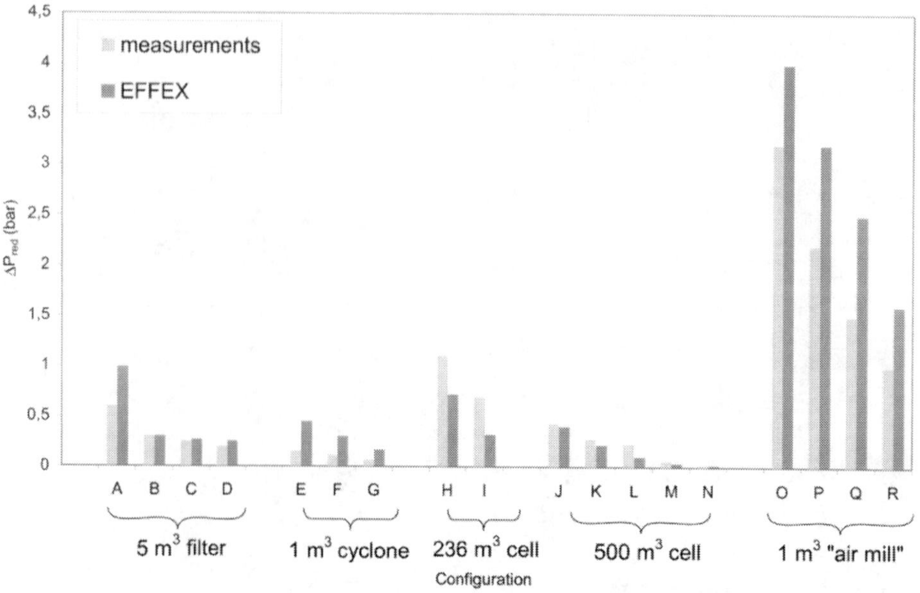

Figure 8. Comparison between predicted and measured overpressure in vented isolated vessels (configurations from Table 1)

A very tricky situation, the simulation of which necessitates to deploy the loop EXPBAT-TURBULENCE-COMBUSTION, appears in interconnected situations when the flame rushes from one vessel to another through an orifice. To test the ability of the code to do this, the excellent data of HSL have been used (Lunn et al., 1996; Holbrow et al., 1996). A standard coal dust explosion is being initiated in the donor vessel, which is simulated first accounting for the initial turbulence level induced by the dispersion of the dust from pressurised reservoirs. Then, the maximum explosion pressure of this first explosion is used as an input to calculate the turbulence in the second vessel with TURBULENCE and subsequent combustion rate (COMBUSTION). Then, EXPBAT is run again for the second vessel. As shown in Figure 9, the level of consistency appears to be satisfactory.

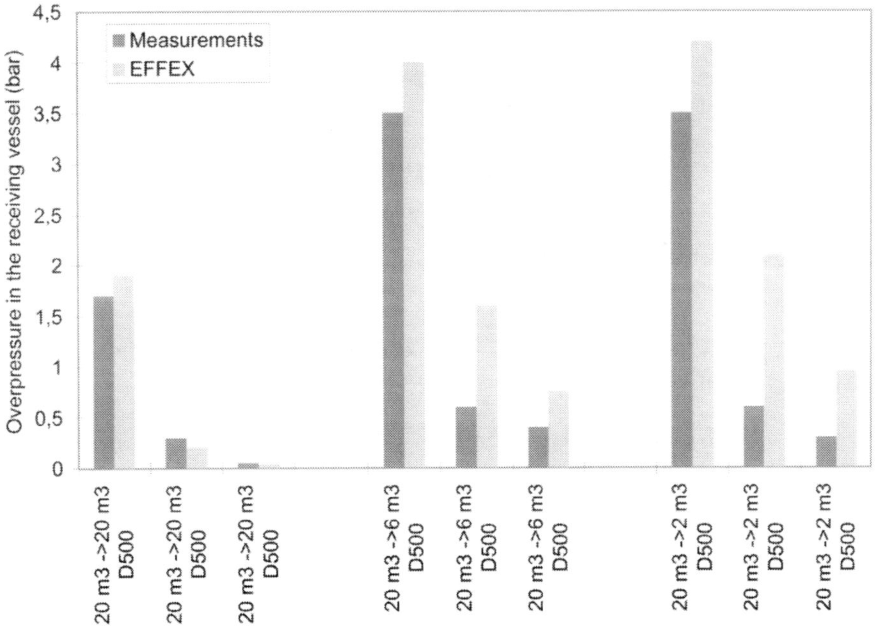

Figure 9. Comparison between predicted and measured overpressure in interconnected vented vessels (coal dust-air clouds, overpressure in the receiving vessel through a 500 mm duct)

Conclusions

Some modelling ways are discussed in the present paper in the context of dust explosions. It is acknowledged that engineering laws may be accurate in their domain of validity and prove to be an excellent mean to train people and promote safety techniques. But usually, they are valid in a very limited range of situations. CFD codes are good at predicting the flow fields and have a broader range of applicability: a large effort has been devoted to their development along the last 25 years and mature, impressive softwares are now available. Their major limitation is certainly that the flame equation is not implemented: combustion is modelled as a kind of heat source. Following several important features of the interaction of the flame with its environment cannot really be tackled like the intrinsic instability. Releasing this shortcoming will undoubtedly require significant changes. A possible promising route has been propose recently (Daubech et al., 2006) in which the flame surface is considered at a 2-D medium separating two fluids with different densities. A "front tracking" method is applied to the flame and jump relationships to connect the flow fields. It seems compatible with existing CFD softwares. Intermediate tools, between "simple", but robust, engineering laws and complex, but somewhat uncontrollable, CFD codes, have been developed like ISOVEX and EFFEX. The example of EFFEX is presented. It is a phenomenological tool able to predict the propagation of a confined dust explosion in a complex network of interconnected volumes. Internal and external pressure effects are modelled together with rupturing and projection of debris. The models are based on precise physical concepts and have been checked against experimental results step by step. The global software performance have been assessed with existing data extracted from well documented dust explosions.

Literature

1. AMYOTTE P.R. (1985), « A review of the effects of turbulence on gas phase, dust and hybrid mixture explosions », Technical report n°004, dept of Chem. Eng., technical University of Nova Scotia, Canada, juin 1985
2. ARNTZEN B.J. (1993), Final Report on modelling in MERGE, CMI report n° CMR-93-F25042
3. BRADLEY D., MITCHESON A. (1976), « Mathematical solutions for explosions in spherical vessels », Comb. and Flame, vol. 26
4. BRADLEY D., MITCHESON A. (1978), « The venting of gaseous explosions in spherical vessels I-theory », Comb. and Flame, vol. 32, pp. 231-236
5. BARTKNECHT W. (1986), « Pressure venting of dust explosions in large vessels », communication au 20th annual Loss Prev. Symp. , New Orleans, avril 1986
6. BARTKNECHT W. (1993), « Explosion-schuzt : Grundlagen und Anwendung », Springer-Verlag, Berlin, ISBN 3-540-55464-5
7. BJERKETVEDT D., BAKKE J.R, Van WINGERDEN K. (1997), « Gas explosion handbook », J. hazardous materials, vol ; 52, pp. 1-150
8. BYCHKOV V.V., LIBERMAN M.A. (2000), "Dynamics and stability of premixed flames", Physics Reports, pp. 115-237BRADLEY B., CRESSWELL T.M., PUTTOCK J.S. (2001), « Flame acceleration due to flame-induced instabilities in large scale explosions », Comb and Flame, vol. 124, pp 551-559
9. BRAY K. (1990), « Studies of turbulent burning velocities », Proceedings of the Royal Societyof London, vol. A431, pp. 315-325
10. Century Dynamics/TNO (1999), AutoReagas user Documentation, version 3.00
11. DAUBECH J., PROUST Ch., SOCHET I. (2006), « Flame front perturbations induced by concentration gradients », communication to the 6th ISHPMIE, Halifax, Canada, sept 2006
12. DONAT K (1971), « Auswal und Bemessung von Druckentlastungseinrichtungen f'r Staubexplosionen », Staub Reinh. Luft, vol. 31, pp.154-160
13. ECKHOFF R.K. (1991), " Dust explosions in the process industries ", Butterworth-Heinemann, Oxford
14. ECKHOFF R.K., ALFERT F., FUHRE K., MILLS J.D., PERDERSEN G.H. (1986), « Maize starch explosions in a 236 m^3 experimenrtal silo with vents in the wall », rapport CMI ref CMI n° 863307-1, decembre 1986
15. ECKHOFF R.K., FUHRE K. (1984), « Dust explosion experiments in a 500 m^3 silo cell », J. of Occupational Accidents, vol. 6
16. GÜLDER Ö.L. (1990), « Turbulent premixed flame propagation models for different combustion regimes », Comptes-rendus du 22nd Symp (Int.) on Combustion

17. HAUERT F., VOGL A., RADANDT S. (1994), « Measurement of turbulence and dust concentration in silos and vessels », communication présentée au 6th Int. Coll. on Dust Explosions, Shenyang, août 1994

18. HINZE J.O. (1975), « Turbulence », 2nd edition, Mc Graw-Hill company, New-York, ISBN 0-07-029037-7

19. HOLBROW P., ANDREWS S., LUNN G.A. (1996), « Dust explosions in vented interconnected vessels », J. Loss Prev. Process Ind, vol. 9

20. JANOVSKY B., PODSTAWKA T., HORKEL J. VEJS L. (2004), « Vented confined explosions : experiments and AutoReagas simulation », communication at the 11th Loss prev. and Safety promotion in the Process Ind., Praha, Czeck republic, June 2004

21. KANSA E.J., PERLEE H.E. (1978), « Constant volume flame propagation : finite sound speed theory », US Bureau of Mines, RI 8163

22. LEWIS B., Von ELBE G. (1987), « Combustion, flames and explosions of gases », 3rd edition, Academic Press, London, ISBN 0-12-446751-2

23. LUNN G.A., HOLBROW P., ANDREWS S., GUMMER J. (1996), « Dust explosions in totally enclosed interconnected vessels », J. Loss Prev. Process Ind, vol. 9

24. MASSON F. (1998), « Explosion d'un silo de céréales », Rapport de synthèse, INERIS EMA-FMs-98-21FP30-07/07/98

25. PINEAU J.P. (1985), »Gas and Dust Explosions in closed and vented Vessels », « A two-day practical approach to gas and dust explosion venting », EUROPEX seminar, Frankfurt, June 1985

26. PROUST Ch., ROUX P., CHUUON B. (2000), "Prévoir les effets des explosions de poussières sur l'environnement : EFFEX un outil de simulation", Rapport final d'une étude réalisée pour le compte du ministère de l'Environnement , www.ineris.com

27. PROUST Ch. (2006), " A few fundamental aspects about ignition and flame propagation in dust clouds ", J. Loss Prev. Process Ind., vol. 19, pp. 104-120

28. PROUST Ch. (2005), " The usefulness of phenomenological tools to simulate the consequences of dust explosions : the experience of EFFEX", communication to the international ESMG symposium, oct 2005, Nürnberg, Germany

29. PROUST Ch. (2004), " Formation, inflammation, combustion des atmosphères explosives (ATEX) et effets associés ", Mémoire d'HdR Institut National Polytechnique de Lorraine.

30. ROUX P., PROUST Ch. (2003), « Méthode de protection contre les explosions : évents - explosions de poussières », Rapport final établi pour le compte du Ministère de l'Environnement et du Développement Durable

31. SCHNEIDER H., PROUST Ch. (2005), "Laminar and turbulent burning velocities of dust clouds", paper presented to the international ESMG symposium, oct 2005, Nürnberg, Germany

32. SKJOLD T., ARNTZEN B.J., HANSEN O.R., STORVIK I.E., ECKHOFF R.K. (2006), « Simulation of dust explosions in complex geometries with experimental input from standardized tests », J. Loss Prevention, Process. Ind. , vol. 19, pp. 210-217

33. TAMANINI F., VALUILIS J.V. (1996), "Improved guidelines for the sizing of vents in dust explosions", J. Loss Prevention, Process. Ind. , vol. 9, pp. 105-118.

34. TAMANINI F. (1998), "The role of turbulence in dust explosions", J. Loss Prevention, Process. Ind. , vol. 11, pp. 1-10.

35. TONKIN P.S., BERLEMONT C.F.J. (1972), « Dust explosions in a large scale cyclone plant », Internal report of the Joint Fire Research Organization, n° 942, July 1972

36. Van WINGERDEN K. (1994), « Measurement of fundamental burning velocity of dust-air mixtures in industrial situations », Interim report of CREDIT project, October 1994

37. WILKENING H., HULD T. (1999), « An adaptative 3-D solver for modelling explosions on large industrial environment scale », Comb ; Sci. Techn., vol. 149, pp. 361-387

LOSS PREVENTION SYMPOSIUM

Simulation and Application of Blast Wave-Target Interaction

Jihui Geng, Ph.D.
Kelly Thomas, Ph.D.
Baker Engineering and Risk Consultants, Inc.
San Antonio, TX 78218 U.S.A.
Phone: 210-824-5960
jgeng@bakerrisk.com

Prepared for Presentation at
American Institute of Chemical Engineers
2007 Spring National Meeting
41st Annual Loss Prevention Symposium
Houston, Texas
April 22–26, 2007

UNPUBLISHED

Simulation and Application of Blast Wave-Target Interaction

Jihui Geng, Ph.D.
Kelly Thomas, Ph.D.
Baker Engineering and Risk Consultants, Inc. (BakerRisk)
San Antonio, TX 78218 U.S.A.
Phone: 210-824-5960
jgeng@bakerrisk.com

Abstract

This paper presents example applications of a computational fluid dynamics (CFD) computer code to the evaluation of blast wave propagation and interaction problems relevant to the chemical processing and petroleum refining industries. The specific phenomena illustrated in these example applications are: clearing of a blast wave over a wall, enhancement of blast loads behind a door blast shield, and the development of an enhanced negative phase on an enclosure wall.

The code used for these evaluations is the Blast Wave-Target Interaction (BWTI ™) simulation package. BWTI ™ is an integrated three-dimensional (3D) CFD code and visualization system developed by BakerRisk to provide the capability to simulate the generation and propagation of blast and shock waves and the interaction of such waves with structures (e.g., buildings, blast walls, etc.). The code is capable of modeling blast waves resulting from vapor cloud explosions (VCE), bursting pressure vessels (BPV), and high explosive (HE) detonations. BWTI™ was developed to run on a personal computer (PC) with acceptable run times while still providing reasonably accurate evaluations. The main application area for this type of tool is an engineering-level analysis involving scenarios where blast shielding and focusing are relevant, such that simplified methods may not yield accurate results.

1. Introduction

There are a variety of simplified engineering analysis methods available to evaluate the generation and propagation of blast and shock waves from VCE, BPV and HE explosion sources. For example, the Baker-Strehlow-Tang (BST) and TNO multi-energy methods are both widely used to estimate the properties of blast waves resulting from a VCE[1]. Blast charts are available that can be used for the analysis of a BPV or HE explosions[2,3]. For many engineering analyses, the application of these simplified methods, coupled with a simplified representation of the blast wave interaction with the structure of interest, is sufficient for the purposes of the required evaluation. However, the application of simplified methods is not always justified. The simplified methods may lead to inaccurate results in cases where blast wave focusing, shielding and reverberations play a large role[4]. The application of a more detailed analysis method, such as CFD, can be justified for such cases in order to ensure a sufficiently accurate evaluation.

2. Overview of CFD Code

BWTI™ is an integrated three-dimensional CFD code and visualization system. An object-oriented interface is incorporated into the system so that models can be rapidly constructed and results can be readily viewed. This allows analyses to be conducted in a timely and cost-effective fashion. Figure 1 illustrates the use of this interface to view a model, the corresponding unstructured mesh, and the results obtained with the model. Model results can also be exported for use with other visualization software.

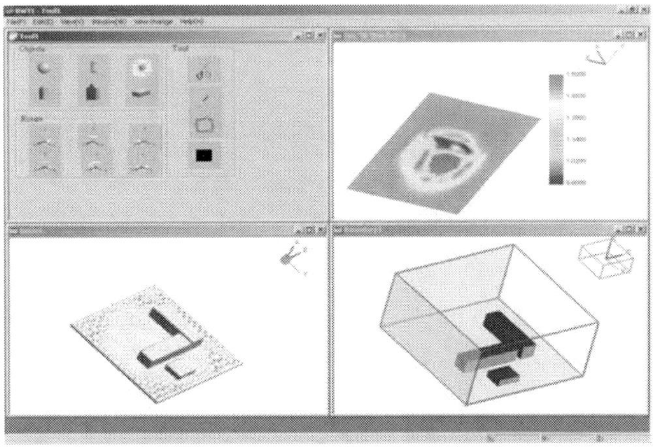

Figure 1. Code Interface Screen

BWTI ™ employs an unstructured adaptive mesh for both 2D and 3D analyses[5]. The use of an unstructured mesh formulation allows irregular geometries to be conveniently represented. An adaptive mesh allows a refined grid to be employed in the vicinity of the blast wave while using a coarser grid away from the wave, allowing an accurate simulation with reduced computational demands. The code is also capable of dynamic mesh generation in the vicinity of moving boundaries for 2D simulations[6].

The code solves the eulerian governing equations. The Finite Volume Method is employed on the unstructured meshes, with application of the PLM method, Godunov and HLLC schemes. A double-gamma treatment is used to simplify the calculation of thermo-physical properties. An energy wave model is employed for subsonic VCEs (i.e., deflagrations), with the flame speed being an input to the model (i.e., the code does not directly model the turbulent combustion and flame acceleration). For gas-phase detonations, a two-step model is used to simulate the detonation wave.

Examples of typical BWTI analyses are shown in Figure 2 through Figure 4. Figure 2 shows an example of VCE blast wave propagation in the vicinity of a vented enclosure. The blast wave exits the enclosure at about 80 ms, and the vented blast wave has propagated over most of surrounding buildings by 230 ms. Figure 3 illustrates a propagation process of a blast wave from a BPV internal to an enclosure. The initial expansion of the blast wave from a tunnel into a chamber can be seen at time of 20 (dimensionless), with propagation throughout the enclosure by a time of 98. Figure 4 illustrates the interaction of an HE blast wave with a set of high-rise buildings.

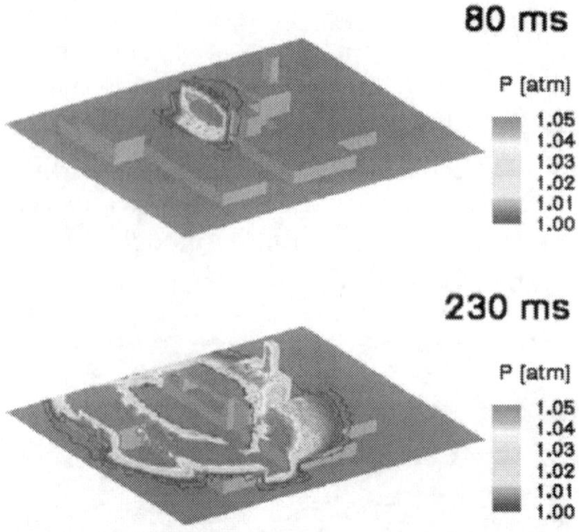

Figure 2. Vented VCE Blast Wave Interaction with Surrounding Buildings

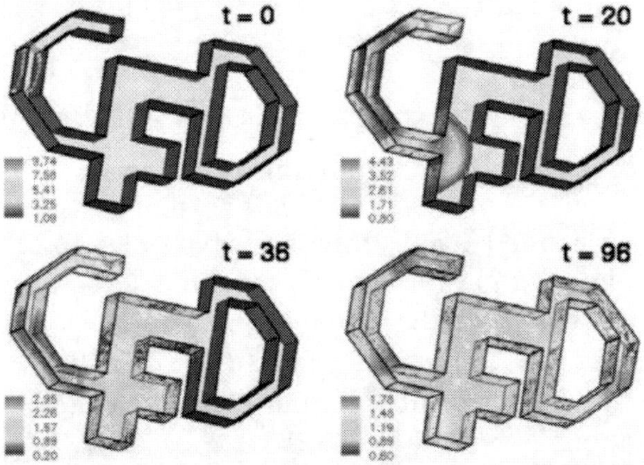

Figure 3. Propagation of BPV Blast Wave Within Enclosure

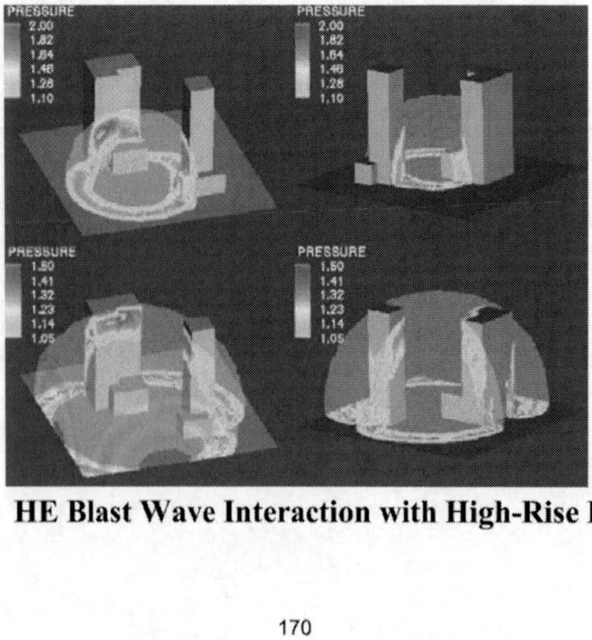

Figure 4. HE Blast Wave Interaction with High-Rise Buildings

3. Example Applications

The specific phenomena illustrated in these example applications are: the clearing of a blast wave over a blast wall, enhancement of blast loads behind a door blast shield, and the development of an enhanced negative phase on an enclosure wall. Each of these examples is discussed separately below.

3.1 Blast Wave Clearing

A typical free-field (side-on or incident) pressure history for a blast wave generated from a vapor cloud explosion with a subsonic flame speed (i.e., a deflagration) is shown in Figure 5. The positive phase has a finite rise time and is followed by a negative phase. As the incident blast wave impinges on an infinite rigid wall, the wall impedes the expansion of the blast wave, halting the forward motion of the wave and causing it to reflect off the wall. The reflected blast wave is also shown in Figure 5. The reflected peak overpressure will be in excess of twice the free-field pressure, depending on the incident blast wave strength.

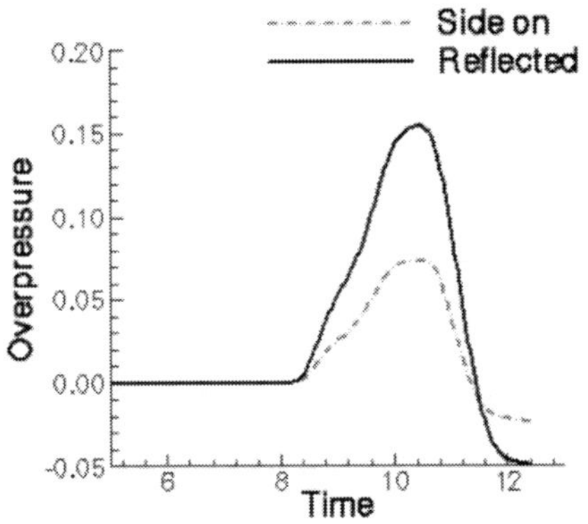

Figure 5. Typical Free-Field and Reflected VCE Blast Waves

When the blast wave reflects from a wall with a finite height (H_0) such as a front surface of a typical structure or building, rarefaction waves are created at the edges of the front wall, with the rarefactions sweeping down from the top and inward from the sides. This process is generally referred to as clearing, and is illustrated in Figure 6. The rarefaction waves may begin propagating along the front of the wall before the peak of the wave reaches the wall, which reduces the pressure applied to the front wall.

Figure 7 shows the ratio of applied blast loads [overpressure (P_A) and impulse (I_A)] to the fully reflected blast loads (P_R and I_R); the fully reflected loads are those that would be developed on an infinite wall. At ground level (H=0), the applied blast loads approach the reflected values, but are still slightly smaller. The applied blast loads at the top edge of the wall are decreased dramatically due to the clearing effect. A localized region exists near the top of the wall where the pressure and impulse are higher than for adjacent locations. This is due to the Mach reflection of the pressure wave over the edge of the wall, as can be seen in Figure 6.

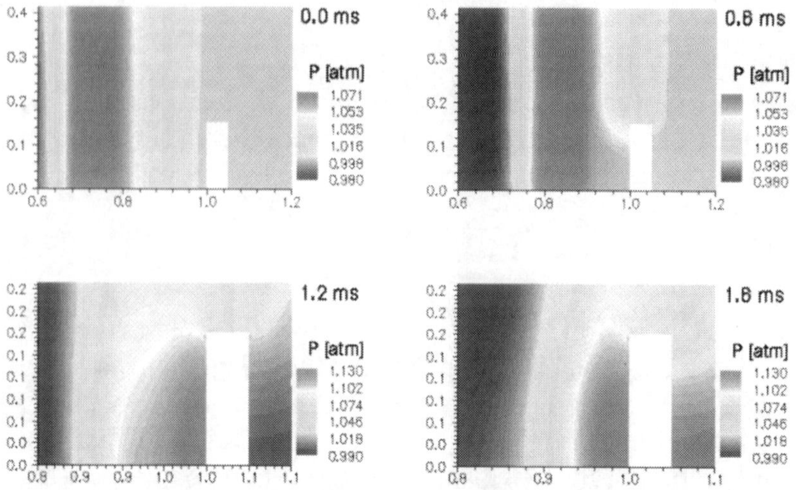

Figure 6. Blast Clearing Over Wall and Mach Reflection

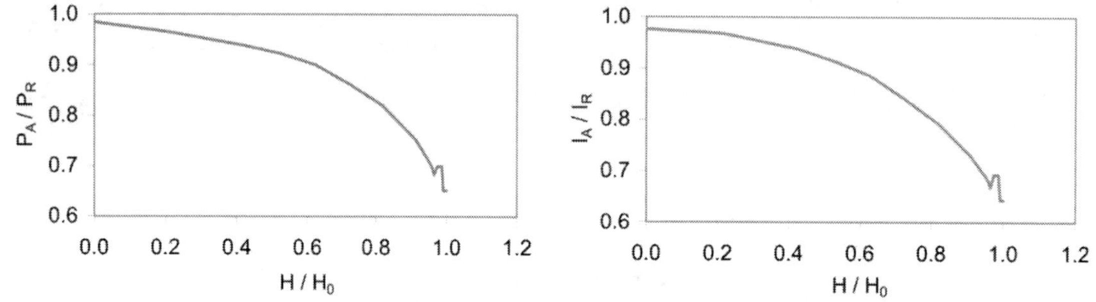

Figure 7. Relative Blast Loading on Wall

3.2 Blast Load Enhancement behind a Door Blast Shield

A door blast shield is a potential mitigation option when the blast load on an entryway to a building is sufficiently large that a blast door may be required. That is, when a conventional door does not offer sufficient blast resistance. Both testing and CFD analysis were carried out to evaluate the performance of a conventional "L-Shaped" door blast shield design[7]; this door blast shield design has been utilized at a variety of commercial and government sites. A series of tests was first performed using a scale model of the shield in a large shock tube fitted with a 16-foot long, open-ended extension section with an 8-foot square cross section. Tests were conducted with side-on pressures ranging from 1 to 4 psig and durations up to approximately 30 milliseconds. The scale model was placed such that the blast wave would strike the model at both a normal (i.e., 90 degrees) orientation and at 45 degrees. The tests showed that the "L-Shaped" shield was not effective at reducing the blast load applied to the door, and actually increased the applied blast load under some conditions. A simplified 2D CFD model was utilized to develop an understanding of the test results, and to explore alternative blast shielding arrangements. A comparison of the predicted and test data is shown in Figure 8. Although a simplified 2D CFD model was used, the agreement with the test data is quite good.

Figure 9 shows pressure contours for a blast wave striking the "L-Shaped" shield at a normal orientation; the pressures shown in this figure are in units of atmospheres (absolute). The initial rise in pressure at the door location is due to the diffraction of the incoming shock wave, the diffraction of the reflection of this shock wave off the building, the coalescence of these waves, and their reflection off the sidewall of the shield. The applied overpressure on the wall is only 9.6 psi, while the applied overpressure at door location behind the shield is 15 psi. That is, the "L-Shaped" door blast shield is not effective at reducing the applied door blast load, and can actually increase the applied load on the door. The use of a CFD code allowed the causes of this behavior to be understood, and supported the development of alternative shield arrangements.

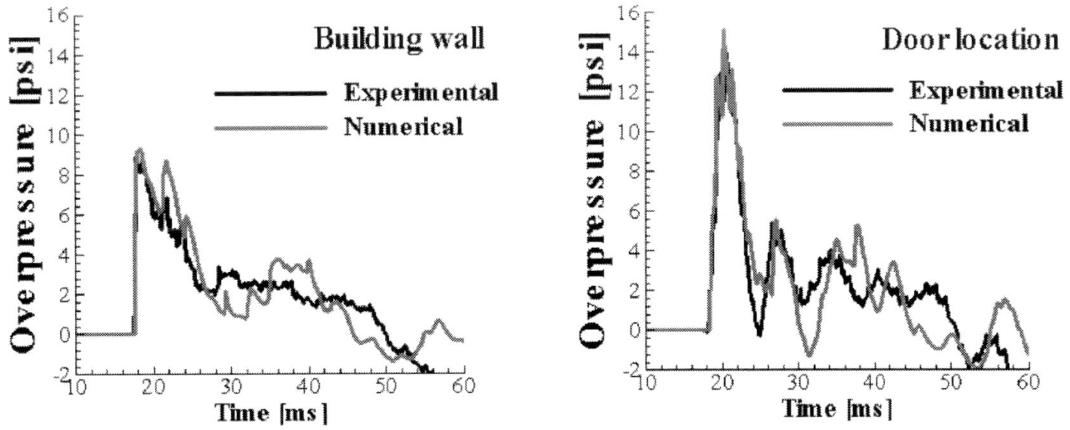

Figure 8. Comparison of Numerical (2D) and Experimental Pressure Histories

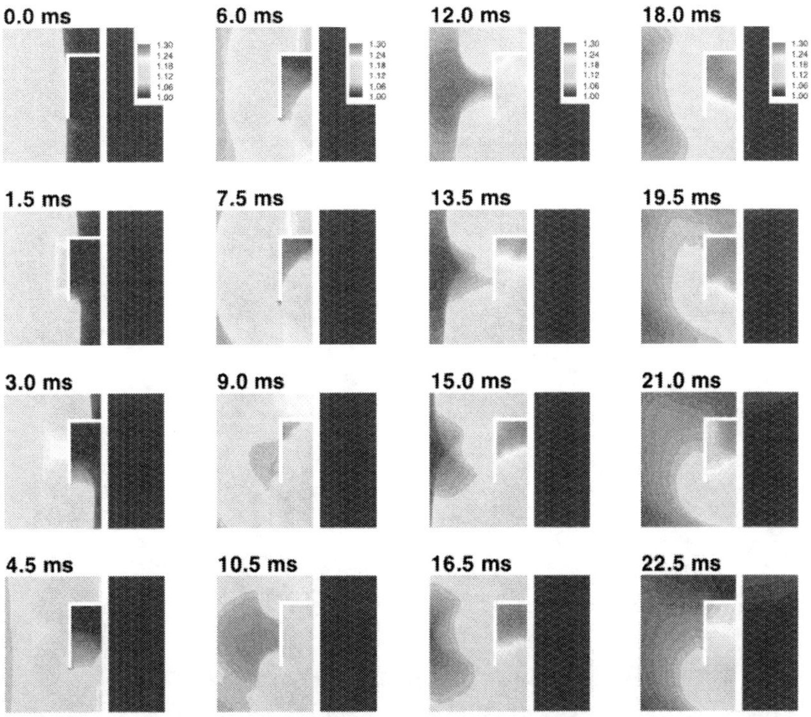

Figure 9. Pressure Contours (normal blast wave orientation)

3.3 Development of an Enhanced Negative Phase

The pressure history for an ideal free-field (side-on or incident) blast wave is shown in Figure 10. The negative phase of the blast wave is ignored in most blast load evaluations, and only blast parameters associated with the positive phase are considered or reported. However, there are scenarios in which the negative phase may become important to the response of the structure being evaluated. A blast wave interacting with a building wall containing a large opening, as shown in Figure 11, is an example of such a scenario. The blast wave passes through the opening and enters the building at 165 ms. The wave propagates along the inside face of the front wall starting at about 170 ms, and a negative phase is developed on the exterior of the wall at approximately 175 ms. Figure 12 shows the pressure histories on the outside and inside face of the wall, along with the net applied load (i.e., the difference between these two). As can be seen, the net loading during the negative phase is significant, and may impact the response of the structure.

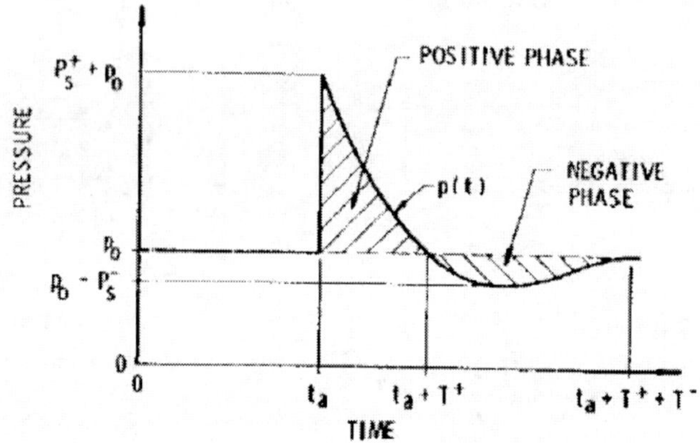

Figure 10. Ideal Blast Wave Pressure History[8]

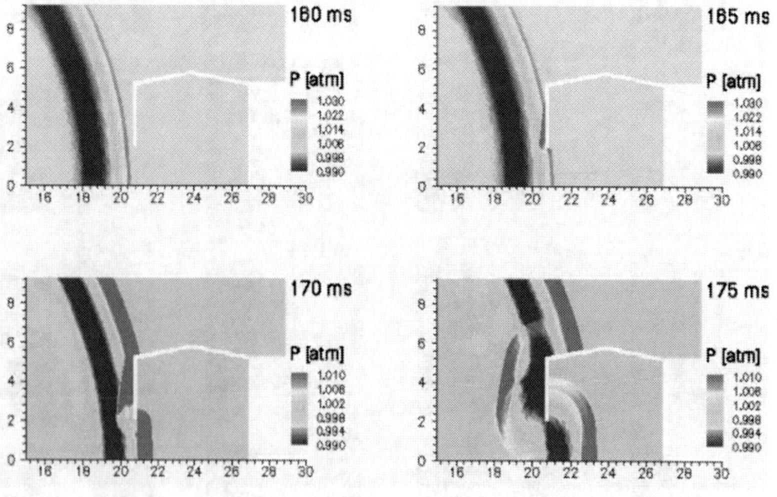

Figure 11. Pressure Contours at Selected Times

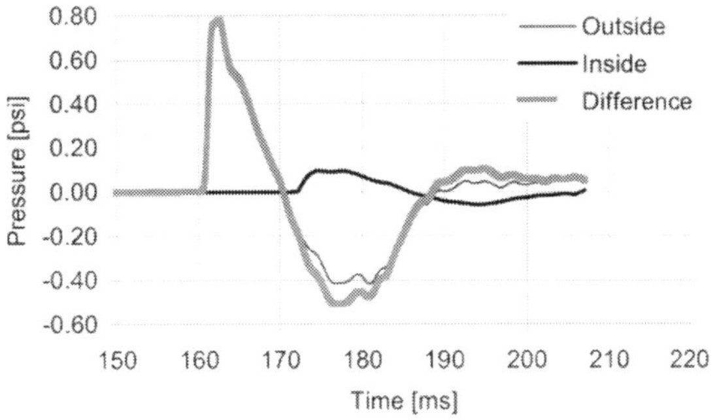

Figure 12. Interior Pressure Histories at Selected Locations

4. Summary

Example applications of a computational fluid dynamics (CFD) computer code to the evaluation of relevant blast wave propagation and interaction problems were provided. These examples illustrate typical situations where the application of simplified methods is not necessarily justified, and the use of a CFD code may be required to capture the relevant phenomena. CFD analysis should be considered in cases where blast wave focusing, shielding and reverberations play a large role. Modern computational resources, combined with computational packages designed to reduce required simulation times, make the application of such codes a reasonable choice for engineering-level analysis.

5. References

1) Pierorazio, A.J., J. K. Thomas, Q.A. Baker and D.E. Ketchum (2005) "An Update to the Baker-Strehlow-Tang Vapor Cloud Explosion Prediction Methodology Flame Speed Table," *Process Safety Progress*, 24(1): 59-65.

2) Baker, Q.A. and M.J. Tang (2004) "Effect of Vessel Temperature on Blast Loads," *Proc. 2004 ASME/JSME Pressure Vessels and Piping Conference*, San Diego, CA.

3) Baker, W.E., P.A. Cox, P.S. Westine, J.J. Kulesz and R.A. Strehlow (1983) Explosion Hazards and Evaluation, Elsevier Scientific, New York, NY.

4) Herrmann, D.D. (2005) "Developing a Sound Basis for the Design of Vented Explosion Barricades," *Process Safety Progress*, 24(1), Date: March 2005, Pages: 52-58

5) Geng, J.H., D.H. Wu and H.Z. Wu (2001) "Generation and application of unstructured adaptive meshes with moving boundaries," *Trans. of NUAA*, 18 (suppl):56-59.

6) Geng, J.H., H.Z. Wu and H.Z. Li (2001) "Numerical simulation of blast wave-induced body motion," *Proc. 23rd Intl. Symp. on Shock Waves*, Fort Worth, TX, pp. 635-640.

7) Thomas, J.K., J. Geng and A.D. Hallgarth (2004) "Evaluation of Door Blast Shields," *31st Department of Defense Explosive Safety Seminar*, San Antonio, TX.

8) Baker, W. E. (1973) *Explosions in Air*, Univ. of Texas Press, Austin, Texas.

The Design of Venting of Gas Explosions in the Presence of a Discharge Duct

Almerinda Di Benedetto
Istituto di Ricerche sulla Combustione
Consiglio Nazionale delle Ricerche
Via Diocleziano 328, 80125 Napoli, Italy
Phone: +39 0817682947
dibenede@irc.cnr.it

Paola Russo
Dipartimento di Ingegneria Chimica e Alimentare, Università di Salerno (I)
Phone: +39 089964090
parusso@unisa.it

Ernesto Salzano
Istituto di Ricerche sulla Combustione
Consiglio Nazionale delle Ricerche
Via Diocleziano 328, 80125 Napoli, Italy
Phone: +39 0817621922
salzano@irc.cnr.it

Prepared for Presentation at
American Institute of Chemical Engineers
2006 Spring National Meeting
41st Annual Loss Prevention Symposium
Houston, TX
April 22-26, 2007

UNPUBLISHED

The Design of Venting Of Gas Explosions in The Presence of a Discharge Duct

Almerinda Di Benedetto
Istituto di Ricerche sulla Combustione
Consiglio Nazionale delle Ricerche
Via Diocleziano 328, 80125 Napoli, Italy
Phone: +39 0817682947
dibenede@irc.cnr.it

Paola Russo
Dipartimento di Ingegneria Chimica e Alimentare, Università di Salerno (I)
Phone: +39 089964090
parusso@unisa.it

Ernesto Salzano
Istituto di Ricerche sulla Combustione
Consiglio Nazionale delle Ricerche
Via Diocleziano 328, 80125 Napoli, Italy
Phone: +39 0817621922
salzano@irc.cnr.it

Abstract

Venting devices are the main solution for the mitigation of accidental explosion. However, relief ducts are mandatory for the discharge of combustion products to safe location. On the other hand, the presence of a duct is likely to increase the severity of the explosion with respect to simply vented vessels. To this regard, several mechanisms have been proposed to account for the enhanced violence of the explosion in such configuration: secondary explosion in the duct (burn-up), frictional drag and inertia of the gas column in the duct, acoustic and Helmholtz oscillations. The relative effects of the cited phenomena and their weight on maximum pressure reached during ducted vented explosion is uncertain. As a consequence, appropriate design of duct-venting configuration is still a matter of debate and it's not surprising that the available guidelines for the design of ducted vents for gas explosions can lead to gross errors.

The authors have recently proposed experimental and numerical analyses (Computational Fluid Dynamics) for the analysis of duct-vented phenomena on the peak pressure measured in either small, medium or large scale equipment. Results of these studies highlighted that the two main phenomena responsible for the increased peak pressure are: i) the explosion occurring in the initial sections of the duct leading to a backflow from the duct towards the main vessel, thus restricting the effective vent section; ii) the turbulization of flame within the combustion chamber, due to the backflow, which enhances the combustion rate and the rate of pressure rise. In this paper, we propose an engineering correlation based on semi-empirical engineering methodologies aiming at quantifying the relation between the geometric properties (venting section, duct length, vessel volume) and the peak pressure occurring in the combustion chamber in the presence of a duct fitted on the vent panel.

1. Introduction

Venting devices are the main solution for the mitigation of accidental explosion in industrial equipment. Relief ducts are mandatory for the discharge of combustion products to safe location as recommended by the international standards NFPA 68 (2002) [1] and VDI 3673 (2002) [2]. The presence of duct is likely to increase the severity of the explosion and the peak pressure with respect to simply vented vessels [3-7].

Up to now, many studies have been performed to the aim of understanding the mechanisms driving the explosion severity when vented through relief ducts for both gas [8-12] and dusts [4,13]. Several mechanisms have been proposed to account for the observed increased peak pressure in such configurations: frictional drag and inertia of the gas column in the duct [4,14,15]; acoustic and Helmholtz oscillations; secondary explosion in the duct (burn-up) [5,6,10-12,16,17].

Although some explanations on the mechanism driving the pressure trend during explosion in ducted vented vessel has been given in the literature, the weight and the role of mixture reactivity, geometrical parameters and operating conditions are not yet quantified. Furthermore, we [19] recently showed that the available guidelines for the design of ducted vents for gas explosions as those proposed by Bartknecht (1981) [18], also reported in NFPA 68 [1], can lead to gross errors.

Recently we performed CFD simulations showing that the main mechanism affecting the pressure trend during gas explosion in ducted-vented vessel is the violent combustion which occurs in the initial sections of the relief tube [12]. This phenomenon is currently referred as "burn-up". On the other hand, we have found that pressure drop and gas column inertia, acoustic and Helmholtz oscillations are negligible for the increase of pressure. Due to the burn-up, the pressure impulse in the duct induces temporarily a flow reversal across the vent thus enhancing the burning rate by means of turbulization of the flame in the main protected vessel [5-7, 10-11]. Besides, an added negative effect on the protection system is given by the reduction of effectiveness of the venting flow rate from the vessel toward the duct [12]. However, CFD results have unambiguously showed that the increase of the burning rate due to turbulization is not a necessary condition for the increase of pressure observed experimentally when ducting the vent [12]. The CFD modeling results may be a useful tool for driving the study of a project correlation able to quantify the link existing between the peak pressure and the geometrical and operating conditions. Indeed, we recently have showed the relevant role of coupling detailed modeling (CFD) to derive empirical correlation when dealing with complex phenomena such as gas explosion [20].

In the present paper, we then start from our CFD results which allow to identify the mechanisms and then the most important parameters affecting explosion in a vessel fitted to a duct. We perform a parametric analysis of the experimental data available on explosion in ducted vented vessel to eventually derive an empirical correlation able to correlate the vent section to the peak pressure when a duct is fitted to the vessel.

2. Background

In previous paper, we have tested all correlation available in the literature to evaluate the experimental peak pressure measured during gas explosion in ducted vented vessel [19]. As expected, NFPA 68 correlation for gas explosion [1] is not adequate as it underestimates the peak pressure and takes into account only the effects related to the duct length. On the contrary, both experimental and theoretical studies reported in the literature clearly show that the duct length can not be assumed as the unique parameter affecting the overpressure with respect to the ductless system [5-6, 10-12, 14]. Indeed, other parameters which may strongly influence the duct burn-up have to be considered In particular, the duct diameter and the vessel volume have been shown to be essential. Moreover, according to the experimental results, the overpressure is also influenced by the ignition position, which is not accounted in any of the proposed correlations.

In the following, the experimental data available in the literature for gas explosion in ducted vented vessel are used to perform a sensitivity analysis and to derive an empirical correlation.

3. Approach description

Our methodology is based on the assumption that the ducted-vented explosion phenomena are well represented by turbulent Bradley number (Br_t) which quantifies the intensification of both turbulization and frictional effects in a simply vented vessel [21,22]. The turbulent Bradley number is different from the Bradley number since it accounts for the turbulization and the frictional effects of the vent section trough two parameters named as the turbulization factor (χ) and the discharge factor (μ).

In the presence of a duct, we assume that the intense burn-up occurring in the initial section of the duct affects both flame turbulization and frictional effects.

More precisely, it gives rise to additional reduction of the vent section, thus affecting the discharge flow and then the parameter, μ. Moreover, the resulting inversion of flow from the duct to the vessel, due to high local values of pressure in the duct entrance, produces turbulization of the mixture in the vessel and then the parameter, χ.

The quantification of the these two effects, the discharge flow reduction and the flame turbulization, is still obtained by means of the evaluation of the "discharge coefficient" (μ) and the "turbulization factor" (χ), respectively.

For the sake of describing duct-vented explosion phenomenon, we have then defined a ducted-turbulent Br number ($Br_{t\ ducted}$) which takes into account the effect of the duct with respect to the conventional turbulent Bradley number defined for un-ducted vented vessels, here referred as $Br_{t\ unducted}$.

The ratio of the two numbers is equal to the ratio between the corresponding DOI numbers as in the following:

$$\frac{Br_{t\,ducted}}{Br_{t\,un-ducted}} = \frac{(\mu/\chi)_{ducted}}{(\mu/\chi)_{un-ducted}} = f(reactivity, geometrical\ properties) \quad (1)$$

The above defined ratio (*f*) is expected to be a function of reactivity and geometrical properties of the ducted vessel configuration so allowing to quantifying the effect of the duct on venting. In order to calculate the *f* function we then used the experimental data of peak pressure reported in the literature and showed here in Table 1, where the data have been obtained by varying the ignition position, the duct diameter and length, the vessel volume and the mixture reactivity (fuel type and composition).

We used the experimental peak overpressure values P_{red} to calculate the turbulent ducted Bradley number ($Br_{t\,ducted}$) by means of the following equations obtained and tested for simply vented vessels [23]:

$$P_M = \frac{\pi_{red}}{\pi_v^{1.5}} = Br_{t\text{-}ducted}^{-2.4} \qquad P_M \le 1;\ Br_{t\,ducted} \ge 1 \quad (2)$$

$$P_M = \frac{\pi_{red}}{\pi_v^{1.5}} = P_M^* - 6\,Br_{t\,ducted}^{0.5} \qquad P_M > 1;\ Br_{t\,ducted} < 1 \quad (3)$$

where P_M is the reduced pressure given by the ratio of π_{red} and π_v, which are, respectively, the dimensionless ratio of the peak overpressure P_{red} and the absolute vent set pressure P_v to the absolute initial pressure P_o; P_M* is the reduced pressure reached when $Br_t = 0$.

The turbulent Bradley number for un-ducted vessels ($Br_{t\,un\text{-}ducted}$) may be calculated as function of the Bradley number and the ratio μ/χ (DOI number) for un-ducted vessel as follows:

$$Br_{t\,un\text{-}ducted} = 0.21\sqrt{\frac{E}{\gamma_u}\left(\frac{\mu}{\chi}\right)_{un-ducted}}\,Br \quad (4)$$

where γ_u is the specific heat ratio for the unburned fuel at initial condition and E is the expansion ratio and Br is the vent ratio coefficient (which was named Bradley number by Molkov) defined by Bradley & Mitcheson in pioneering works [21,22].

The DOI number for simple vented vessel may be calculated by means of the empirical correlation proposed by Molkov [25]:

$$\left(\frac{\chi}{\mu}\right)_{unducted} = \alpha\left[\frac{\left(1+10\sqrt[3]{V_\#}\right)\left(1+0.5Br^{\beta}\right)}{1+\pi_v}\right]^{0.4}\pi_{1,\#}^{0.6} \quad (5)$$

Here, α and β are empirical coefficients and their values are those suggested by the same author ($\alpha = 1.75$, $\beta = 0.5$), $V_{\#}$ is a dimensionless volume (ratio of the actual volume of the vented vessel to a vessel of $1m^3$), and $\pi_{1,\#}$ is the dimensionless initial pressure expressed as the ratio of absolute initial pressure to the pressure of 1 bar.

By applying Eqs 2,3) to the literature experimental data, $Br_{t\ ducted}$ can be calculated, whereas $Br_{t\ un\text{-}ducted}$ was calculated by means of Eqs. (4-5). Hence, the *f* function was calculated as the ratio of the $Br_{t\ ducted}$ to the $Br_{t\ un\text{-}ducted}$ number by means of Eq. 1). Once the *f* function was evaluated from the experiments, we tried to correlate its value as function of the geometrical and reactivity parameters in order to quantify the effect of the duct on the turbulent Bradley number.

To this end, we adopted the following power law equation:

$$f = a \cdot \left(P_M^*\right)^b \cdot \left(S_u\right)^c \cdot \left(V\right)^e \cdot \left(L_t\right)^g \cdot \left(D_t\right)^h \qquad (6)$$

where *a, b, c, e ,g* and *h* are parameters, S_u is the laminar burning velocity, V is the vessel volume, and L_t and D_t are the tube length and diameter, respectively. A regression analysis and a sensitivity analysis have been then carried out with the aims of finding the parameters of Eq. 6). In Table 1 the available experimental data of peak pressure as function of the parameters investigated are given. It is found that the effect of the duct length and diameter, the fuel type and composition, the vessel volume and the vent pressure have been investigated.

Table 1 - Experimental peak overpressures P_{red} reported in literature for duct-venting explosions in vessel with volume V, by varying duct diameter D_t, length L_t, and ignition position. See text for definition of P_o, P_v and P_{red}.

Gas	Concentration	Ignition	L_t [m]	D_t[m]	V [m^3]	P_v [bar]	P_o [bar]	P_{red} [barg]	Reference
Propane	stoichiometric	rear	0.6	0.016	0.00366	1.01	1.01	1.45	[6]
Propane	stoichiometric	rear	0.6	0.021	0.00366	1.01	1.01	1.17	[6]
Propane	stoichiometric	rear	0.6	0.036	0.00366	1.01	1.01	1.27	[6]
Propane	stoichiometric	rear	1.1	0.016	0.00366	1.01	1.01	1.80	[6]
Propane	stoichiometric	rear	1.1	0.021	0.00366	1.01	1.01	1.45	[6]
Propane	stoichiometric	rear	1.1	0.036	0.00366	1.01	1.01	1.92	[6]
Propane	stoichiometric	rear	2.6	0.016	0.00366	1.01	1.01	1.92	[6]
Propane	stoichiometric	rear	2.6	0.021	0.00366	1.01	1.01	1.55	[6]
Propane	stoichiometric	rear	2.6	0.036	0.00366	1.01	1.01	1.92	[6]
Propane	stoichiometric	rear	2.6	0.053	0.00366	1.01	1.01	2.11	[6]
Propane	stoichiometric	central	1.7	0.036	0.00366	1.01	1.01	2.01	[7]
Propane	stoichiometric	central	1.7	0.036	0.00366	1.31	1.01	2.16	[7]
Propane	stoichiometric	central	1.7	0.036	0.00366	1.92	1.01	2.66	[7]
Propane	stoichiometric	central	1.7	0.036	0.00366	3.31	1.01	3.37	[7]
Propane	stoichiometric	rear	1.7	0.036	0.00366	1.01	1.01	1.76	[7]
Propane	stoichiometric	rear	1.7	0.036	0.00366	1.33	1.01	1.88	[7]
Propane	stoichiometric	rear	1.7	0.036	0.00366	1.84	1.01	1.81	[7]
Propane	stoichiometric	near the vent	1.7	0.036	0.00366	2.12	1.01	1.27	[7]
Propane	stoichiometric	near the vent	1.7	0.036	0.00366	3.25	1.01	2.24	[7]
Propane	5%vol	centre	1	0.8446	2.6	1.11	1.01	0.19	[24]
Propane	5%vol	centre	2	0.8446	2.6	1.11	1.01	0.30	[24]
Propane	5%vol	centre	3	0.8446	2.6	1.11	1.01	0.39	[24]
Propane	5%vol	bottom	3	0.8446	2.6	1.11	1.01	1.01	[24]
Acetone	stoichiometric	central	25	0.50	10	1.11	1.01	4.10	[25]
Acetone	stoichiometric	central	25	0.50	10	1.06	1.01	2.80	[25]
Acetone	stoichiometric	central	4	0.20	2	1.16	1.01	4.30	[25]
Acetone	stoichiometric	central	10	0.20	2	1.16	1.01	5.20	[25]
Acetone	stoichiometric	central	10	0.38	2	1.16	1.01	2.15	[25]
Acetone	stoichiometric	central	1.83	0.05	0.027	1.21	1.01	5.00	[25]
Acetone	stoichiometric	central	2.35	0.05	0.027	1.26	1.01	4.40	[25]
Acetone	stoichiometric	central	2.35	0.05	0.027	1.26	1.01	3.50	[25]
Acetone	stoichiometric	central	2.35	0.05	0.027	2.66	1.01	1.90	[25]
Acetone	stoichiometric	central	1.83	0.05	0.027	2.43	1.01	4.40	[25]
Town gas	18%vol	central	0.16	0.035	0.022	1.01	1.01	3.00	[11]
Town gas	18%vol	central	0.32	0.035	0.022	1.01	1.01	4.82	[11]
Town gas	18%vol	central	0.54	0.035	0.022	1.01	1.01	5.65	[11]
Town gas	18%vol	central	0.80	0.035	0.022	1.01	1.01	4.82	[11]
Town gas	18%vol	central	1.40	0.035	0.022	1.01	1.01	5.13	[11]
Town gas	18%vol	central	1.75	0.035	0.022	1.01	1.01	5.18	[11]
Town gas	18%vol	central	2.80	0.035	0.022	1.01	1.01	2.14	[11]
Town gas	18%vol	central	3.50	0.035	0.022	1.01	1.01	4.64	[11]
Town gas	18%vol	central	4.91	0.035	0.022	1.01	1.01	3.57	[11]
Town gas	18%vol	central	6.14	0.035	0.022	1.01	1.01	3.75	[11]
Town gas	18%vol	central	6.75	0.035	0.022	1.01	1.01	3.39	[11]
Town gas	18%vol	central	2.50	0.025	0.022	1.01	1.01	5.00	[11]
Town gas	18%vol	central	2.50	0.025	0.022	1.01	1.01	4.73	[11]
Town gas	18%vol	central	2.50	0.025	0.022	1.01	1.01	4.20	[11]
Town gas	10%vol	central	2.50	0.025	0.020	1.01	1.01	0.82	[10]
Town gas	12%vol	central	2.50	0.025	0.020	1.01	1.01	2.38	[10]
Town gas	14%vol	central	2.50	0.025	0.020	1.01	1.01	2.91	[10]
Town gas	16%vol	central	2.50	0.025	0.020	1.01	1.01	3.47	[10]
Town gas	18%vol	central	2.50	0.025	0.020	1.01	1.01	4.00	[10]
Town gas	20%vol	central	2.50	0.025	0.020	1.01	1.01	4.30	[10]
Town gas	22%vol	central	2.50	0.025	0.020	1.01	1.01	4.82	[10]
Town gas	25%vol	central	2.50	0.025	0.020	1.01	1.01	5.00	[10]
Town gas	30%vol	central	2.50	0.025	0.020	1.01	1.01	0.82	[10]
Town gas	20%vol	central	0.04	0.025	0.020	1.01	1.01	3.68	[10]
Town gas	20%vol	central	0.17	0.025	0.020	1.01	1.01	3.68	[10]
Town gas	20%vol	central	0.30	0.025	0.020	1.01	1.01	6.71	[10]
Town gas	20%vol	central	0.61	0.025	0.020	1.01	1.01	6.36	[10]
Town gas	20%vol	central	1.26	0.025	0.020	1.01	1.01	4.57	[10]
Town gas	20%vol	central	2.50	0.025	0.020	1.01	1.01	4.00	[10]

4. Results and Discussion

Fig. 1 reports the *f* function obtained by Eq.6) by best fitting of experimental data reported in Table 1 vs the *f* function calculated by direct fitting of available experimental data for the central ignition position cases. Table 2 reports the parameters of Eq. 6) for the best fitting.

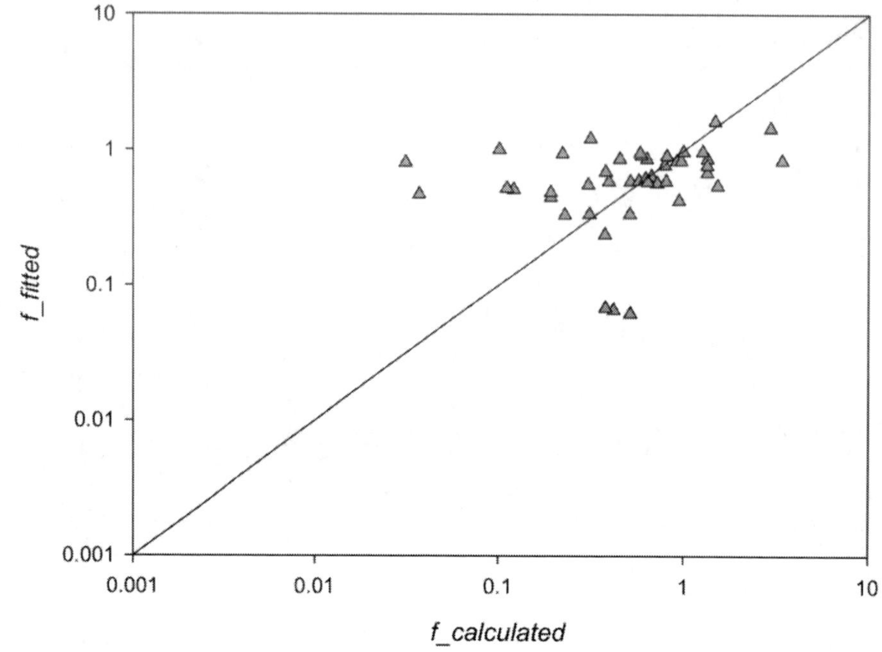

Figure 1. - *f* obtained by best fitting of experimental data (Eq. 6) vs. *f* calculated by direct fitting of experiments, for centrally ignited ducted-vented explosion (Table 1).

Table 2 – Values of the parameters obtained by fitting the experimental values (Eq. 6).

Parameter	
R^2	0.42
a	$3.0\ 10^5$
b	-4.0
c	-0.1
e	-0.4
g	-1.6
h	3.7

Starting from Eq. 6) and the parameter values reported in Table 2, we can now calculate $Br_{t\ ducted}$ and plot the experimental P_M values vs. the $Br_{t\ ducted}$. Results are reported in Figure 2, which shows a satisfactory agreement for most of data.

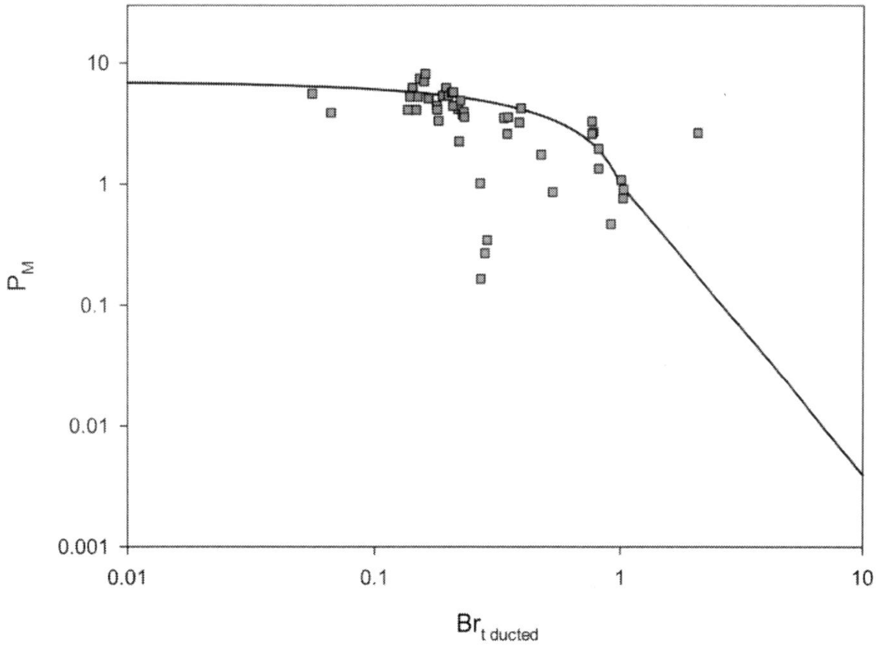

Figure 2 – Maximum pressure (P_M) vs. $Br_{t,ducted}$ as calculated by means of the Eq. 6.

At this point, in order to quantify the weight of each parameter in affecting the f ratio, a sensitivity analysis has been carried out. A sensitivity coefficient Φ_i was then calculated for each variable x_i defined in Eq.6) in the range of their experimental values, according to the following formula:

$$\Phi_i = x_i \frac{\partial f}{\partial x_i} \tag{7}$$

The sensitivity factors increase on decreasing the adiabatic pressure (P_M^*), the burning velocity (S_u), the vessel volume (V) while increase on increasing the duct diameter (D_t).
This behaviour has to be addressed to the mechanisms controlling the peak pressure in the presence of a duct. In our previous paper we showed by means of CFD simulations that pressure trend is affected by the violent combustion occurring in the initial sections of the duct which causes blockage of the venting, consequently burning residual unburnt in a closed-like vessel (Ferrara et al., 2006).

At burn-up the combustion of the unburnt mixture remaining in the main vessel is controlled by the competition between the turbulised flame propagation and the reduced effectiveness of venting.

The sensitivity to the adiabatic pressure (P_M^*) quantifies the intensity of burn-up, while the sensitivity to D_t quantifies the venting rate trough the duct, thus explaining the sensitivity to these parameters.

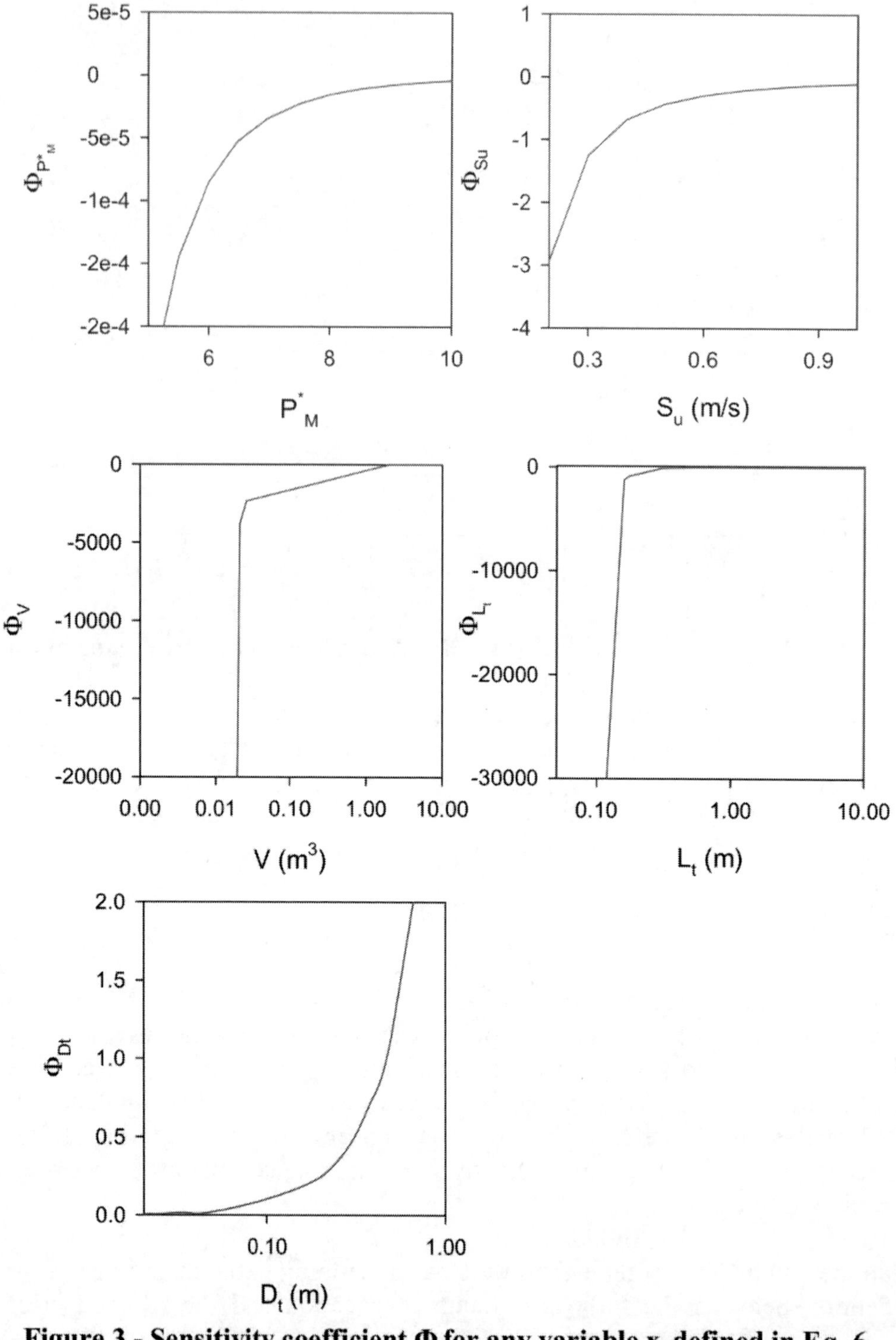

Figure 3 - Sensitivity coefficient Φ for any variable x_i defined in Eq. 6.

4.3 The engineering correlation for duct venting design

According to the results reported above, it turns out that the following relation may reproduce the duct-vented explosion phenomena in equipment:

$$\frac{Br_{t\text{-}ducted}}{Br_{t\text{-}un-ducted}} = \frac{\left(\frac{\mu}{\chi}\right)_{ducted}}{\left(\frac{\mu}{\chi}\right)_{un-ducted}} = f\left(D_t, L_t, S_u, P_M^*, V, \rho\right) \tag{8}$$

That is, we have six variables and three dimensions (length, time, mass). Hence, according to *Buckingham π theorem* we may derive three dimensionless groups, thus obtaining:

$$\frac{Br_{t\text{-}ducted}}{Br_{t\text{-}un-ducted}} = \frac{\left(\frac{\mu}{\chi}\right)_{ducted}}{\left(\frac{\mu}{\chi}\right)_{un-ducted}} = f\left(Re, \frac{P_M^*}{P_o}, \frac{L_t}{D_t}\right) \tag{9}$$

The Reynolds number (Re), is defined as the flame Reynolds number. In their previous paper Rota et al. (1991) used this number to empirically calculate the flame acceleration due to flame instabilities in their model of vented gas deflagration. In their correlation the characteristic length is the vessel volume which accounts for the dimension of the vortexes generated by the back-flow at the moment of burn-up.

$$\mathrm{Re}_F = \frac{S_u V^{1/3} \rho}{\mu} \tag{10}$$

The ratio P_M^*/P_o accounts for the entity of back-flow induced by the duct burn-up which causes a sudden increase of the pressure thuds inducing back-flow.

The ratio L_t/D_t accounts for the effect to the burn-up which causes the reduction of the vent area which is relevant only when L_t is at least of the order of the duct diameter.

The Re number is the flame Reynolds number and it accounts for flame acceleration due to the presence of the venting section.

By rearranging the data we obtained the following correlation:

$$f = a_1 (\mathrm{Re}_F)^{a_2} \left(\frac{P_M^*}{P_o}\right)^{a_3} \left(\frac{L_t}{D_t}\right)^{a_4} \tag{11}$$

The Marquardt-Levenberg algorithm [26] was used to find the parameters of the independent variables (Re, P_M^*/P_o, L_t/D_t) that give the *best fit* between the *f* function Eq. 11) and the *f* function calculated from experimental data. This algorithm, used as standard of nonlinear least-squares routines, seeks the values of the parameters that minimize the sum of the squared differences between the values of the observed and predicted values of the dependent variable (*f* function)

$$SS = \sum_{i=1}^{n} \frac{\left(y_i - y_i^{'}\right)^2}{y_i} \qquad (12)$$

where y_i is the observed and y'_i is the predicted value of the dependent variable.
The parameter values in Eq. 11) are given in Table 3.

We found that only the data of Table 1 with $L_t > 1$ m may be regressed by this formula as at $L_t \leq 1$ m the ducted venting behaves as a simply vented vessel and the $f=1$.

Table 3 – Values of the empirical constants of Eq. 11)

Parameter	Value
a_1	5
a_2	-0.1
a_3	-0.1
a_4	-0.5

The comparison between the *f* function obtained by best fitting of experimental data (Eq. 11) vs. *f* calculated by direct fitting of experiments is shown in Figure 4.

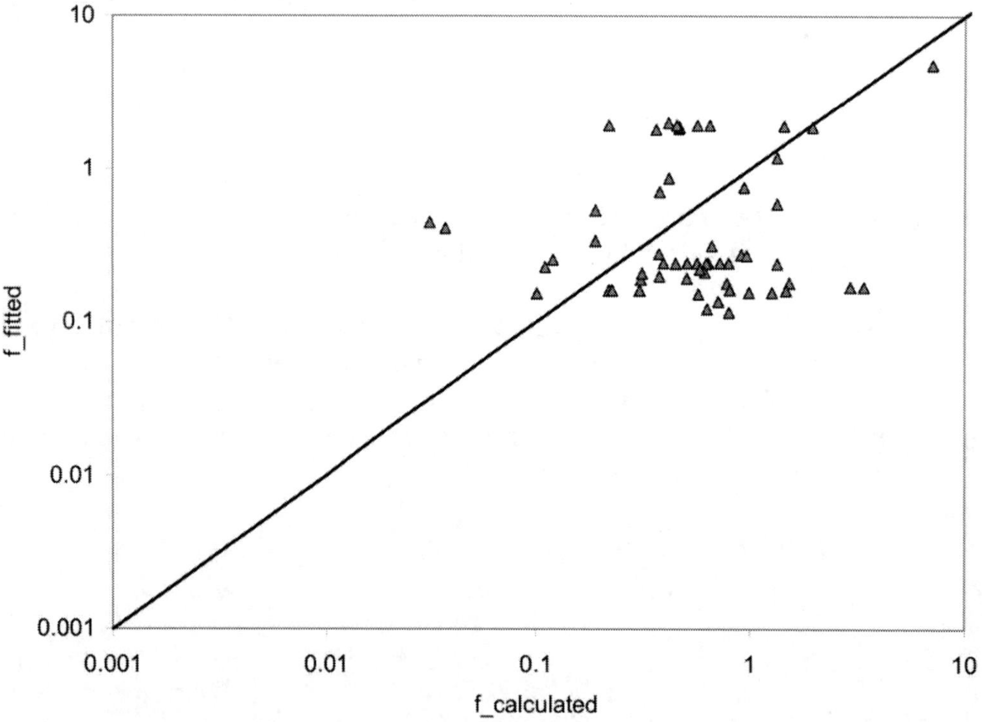

Figure 4. - *f* obtained by best fitting of experimental data (Eq. 11) vs. *f* calculated by direct fitting of experiments, for rear and centrally ignited ducted-vented explosion (Table 1).

The result is quite satisfactory considering that we have described the whole set of experimental data (central and rear ignition) by means of a single correlation which utilizes only dimensionless numbers.

The accuracy of predictions made by means of the proposed correlation has been estimated by comparing the predicted values of the reduced pressure P_M with the corresponding experimental data, as reported in Figure 5 at both Lt≤ 1 m and Lt> 1 m.

Results of Figure 5 shows that both rear and central ignition experimental data are well reproduced, confirming the validity of the *f* function correlation (Eq. 11).

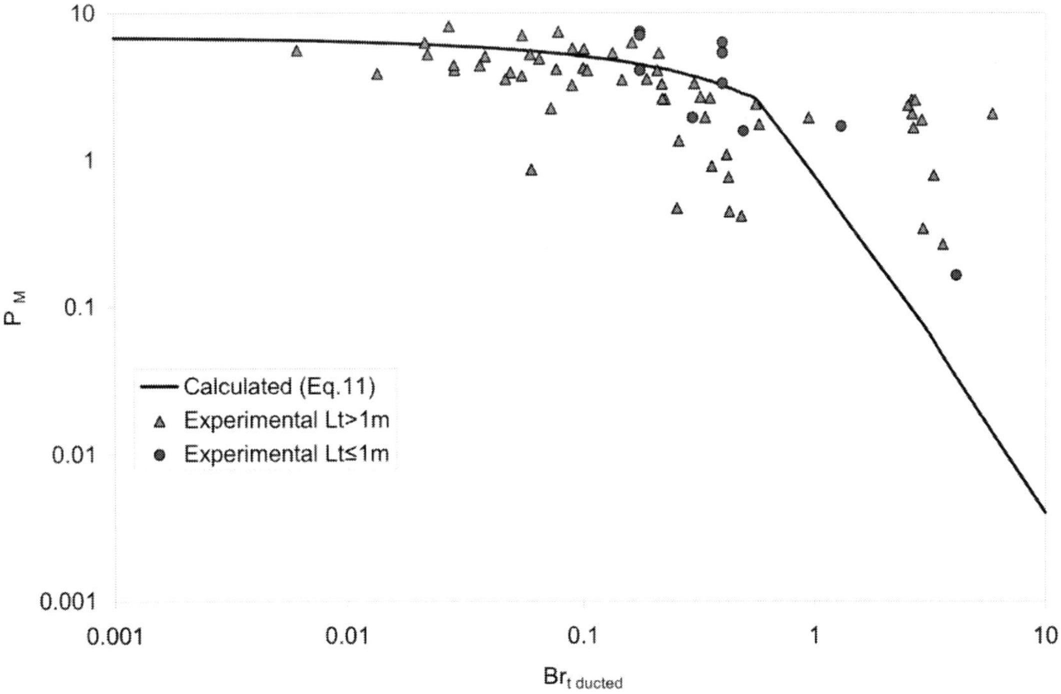

Figure 5 - Maximum pressure (P_M) vs. Br_t as calculated by means of the Eq. 11.

6. Conclusions

The effect of the duct on explosions in vented vessel has been parametrically studied by using the experimental results available in the literature. We started from the correlation and the number which characterize a simply vented vessel and addressed the effect of the duct to an enhancing function (*f*). The *f* function accounts for the increase of turbulization and frictional effects due to the presence of the duct. We then quantify this function as function of the geometrical and reactivity properties. A sensitivity analysis has been performed showing that the most affecting parameters are the duct diameter and the adiabatic pressure which is a measure of the burn-up intensity. A design correlation is also proposed to quantify the link existing between the peak pressure enhancement due to the presence of the duct with respect to a simply vented vessel and the geometrical properties and mixture reactivity. The proposed correlation is conservative since it is derived from the central ignition cases which give the most severe peak pressures.

7. References

[1] NFPA 68, 2002, Guide for Venting of Deflagration, National Fire Protection Association, Quincy, MA, USA.

[2] VDI-Guideline 3673, 2002, Verein Deutscher Ingenieure, Pressure venting of dust explosions, Beuth, Verlag GmbH, Berlin.

[3] Cooper M.G., Fairweather M., Tite J.P., 1986, On the mechanisms of pressure generation in vented explosions, Combustion & Flame, 65, 1.

[4] Lunn G., Crowhurst D., Hey M., 1988, The effect of vent ducts on the reduced explosion pressures of vented dust explosions, Journal of Loss Prevention in the Process Industries, 1, 182-196.

[5] Molkov V.V., Venting of deflagrations: dynamic of the process in systems with a duct and receiver, Proc. 4th Int. Symp. on Fire Safety Science, Ottawa, Canada, 1994..

[6] Ponizy, B., Leyer, J.C., 1999, Flame Dynamics in a Vented Vessel Connected to a Duct: 1. Mechanism of Vessel-Duct Interaction, Combustion and Flame, 116, 259-271.

[7] Ponizy, B., Leyer, J.C., 1999, Flame Dynamics in a Vented Vessel Connected to a Duct: 2. Influence of Ignition Site, Membrane Rupture, and Turbulence, Combustion and Flame, 116, 272-281.

[8] Wiekema B.J., Pasman H.J., Groothuizen Th. M., Wiekema B.J., Pasman H.J., Groothuizen Th.M., 1977, The effect of tubes connected with pressure relief vents, 2nd Int. Symp. on Loss Prevention and Safety Promotion in the Proc. Ind., 223-231..

[9] McCann, D.P.J., Thomas, G.O., Edwards, D.M., 1985, Gasdynamics of vented explosion Part I: experimental studies, Combustion and Flame, 59, 1985, 233-250.

[10] Kordylewski, W., Wach, J., 1986, Influence of Ducting on Explosion Pressure, Combustion and Flame 66, 77-79.

[11] Kordylewski, W., Wach, J., 1988, Influence of Ducting on Explosion Pressure: Small Scale Experiments, Combustion and Flame 71, 51-61.

[12] Ferrara G., Di Benedetto A., Salzano E., Russo G., 2006, CFD analysis of gas explosions vented through relief pipes, Journal of Hazardous Materials 137, 654-665.

[13] Pineau J., Giltaire M., Dangreaux J., Propagation d'explosions de poussieres dans les canalisations, Cahiers de Notes Documentaires, No. 98, 1er Trimestre 1980.

[14] Ural E.A., 1993, A simplified method for predicting the effect of ducts connected to explosion vents, Journal of Loss Prevention in the Process Industries, 6, 3-10.

[15] Tamanini F., An improved correlation of experimental data on the effects of ducts in vented and dust explosions, 8th International Symposium on Loss Prevention and Safety Promotion in the Process Industries, vol. 1, 243-253, 1995.

[16] Ferrara G., Di Benedetto A., Salzano E. Russo G., Analysis of venting through relief pipes by means of CFD model, 4th International Seminar on Fire and Explosion Hazards, September 2003, Londonderry, 2003.

[17] Ferrara G., Willacy S.K., Phylaktou H.N., Andrews G.E., Di Benedetto A., Salzano E., Venting of premixed gas explosions with a relief pipe of the same area as the vent, Proceedings of the 2nd European Combustion Meeting, Louvain-la-Neuve, Belgium, 2005.

[18] Bartknecht W., 1981, Explosions, Springer-Verlag, Berlin Heidelberg New York, 1981.

[19] Russo, P., Di Benedetto, A., 2006, The Effects of duct on venting of explosions: critical review, Process Safety and Environmental Protection, 84(B6) 1-14.

[20] Di Sarli V., Di Benedetto A., Salzano E., Ferrara G., Russo, G., Mitigation of Gas Explosions in Industrial Equipment by Means of Venting Systems, in: New Research on Hazardous Materials, Nova Science Publishers, Editor: Leonora G. Mason, 2007, in print.

[21] Bradley, D., Mitchelson, A., 1978, The venting of gaseous explosions in spherical vessels, Combustion and Flame, 32, 221– 236.

[22] Bradley, D., Mitchelson, A., 1978, The venting of gaseous explosions in spherical vessels. II – Experimental, Combustion and Flame, 32, 236-255.

[23] Molkov, V.V., 2001, Unified correlations for vent sizing of enclosures at atmospheric and elevated pressures, Journal of Loss Prevention in the Process Industries, 14, 567-574.

[24] De Good R., and Chatrathi K.J., 1991, Comparative analysis of test work studying factors influencing pressures developed in vented deflagrations, Journal of Loss Prevention in the Process Industries, 4, 297-304.

[25] Molkov V.V., Baratov A.N., and Korolchenko A.Y., 1993, Dynamics of gas explosions in vented vessels: a critical review and progress, Progress in Astronautics and Aeronautics, 154, 117-131.

[26] Press, W.H., Flannery, B.P., Teukolsky, S.A., and Vetterling, W.T., 1986, Numerical Recipes. Cambridge: Cambridge University Press.

LOSS PREVENTION SYMPOSIUM

CFD-based Risk Assessment for Hydrogen Applications

Olav R. Hansen and Prankul Middha

GexCon AS
P. O. Box 6015 Postterminalen
NO-5892 Bergen Norway
Phone: +47-55-57-4318
olav@gexcon.com

Prepared for Presentation at
American Institute of Chemical Engineers
2007 Spring National Meeting
41st Annual Loss Prevention Symposium
Houston, Texas
April 22–26, 2007

UNPUBLISHED

CFD-based Risk Assessment for Hydrogen Applications

Olav R. Hansen and Prankul Middha
GexCon AS, P.O. Box 6015 Postterminalen, Bergen, N-5892, Norway

Abstract

Computational Fluid Dynamics (CFD) calculations for gas explosion safety have been widely used for doing risk assessments within the oil and gas industry for more than a decade. Based on predicted consequences of a range of potential accident scenarios a risk level is predicted. The development of applications using hydrogen as a clean energy carrier has accelerated in recent years, and hydrogen may be used widely in future. Due to the very high reactivity of hydrogen, safe handling is critical. For most applications it is not realistic to perform an extensive risk assessment similar to what is done for large petrochemical installations. On the other hand, simplified methods, like venting guidelines, may have a questionable validity for hydrogen. The use of simple methods, if these actually are conservative, will in general predict too high consequences for the majority of scenarios, as these are not able to represent actual geometry and physics of the explosion.

In this paper a 3-step approach is proposed, in which the CFD-tool FLACS is used to estimate the risk. The initial approach will be to carry out a "worst-case" calculation evaluating the consequences if a full stoichiometric gas cloud is ignited. Mitigation measures can also be considered. As a second step, if potential consequences of the initial approach are not acceptable, the assumptions are refined and more calculations are performed to make the evaluations more realistic and reduce unnecessary conservatism of the chosen worst-case scenarios. Typically a number of dispersion calculations will be performed to generate likely gas clouds, which are subsequently ignited. If estimated consequences are still not acceptable, a more comprehensive study, including ventilation, dispersion and explosion, is performed to evaluate the probability for unacceptable events. Calculation examples will be used to illustrate the different approaches. The proposed approach is thus very flexible, and can be tailored to the scenario under consideration.

1. Introduction

The possibility of using hydrogen as an energy carrier has increasingly caught interest of both public and government policy makers in recent times due to the growth in global awareness about the impact of greenhouse gases and the finite nature of fossil fuel reserves. The expected scarcity of fossil fuels and the carbon induced climate changes being discovered on a daily basis makes the deployment of hydrogen in combination with renewable energy sources and possibly nuclear energy an interesting alternative. However, the safety of hydrogen during production, and in particular application remains a significant concern. The nuclear industry has been particularly interested in evaluating hydrogen safety, especially due to accidents like Chernobyl and Three Mile Island, and the potential increase in use of nuclear power. The hazards from hydrogen primarily stem from its wide flammability range, extremely fast burning rate (order of magnitude larger compared to natural gas), and the considerable amount of energy released when it burns or explodes. Hydrogen is also quite different from natural gas in certain other ways, some of which actually help to reduce the risk of using the gas. Hydrogen is much lighter than air and therefore, has very strong buoyancy that will quickly remove the gas in an unconfined situation. However, much lower energies are needed to ignite hydrogen and mitigation methods traditionally used for natural gas seldom work in case of H_2.

In any future hydrogen economy, two principal applications can be envisioned: (1) Cars or other vehicles using hydrogen either in fuel cells or in a combustion engine and (2) Fuel cells providing heating and electricity to households. A pre-requisite for any such commercial application of hydrogen is that the safety of the required infrastructure is investigated and that its design is made such that the associated risk is at least not higher than that of existing supplies. The infrastructure required for these applications can be schematically represented in Figure 1. The figure shows that hydrogen production, storage, distribution, and refueling units each form a key component of any large-scale consumer use in the coming years. Each of these activities comes with their own inherent risk. However, all of them cannot be analyzed in detail, due to time and resource limitations and the fact that many of the details about the design features of these installations are not in public domain. It is possible to choose the objects with a representative hazard for the entire system, such as storage tanks and filling stations, but it must be done in a reasonably conservative fashion.

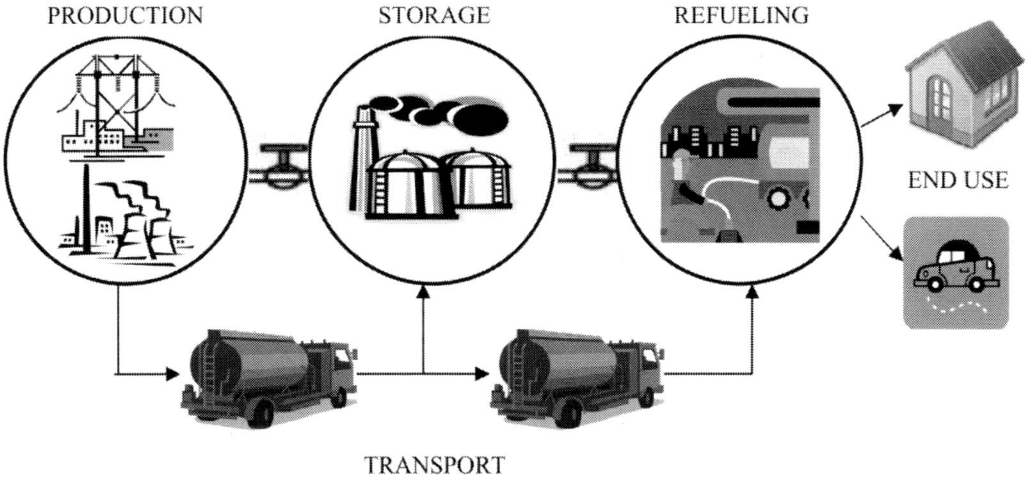

Figure 1. Hydrogen production, storage, distribution, and end-use pathways for a possible future application (based on [1,2])

In recent years, CFD has been increasingly used to perform quantitative risk assessments, especially in the oil and gas industry. As guidance to the NORSOK standard [3] it is described how a quantitative explosion risk assessment should preferably be carried out with the use of CFD for ventilation, dispersion and explosion calculations. Based on predicted consequences of a range of potential accident scenarios a risk level is predicted. However, hydrogen risk is handled very differently at the present time. There is a significant focus on regulations, codes, and standards (RCS) e.g. safety distance rules. CFD is used only occasionally, and mainly for worst-case evaluations. We must accept that it is not realistic to perform an extensive risk assessment similar to what is done for large petrochemical installations for most applications. On the other hand, simplified methods, like venting guidelines, may have a questionable validity for hydrogen. Therefore, a comprehensive procedure for risk assessment of hydrogen applications considering its most important objects is needed and experiences from the oil and gas industry must be used. This article describes such an approach, in a fairly detailed manner with calculation examples, where appropriate. Various levels of treatment, ranging from analyzing the worst-case scenario to a comprehensive study including ventilation, dispersion and explosion, to evaluate the probability for unacceptable events is considered.

However, we need to be very careful before applying a CFD tool to carry out such risk assessments. The tool needs to be well validated against a range of relevant experiments (with studies on variations of various important parameters that may affect explosion loads and hence risk). The validation should be an integrated part of development, and not be the responsibility of the end user. The CFD-tool should have no unphysical calibration parameters. Further, tools that have 5-7 turbulence models and their associated constants have limited value for the risk analyses as the results are bound to be model dependent. Clear user guidelines must exist and different users should be able to achieve similar results, even when predicting blind. The document is based on use of such a validated CFD-tool. Empirical methods and guidelines can also be acceptable to replace certain calculations, provided these are consistently conservative compared to the validated CFD-tool. Another requirement for carrying out accurate risk assessments is the need to model numerous scenarios. Therefore, the CFD tool must enable efficient definition of scenarios and have moderate calculation times. Based on CFD calculations, a typical risk assessment procedure can be summarized as follows:

Step 1: Worst-case assessment (3-5 calculations)

- Explode the worst credible scenario (full stoichiometric cloud?); evaluate the consequences
- If the consequences are unacceptable, evaluate modifications of design or mitigation methods (or go to Step 2)

Step 2: "Realistic worst-case" assessment (20-30 calculations)

- Simulate releases with ventilation and estimate worst-case "realistic" flammable cloud
- Explode this scenario and evaluate the consequences
- If the consequences are unacceptable, evaluate modifications of design or mitigation methods (or go to Step 3)

Step 3: Probabilistic risk assessment (100-200 calculations)

- Simulate a range of releases and ventilation conditions, and establish cloud size distribution
- Simulate explosions with the various cloud sizes, and evaluate risk versus acceptance criteria
- Evaluate how risk can be reduced by modifications or mitigation

Section 3 describes the different steps involved in carrying out these calculations in more detail using a validated CFD tool. The next section describes why FLACS is suitable for performing these studies.

2. The CFD software FLACS

FLACS has become increasingly accepted for risk assessment and explosion safety studies in offshore oil platforms worldwide. This is due to accurate and fast numerics and very extensive validation against available experimental data. As described above, validation of the CFD tool is one of the most critical requirements for its use in quantitative risk assessments. For natural gas deflagration calculations several hundred experiments have been simulated using FLACS as part of validation exercises in the 1990s [4]. Grid independency studies as well as more basic investigations (shock-tube, transport of passive scalar, turbulence generation behind object with different grids, etc.) were important parts of a validation study. Similar studies have been carried out to ensure good performance of FLACS in other areas, including far-field blast propagation from deflagrations [5,6], dispersion of flammable gas and ignition of non-homogeneous gas cloud relevant for QRA-studies towards offshore oil and gas industry [7], dispersion of tracer gas [8] and dust explosion simulations with DESC, a tool based on FLACS [9].

Despite some sporadic explosion tests at GexCon (CMI) with hydrogen before 1995, the main focus was natural gas explosions. Through the 1990s when the validity of FLACS for natural gas dispersion and explosion predictions improved, the limitations when simulating hydrogen became more visible. With increasing interest in hydrogen safety in recent years as described above, a strong effort has been made in recent years to learn more about hydrogen explosions and improve FLACS in that area. In 1998, a project funded by oil companies was carried out to study the effect of inert gas dilution on deflagrations, which also included some small-scale hydrogen experiments in a channel with obstacles [10]. In 1999, more hydrogen explosion experiments were carried out with partial funding from the Norwegian Research Council. The effect of various parameters such as gas concentration, obstacle height and number, ignition position, and confinement was studied. In 2001, a 3-year research project was established to improve the validity of FLACS when simulating hydrogen with support from Statoil, Norsk Hydro and IHI. A significant improvement in performance was obtained through a large number of varied small-scale experiments combined with simulations and model improvements. More details are presented in Ref. [11]. A brief list of important validation exercises involving hydrogen is presented below:

- Ventilation
  - Studies at oil platforms and full scale test site (with many different wind strengths, directions, etc.) including Beryl Bravo (Mobil), Oseberg A (Hydro), Spadeadam JIP, Claymore

- Dispersion
  - Subsonic jets (GexCon experiments [12], Swain tests [13], etc.)
  - High-pressure jets (HSL experiments [14], INERIS tests [15])
  - More recent experiments from INERIS and FZK (unpublished)

- Explosion
  - Large-scale: Sandia FLAME facility (varying concentration, confinement, and congestion) [11,16]; Fh-ICT 20m balloon (low confinement) [17,18], SRI Tunnel [19]
  - Small-scale: GexCon tests (varying congestion, confinement, concentration, and ignition point) [20]

- Flame acceleration in pipes: McGill detonation tubes [21,22], SRI tubes [23,24], various experiments under the HYCOM project [25]
- DDT in various geometries: Fh-ICT lane [26,27], KOPER facility [22,28]

With the extensive work carried out in recent years and being continued at present, the validation status of FLACS in simulating scenarios relevant for hydrogen safety can be considered good. There is still a major focus on hydrogen safety at GexCon, with involvement in the EU-sponsored HySafe network of excellence and IEA task 19 on hydrogen safety where we are focused on "applied CFD" for risk assessments.

3. Risk Assessment Procedure

In the following, a procedure for the complete calculation of explosion risk from a leak that may lead to unacceptable structural response is described. The purpose of the procedure is to standardize the analyses so that the risk of explosions can be compared between different areas, installations and concepts, even if the analyses are performed in different circumstances and by different personnel. Due to a transparency in the assumptions and intermediate results, the use of such a method will encourage improvement of the method as some conservatism should be included in assumptions with less precision. This methodology covers explosion and dispersion calculations, but it should be possible to combine the methodology with analysis of fire in the risk analysis in a consistent way. The purpose of the dispersion analysis is to relate the frequencies of the different leak scenarios through the ignition probability to the frequencies for the different standard ignited stoichiometric clouds. The frequencies for the different ignited homogenous clouds are combined with the explosion simulations in order to estimate the explosion risk including design loads when necessary.

In order to facilitate the evaluation and the comparison between different studies, presentation of results from the different intermediate steps will be required. This can include:

- Leak frequencies for the different leak sources
- Typical leak durations or profiles (pressure, hole size and inventory)
- Assumptions on ignition probability intensities for spontaneous ignition, and intermittent and continuous ignition sources and how these are influenced by detection/shut-down
- Ventilation conditions (forced or natural ventilation distribution, emergency ventilation)
- Frequencies (exposure time) for different gas cloud sizes when assuming no ignition.
- Frequencies for ignited gas cloud sizes and total ignition frequency.
- Sketches showing sizes and location of gas clouds and ignition points used in the explosion simulations
- Explosion pressures (and impulses) as function of gas cloud sizes and frequency

3.1 Simplifications in modelling

Before describing the procedure in detail, it must be stressed that various steps of the procedure can be simplified by replacing some of the calculations by using a less precise, conservative assumption. The procedure has quite extensive requirements to the factors that shall be considered and the number of parameter variations. It is not necessary to follow this rigorously if it can be documented that simplifications do not influence the results significantly. If there are steps of the procedure that consume significantly more resources than the other, it is recommended to simplify this step initially (e.g. detailed geometry building, or very detailed dispersion simulation study), and rather refine at a later stage if necessary to comply with acceptance criteria. Examples of such conservative simplifications are:

- Skip ventilation and leak calculations and assume stoichiometric concentration everywhere with high probability (worst-case; step 1 of the proposed RA procedure). If a situation normally containing flammable hydrogen-air mixtures is studied, such an approach with the actual concentration (or a distribution of concentrations with associated probability) should be studied.
- Perform a certain number of realistic leak calculations (10-15) with the aim to generate as dangerous as possible a gas cloud, pick most dangerous cloud and assume high probability (realistic worst-case; step 2 of the proposed RA procedure).
- If ventilation conditions are properly specified (forced or no ventilation), this will limit the number of scenarios that have to be studied (only one or a few ventilation conditions must be considered). For hydrogen jets from high-pressure systems the momentum from a hydrogen jet combined with strong buoyancy may dominate the mixing at flammable concentrations, so that changes in natural ventilation will only marginally influence the calculated risk.
- If the geometry is not 100% known, one should estimate this using conservative assumption for congestion (correct amount or more) and vent opening (correct size or smaller, or alternatively also larger vent openings if external effects are studied).

When simplified relations between input parameters and the results from the CFD calculations are used both for gas dispersion and explosion, the applicability of these relations should be documented. Further, if there is a doubt whether a simplification is conservative or not, sensitivity calculations should be performed to check this.

For situations in which the complexity of the study is limited, one could choose to ignite the realistic non-homogenous cloud. This should also be done for selected cases if there is a possibility that strong pre-ignition turbulence (e.g. from a high-pressure jet) may strongly accelerate the flames. Symmetry considerations, reasoning and simplifications based on understanding of the physics may be used to reduce the number of scenarios to be considered for simulation by a CFD-tool. The size distribution for equivalent clouds is established on this basis. Information about the most probable cloud locations may also be used if available (usually towards the ceiling for hydrogen).

3.2 The sequence of events

3.2.1 Leakage

3.2.1.1 Frequency of classes/distribution

The starting point is a distribution of leak sizes. Based upon the pressure in the system, initial leak rates are calculated and classified according to a distribution with narrow ranges. Based on the distribution of estimated leak sizes and durations, a set of representative leak sizes and durations is defined. If a conservative approach is used, the highest leak rate and longest duration should generally be used to represent each class (however, total release mass should not exceed inventory). If this gives unnecessary conservatism, another class with lower release rate or shorter duration should be defined. In cases where the resulting gas clouds end up rich (most of cloud above 30% H_2), one should consider using an average or low-end leak rate/duration to represent the leak size for the class. Each leak scenario considered must be assigned a frequency; the distribution of scenarios and frequencies must be realistic or conservative relative to reality.

3.2.1.2 Leak profiles

Time dependent leak rates should be used, and the ignition model must be time dependent. Deviations from this must be proven to be conservative.

3.2.1.3 Shape of equivalent stoichiometric cloud

For hydrogen, the buoyancy is so strong that one could always assume that a confined and semi-confined cloud is located near the ceiling or below any other horizontal confinement (a possible deviation from this may be large liquid hydrogen releases in hot and dry surroundings). For a free jet in a less confined situation, the jet momentum will usually dominate the mixing of the jet until the hydrogen concentration has become lean unless the wind is very strong. The cloud should be assumed to be located a small distance downwind of the jet, if possible conservatively towards obstructions/walls. The cloud should be made as a cubical rectangular box with assumed "planes of symmetry" towards confinement. The aspect ratio for a free cloud should be 1:1:1, for a cloud towards the ceiling 2:2:1, towards ceiling with one sidewall 2:1:1, etc. Alternatively, for a low momentum release in a confined situation with strong stratification, the cloud should be assumed to cover full ceiling area (or area between beams).

Sufficient grid resolution must be chosen to represent the gas cloud accurately, based on experience and validation. In case of FLACS, the grid resolution resolving the cloud must always be a minimum of 10 cells through a stratified layer near the ceiling, or at least 13 grid cells for directions of the cloud where no confinement exist. If the stratified layer has a very long aspect ratio, one could consider to reduce the length/width and to increase thickness (keep volume constant), in order to fulfil grid requirements. This is assumed to be conservative with regard to explosion strength, but may change the position of the explosion somewhat. The explosion calculations of equivalent homogeneous clouds can be carried out prior to the dispersion calculations if that is found useful, as this will give an early indication of what explosion loads that can be expected.

3.2.1.4 Location and direction of leak

In order to obtain a representative distribution more than one leak point should normally be used. For each leak location, one should consider leaks along all six directions and also a diffuse leak (to represent jet impinging onto something and quickly losing momentum). If there are good reasons not to expect leaks in several directions, or impingement is unlikely, these scenarios can be skipped. Symmetry considerations and evaluations based on the understanding of physics, geometry, wind and ventilation directions may be used in order to limit the number of scenarios that need explicit simulation.
Validated grid guidelines with local grid refinement near leak, must be applied. For low momentum leaks it may also be necessary to refine grid along jet for horizontal jets.

3.2.1.5 Medium

Liquid hydrogen releases can be represented by diffusive (no momentum) releases spread around the assumed pool area, with release of gaseous hydrogen at its boiling point (approx. 20 K). The methodology can also be applied for synthesis gas and other hydrogen mixtures that are highly buoyant in air.

3.2.2 Cloud formation

3.2.2.1 Ventilation, wind direction and strength

For forced ventilation the relevant ventilation system should be simulated during the calculations. If there are alternative ventilation intensities, or if natural ventilation also will influence the ventilation, one should consider how wind with various strengths from a number of wind directions would influence the ventilation efficiency in the area considered. One should aim for simplifications by sorting the result into scenarios with certain ventilation efficiency. If natural ventilation alone decides the ventilation, one may only have to simulate a single wind strength for each wind direction and assume that ventilation scales with wind speed. However, if there is a combination between forced and natural ventilation, one must select more wind speeds for each wind direction.

The number of ventilation rates to be simulated should depend on how much impact a change in ventilation has on the results. If only a marginal change of results is expected with increasing wind it may be sufficient to simulate 2-3 wind different conditions and let these represent the conditions that were not modelled. If ventilation is very important for the results, the number of ventilation regimes shall be of the same order as the number of leak rates. For dominating forced ventilation, one can reduce the number of ventilation scenarios. It should be kept in mind that for a H_2 release, there is only minimal effect of wind, unless it is very strong. One may though have to consider failure of the ventilation system etc. If a system with emergency ventilation exists at gas detection, this should be modelled.

3.2.2.2 Calculation of equivalent stoichiometric cloud

In the recommended approach the inhomogeneous clouds are exploded through the establishment of an equivalent homogenous cloud. This can be an equivalent stoichiometric cloud, or alternatively a homogeneous gas cloud with a lower or higher concentration with the same equivalent stoichiometric cloud as the dispersed gas cloud. The size of the equivalent stoichiometric cloud at the time of ignition should normally be calculated as the amount of gas in the explosive region, weighted by the concentration dependency of the flame speed and expansion. For a confined scenario, or a scenario where very high flame speeds are expected (see comment below), one should use only expansion based weighting (Q8-volume output from FLACS). For unconfined situations one could reduce the conservatism and use a weighting of reactivity and expansion (Q9-volume in FLACS). For vented rooms and other situations with a significant confinement, a weighting between Q8 and Q9-volumes is suggested. This proposal is based on scaling and dimensional arguments and more validation is therefore needed.

Equivalent volume $Q = (T_b \times Q8 + T_v \times Q9) / (T_b + T_v)$

$$\text{with } T_v = \sum_{i=x,y,z} \frac{\dim(i)}{(\mathrm{por}(i,\min) + \mathrm{por}(i,\max))} \times \frac{1}{c_{sa}} \times \sqrt{\frac{Q8}{V}} \times E$$

$$\text{and } T_b = 0.5 \times \frac{\sqrt{(G_x^2 + G_y^2 + G_z^2)}}{S_{ql} \times (E+1)}$$

Here,

T_v = estimated time to vent overpressures

T_b = estimated time to burn cloud with central ignition and quasilaminar flame speed

c_{sa} = speed of sound in cold air ($\approx$ 340 m/s)

V = effective room volume where air and gas may flow

E = relative volume increase for stoichiometric combustion (6.5 for ambient conditions)

por(x,min) = porosity (area fraction vented) of wall/boundary in X direction (with low X-coordinate)

dim(x) = room length (m) in x-direction

G_x = gas cloud length in x-direction

S_{ql} = quasilaminar burning velocity, approx. 3.5 (wrinkling) $\times$ 2.7 m/s

As described above, for scenarios (cloud/congestion combinations) with low or moderate congestion (or cloud sizes much smaller than room dimension, e.g. < 10%) where very high flame speeds can be achieved, one should use Q8 value instead. To evaluate this, a number of cloud sizes can be simulated to identify a critical cloud size, Q_{crit}, for which flame speeds exceed e.g. 200 m/s. For clouds with Q8 < Q_{crit}, the weighting procedure above can be applied, for clouds with Q8 > Q_{crit}, one should apply the Q8 cloud as representative cloud size.

If the ventilation or a high-momentum leak (jet) creates a significant turbulence in the region where ignition is expected, this turbulence should be defined as an initial condition for the CFD solver. The main parameter that should be defined should be turbulent fluctuations (product of characteristic velocity and turbulence intensity). The turbulent length scale should not be assigned a too large value (As a guideline, maximum 10% of smallest cloud dimension and maximum 50% of grid cell size should be used). For hydrogen it is expected that the flame interaction with the turbulence induced by a jet becomes important in many cases, and one should also consider to model explosion in a range of realistic, but more hazardous leak scenarios to evaluate whether Q9 or Q8 should be used. If explosion pressures from realistic calculations become higher than Q8 pressures, one may have to include the effect of initial turbulence in the explosion calculations.

Due to the high reactivity of hydrogen, a transition from deflagration to detonation (DDT) may happen in a variety of scenarios involving H_2-air mixtures and can be a significant threat and result in large-scale damage. DDT is characterized by very high pressures (~ 20 barg) and rapid flame fronts (~ 1800 m/s). If a worst-case realistic cloud with ignition near jet or a cloud with Q8 volume shows indications of a possibility of DDT, the risk for this should be evaluated in detail. GexCon has developed a methodology to evaluate the risk of DDT in a given scenario in terms of the spatial pressure gradient across the flame front (see references

21 and 29 for more information). Since DDT leads to consequences that are in general much more severe than a deflagration, a much more detailed probabilistic study to characterize the possibility of particular scenarios that have an increased likelihood of a transition to detonation is recommended. This can include the probability of certain leak directions, momentums, locations, sizes, etc. which may lead to the "worst-case" gas clouds which can result in very strong flame acceleration. In general, for systems that are deemed dangerous and susceptible to DDT, the size of the gas cloud with concentrations that lie in the detonable range should be included in the reporting. Traditionally, the detonability limits of hydrogen are between 18 and 59% H_2. However, some recently reported experiments in tubes have found indications for DDT for H_2 concentrations as low as 13%.

3.2.2.3 Dispersion, selection of models

Well-validated CFD models should be used for dispersion calculations. In order to include the large number of parameter variations that the procedure implies, it may be necessary to use correlations based on these dispersion calculations. The validity of these correlations must be documented by independent calculations. Typical number of CFD dispersion calculations could be 10-100. In a simplified approach, it is possible to replace the CFD simulations by simpler models or assumptions, but it must be justified that the simpler method is consistently conservative.

3.2.3 Ignition

For a quantitative explosion risk assessment the explosion simulations should be performed with various idealized clouds of variable size and typically using stoichiometric concentration. In practice, the explosion simulations can therefore be performed before a probability distribution of cloud sizes is established through a dispersion study.

3.2.3.1 Location of gas cloud and point of ignition

The required ignition energy is very low for reactive mixtures of hydrogen, but increases for lean mixtures as well as for regions with high turbulence. Hence, ignition of a flammable gas cloud is likely. Self-ignition of clouds formed from a high-pressure release has been observed experimentally several times. For the purpose of RA, the distribution of ignition locations should be chosen to represent reality. If there are one or more highly likely ignition locations that dominate the ignition frequencies, these may be used. Otherwise, it should be assumed that a constant ignition source might lead to end ignition (where concentration reaches 6-10% H_2), whereas intermittent ignition sources will be more arbitrarily distributed (with higher likelihood centrally in the cloud where concentrations are above 6-10% H_2). For stratified clouds, end ignition will mean ignition in the lower end of the cloud.

For smaller clouds (flame travel less than 1m to open boundary), ignition with maximum distance to edges of gas cloud is normally a conservative choice (for a cloud located in a corner, this means corner ignition, not central ignition). This could be used to represent all scenarios. Alternatively a distribution of ignition points could be applied. For larger clouds, a homogenous distribution of ignition positions should be applied. It should be kept in mind that the gas clouds with the possibility of the longest flame travel are often the most dangerous ones.

3.2.3.2 Exposure of ignition source to the cloud

Algorithms for determining ignition probabilities should be established. These may be based on methods similar to those used according to Norsok guidelines (TDIIM), but should be modified to reflect a lower ignition energy for hydrogen compared to natural gas. For hydrogen, spontaneous ignition has also been observed in experiments. This also needs to be considered when establishing the likelihood of ignition in various scenarios. Therefore, three ignition probabilities should be established:

PI_{const} = ignition probability from constant ignition sources (per m^3 that is exposed to flammable gas for the first time in the last 1 second)

$PI_{intermittent}$ = ignition probability for intermittent ignition sources (per m^3 flammable volume and second)

$PI_{spontaneous}$ = ignition probability for spontaneous ignition sources (early ignition)

The ignition intensities can also be time independent reflecting the effect of measures like switching off power in the exposed volume. In that case algorithms for that should be established.

From the CFD calculations the volume of the gas cloud with concentration between LFL and UFL will give the volume that may be ignited if exposed to an intermittent ignition source. This information should thus be related to the intermittent ignition frequencies defined. In the case of constant ignition source, the Q6 output from FLACS (or similar from another CFD tool) gives the cloud volume that was exposed to flammable gas concentrations for the first time last second. This information should be combined with the probabilities for ignition by constant ignition sources. Thus for each time step (or each 1s) for each dispersion calculation, the probability for ignition from spontaneous, intermittent and constant ignition sources should be established, and this probability should be added to a gas cloud size class (based on corresponding Q8, Q9 or a combination).

If gas cloud becomes rich (in a well mixed state, but more gas available than needed to fill room with stoichiometric concentration) this may be represented either as a stoichiometric cloud or as a rich cloud with a slightly higher reactivity than observed. This can be done by using the flammable volume to establish the volume of the cloud and the flammable mass to establish its concentration.

3.2.4 Explosion Load

3.2.4.1 Definition of load

In order to determine the structural response and safety margin of safety for critical structural elements of an installation, a time dependent gas explosion overpressure and drag force must be applied. Design values of load for structural design should be established based on the relation between pressure and impulse (*P-I*) of a predicted load, overpressure/drag and the frequency of exceedance of the load.

The *P-I* diagram is created in the following way for all observations i:

$Z(P_i, I_i)$ = SUM (probability for all points where $P > P_i$ or $I > I_i$)

Thus the load to be used for dimensioning will not be a single pressure value, but the collection of pressure and impulse (time) combinations with a given frequency for exceeding either pressure or impulse. If this *P-I* surface is compared to given *P-I* diagrams for wall strength, it will be possible to establish frequency for damage. A relation between maximum pressure and duration may be used if the simulations show this to be a good simplification, but then the relation has to be documented. The *P-I* surface should be established for the average pressure over the predefined area. In cases where there is a considerable variation in the overpressure over the exposed area, an additional *P-I* relation for the local area with the highest local overpressure should be established. A drag load must be assessed for larger piping and equipment, in particular in connection to equipment near vents.

3.2.4.2 Selection of standard for modelling

The standard for modelling should be selected according to requirements and guidelines from the software supplier. Experience shows that it is vital to use a sufficiently detailed representation of the geometry in the explosion simulations, and also that grid guidelines must be followed.

3.2.4.3 External pressure

Propagation of explosions from a room into gas blown out of the room produces an external pressure field that may create considerable pressure loads on external targets. In some cases it will be more time consuming to calculate far-field effects, and as a first approach one could calculate the loads for worst-case and dimensioning scenarios. If necessary, one should include this output for all explosion calculations.

3.2.5 Calculations of response

3.2.5.1 Relation of response analysis to risk acceptance criteria (RAC)

In case the explosion loads may threaten the integrity of structures, structural response calculations could be performed. If the acceptance criteria for explosions relate to the frequency of the explosion load, the purpose of the response analysis is to show compliance of the response with the strength and functionality (limit state) requirements. The purpose of the response analysis is to show that the frequency of exceedance of the limit state requirements is within the acceptance criteria. This basically involves the comparison of the *P-I* diagram to RAC. Examples of strength and functionality requirements that should be considered include: a) Global structural collapse, b) Damage to equipment and piping resulting in unacceptable escalation of events, c) Unacceptable damage to safety critical equipment after the explosion, d) Explosion loads potentially killing people (small scale hydrogen facilities), etc.

3.2.5.2 Selection of principles for analysis

Prediction of responses should be carried out using appropriate analytical models. SDOF and finite element analysis modelling technique with dynamic and non-linear capabilities (NLFEM) may be utilised. Dynamic and deformation characteristics of the structure should be adequately accounted for in the selected analytical procedure. More details can be found in the NORSOK standard Annex G [2]. The description of explosion load needed for structural analysis can be simplified as:

1. A triangular, time dependent overpressure/drag given by the dimensioning incidents from the pressure-impulse surface.
2. Direct linearization of the overpressure-time curve resulting from each CFD simulation (However, modern NLFEM programs have the capability to use complete overpressure-time descriptions and thus no load linearization is required).

Subsequently, response analysis of each CFD simulation can be carried out. Each response is evaluated as acceptable or unacceptable based on the limit state criteria. Finally the total frequency for unacceptable response is obtained as the sum of the frequencies for the single scenarios with unacceptable response. Thus, the load for each scenario is transferred directly to a response.

3.2.6 Calculation of mitigating measures

The mitigation of a hydrogen explosion is very difficult once it has been initiated, but a lot of "mitigation measures" can be applied beforehand by changing the design thus influencing the dispersion phase. The effect of such measures can be evaluated by repeated simulations. The number of simulations that need to be redone will depend on whether the measure has effect on dispersion, explosion, or both. Current guidelines for water deluge must be modified for hydrogen, and experimental validation is needed before method can be applied. A few steps that may be taken to reduce the mitigate the consequences of any accident could be:

1. Reducing leak frequency or design to limit dangerous leak types
2. Keeping people away from most hazardous areas by establishing good safety distances

4. Illustration of an example case

In this section, an example case is presented to illustrate the methods that are described in the previous section. The scenario chosen is a possible dangerous situation in a hydrogen workshop caused by a small release of hydrogen (of the order of 10g), which may get ignited subsequently. The goal is evaluate the risk if this happens in terms of pressure loads. The schematic views of two different geometrical configurations that are considered are presented in Figure 2. The two geometries correspond to two different confinement situations for the leaked gas, a) one horizontal plate (1.00m x 1.00m) and b) one horizontal plate with 4 vertical sidewalls (1.00m x 0.50m). The second configuration represents a "hood" scenario. Thus, the second scenario involves a partially enclosed volume of $0.5 m^3$ (open downwards). In this exercise, only the geometrical details presented in Fig. 2 are considered for the purpose of risk evaluation.

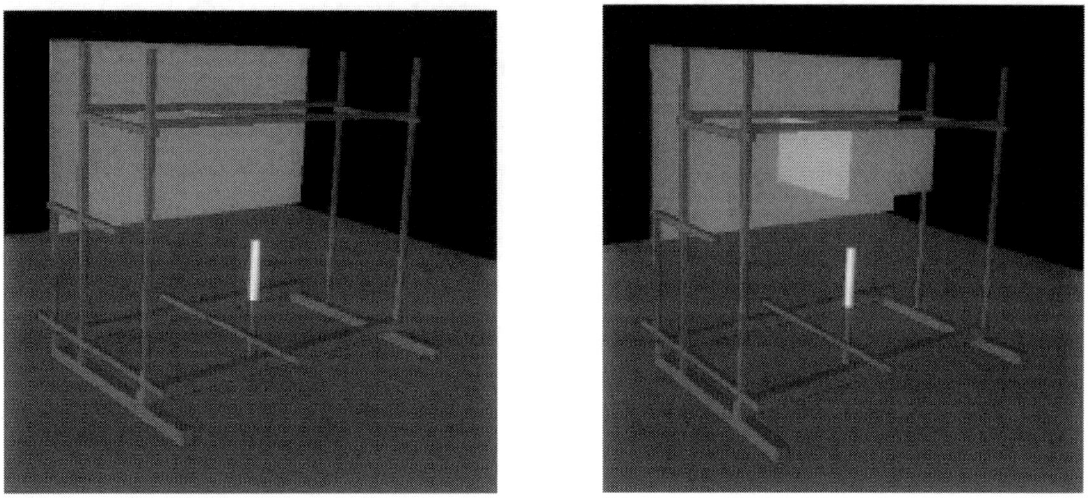

Figure 2. Geometrical configurations of the "workshop" scenario considered for risk analysis, plate only (left) and plate with sidewalls (right).

The worst-case scenario entails the leaking gas accumulating in the 0.5 m^3 box. Therefore, explosion calculations with a 500-litre stoichiometric H_2-air cloud filling the hood are carried out using FLACS. The results are presented in Figure 3. Clearly, it is seen that the worst-case scenario leads to some unacceptable consequences for the hood geometry, especially in case pre-ignition turbulence is considered (which is likely to be present if there is a high-momentum hydrogen release which then gets ignited). The pressure in this case is predicted to be around 500 mbar. A typical mitigation measure in this case could be the removal of the sidewalls. This leads to a considerable reduction in predicted overpressures (up to a factor of 5). However, the danger may still be significant.

Next, step 2 of the proposed risk assessment method is considered which involves realistic "worst-case" assessment. In this, the worst possible "credible" gas cloud is evaluated. A range of simulations is performed to investigate the size of the gas cloud that may accumulate in the 0.5 m^3 box. Several different realistic releases are modelled by varying the leak velocity and nozzle size. Three different nozzle sizes and three different leak velocities are chosen. The total hydrogen release amount was kept constant at 10g. The results are presented in Figure 4 where concentration contours between 4% H2 (LFL) and 30% H2 (stoichiometric concentration) are plotted. It can be seen that the smallest nozzle size leads to very small gas clouds. For the other two nozzle sizes, significant gas cloud sizes are seen, especially for the medium and large velocities. It is possible to study the simulation results in a more detailed manner by considering the transient profile of the flammable volume as a function of time. The gas clouds are seen to occupy a volume of larger than 1 m^3 for some cases for a short time before assuming a steady state value of around 0.5 m^3. However, it can be concluded that considering the concentration of the entire cloud to be stoichiometric (as assumed in the first step) is very far from the truth.

The sensitivity of the gas cloud sizes in all 9 scenarios to the geometric configuration was also studied. Figure 5 presents the flammable gas cloud size and shape for the same scenarios as in Figure 4 for the plate only geometry. Again the smallest nozzle size is found to result in very insignificant gas clouds. For the two other nozzle sizes, the resulting gas clouds are found to be smaller than the hood geometry, and after a certain time, no accumulation of hydrogen is observed (as opposed to the hood scenario where 500 litre gas clouds are seen to last for a significant length of time). The maximum flammable volume is around 1500 litre for the hood geometry, and around 1000 litre for the plate geometry. This is significantly smaller than the assumed flammable volume in step 1, and also of much lower average concentration. However, the values of the flammable volume are still large in both cases.

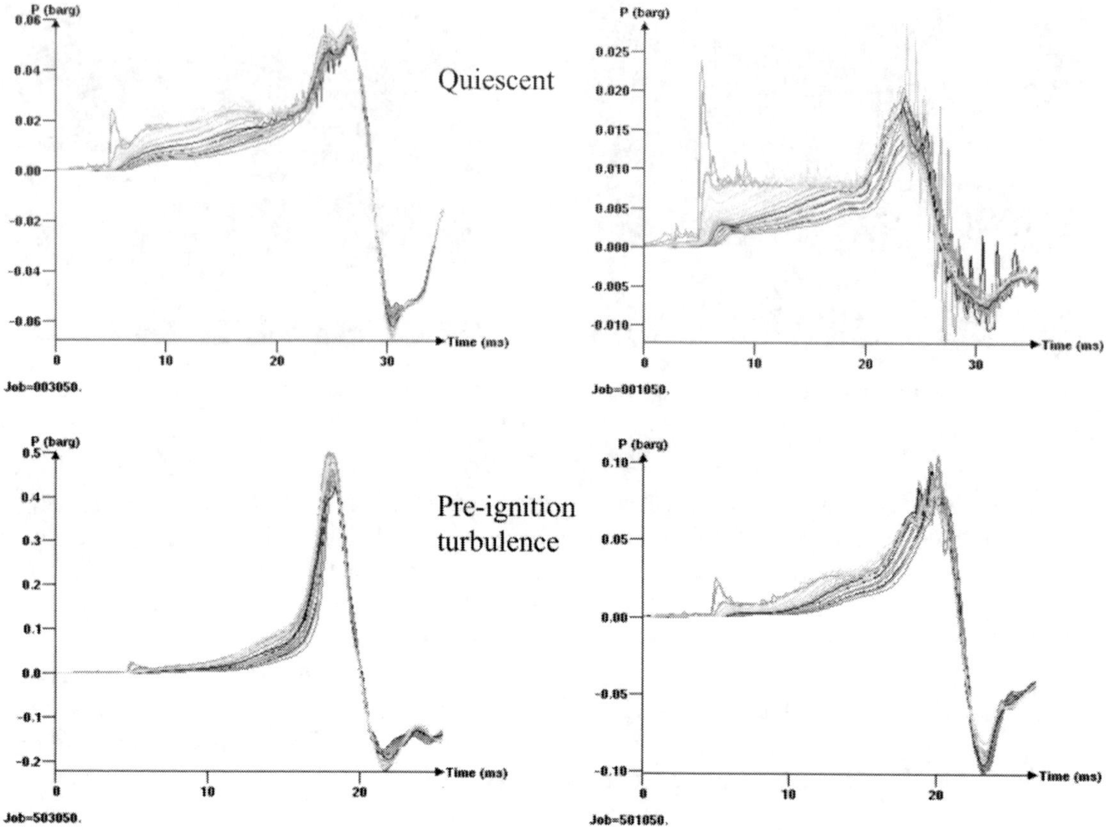

Figure 3. Pressure loads from explosion calculations of a 0.5 m3 stoichiometric H2-air cloud in the "workshop" scenario, hood geometry (left figures), plate geometry (right figures).

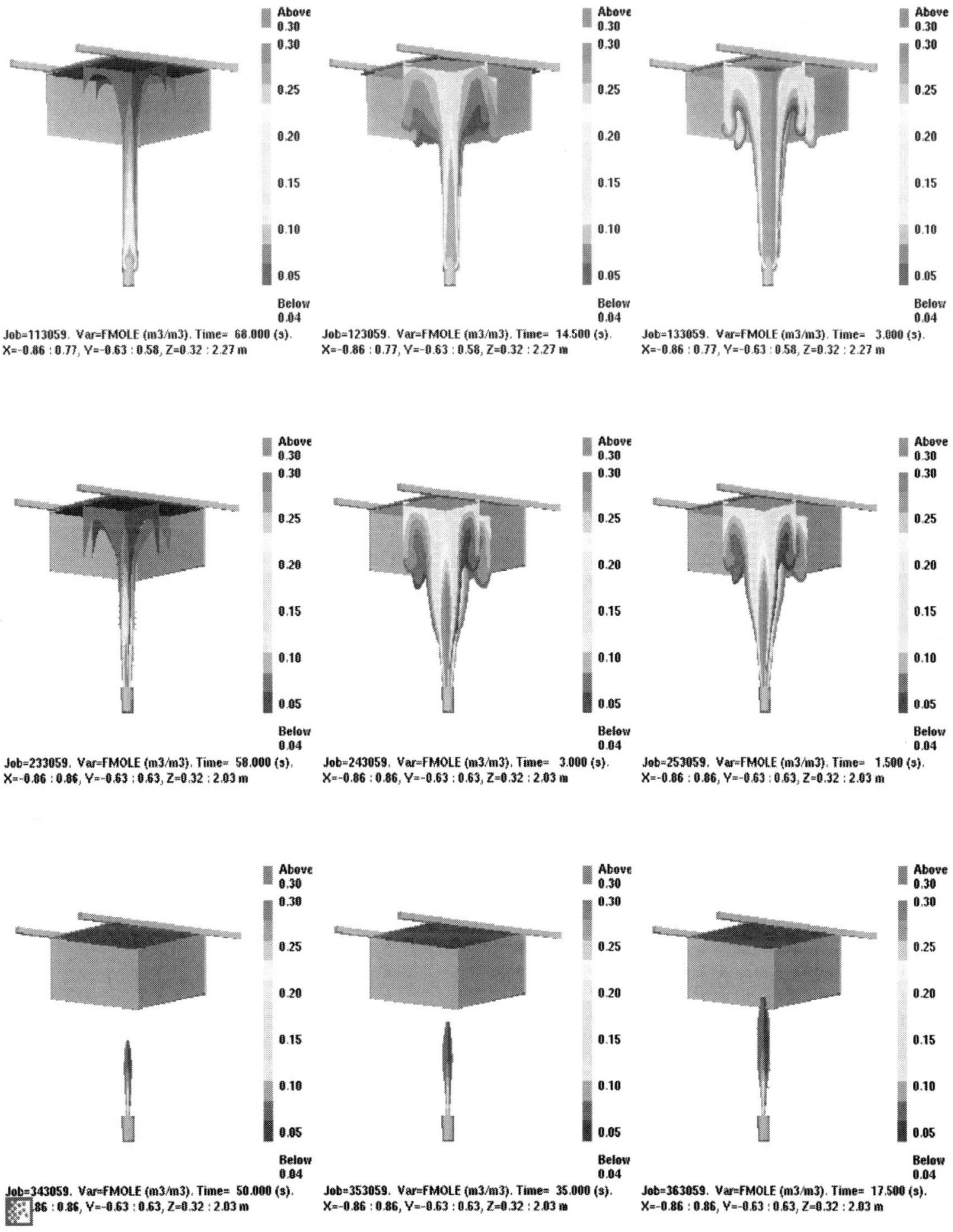

Figure 4. Flammable cloud size and shape towards the end of the release for 9 scenarios in the hood geometry. Large sized hole in left column, medium sized hole in middle column and small sized hole in right column. Low release velocity in upper plot, medium in the center, and high velocity in lower plot.

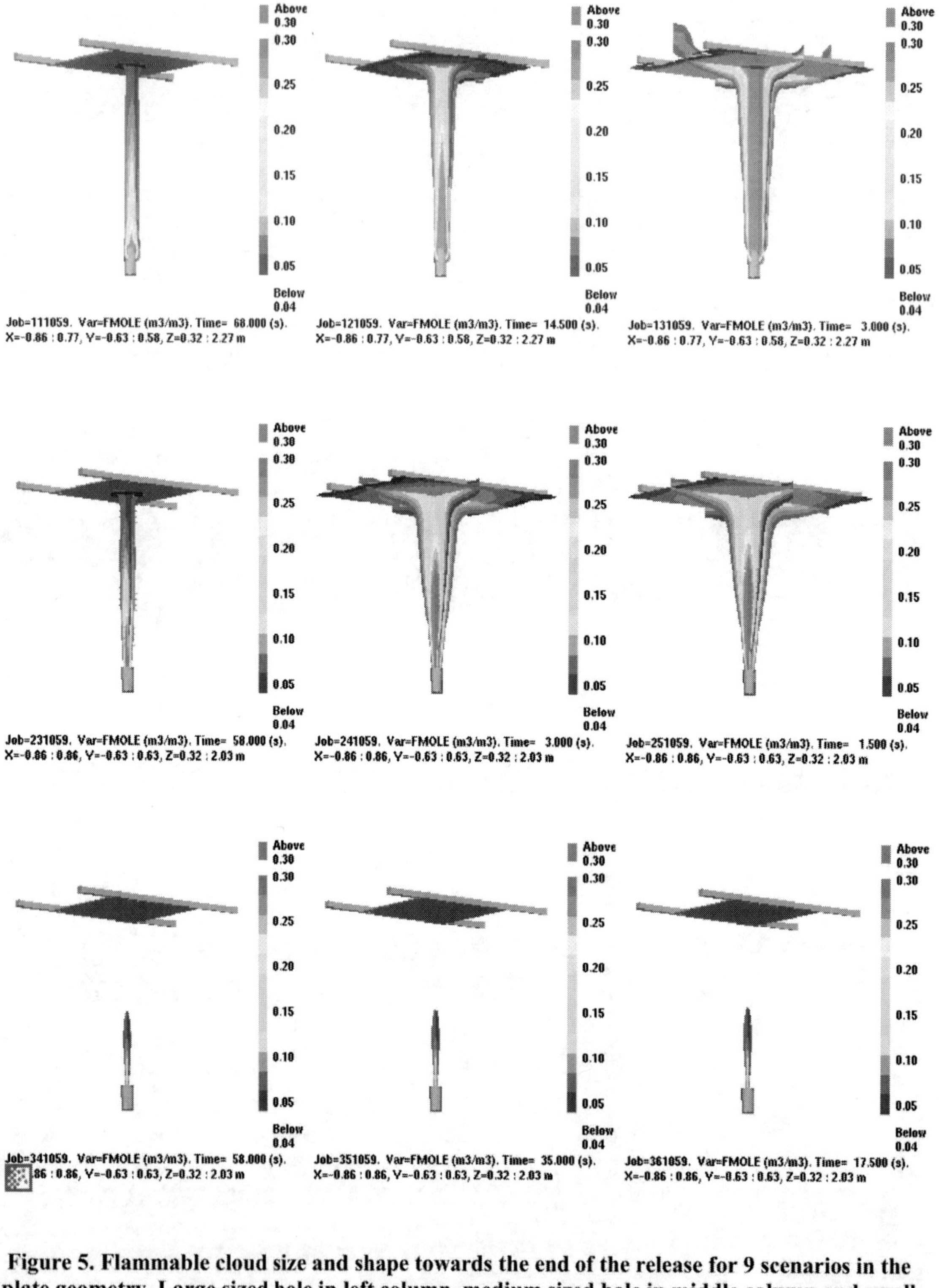

Figure 5. Flammable cloud size and shape towards the end of the release for 9 scenarios in the plate geometry. Large sized hole in left column, medium sized hole in middle column and small sized hole in right column. Low release velocity in upper plot, medium in the center, and high velocity in lower plot.

In order to get a better quantitative estimate of the danger of a given gas cloud, GexCon has developed the concept of an equivalent stoichiometric volume (briefly described in Section 3). This equivalent volume (Q9 output in FLACS) is defined as $Q9 = \sum V_{flam} \times (S \times E)/(S \times E)_{\max}$ where V_{flam} is the flammable volume (of a given concentration), S is the burning velocity, and E is the expansion ratio after combustion. The calculation of the equivalent stoichiometric volume resulted in a maximum value of around 250 litre for the geometry with walls and 80 litre for that without walls. This is much lesser than the 500 litre stoichiometric volume used in the previous step, and the explosion overpressures are expected to be reduced substantially.

The risk can be estimated in a much more rigorous manner by performing a probabilistic risk assessment. This typically involves 100-200 calculations. The study entails the dispersion study to estimate the frequency of occurrence of clouds of different sizes. A much wider range of release scenarios are considered in this kind of study e.g. two different release locations, 6 different leak directions, 9 combinations of leak rates and nozzle sizes (momentum), variable ventilation (not applicable in this situation), total inventory (fixed in this case), etc. A leak frequency is assumed (around 10 incidents/year in this case). Conservative assumptions are used to reduce the number of scenarios. 75% of jets are free jets and may be ignored (as they will not lead to any accumulation of gas). The probability of 2 release locations and 9 leak rate/momentum scenarios can be assumed to be the same.

From these calculations, the annual "exposure time" for different cloud sizes can be estimated. This represents the time that a certain gas cloud volume exists in the "workshop". It can be represented as the following table:

Cloud size (Q9), no ignition	< 1 litre	5 litre	10 litre	25 litre	50 litre	125 litre	250 litre
Exposure time / year	31 s	33 s	8.7 s	8.5 s	30 s	7.1 s	14.4 s

Next, the ignition probabilities are estimated. For this purpose, three contributions are considered as described in Section 3. The contributions from spontaneous, constant, and intermittent ignition sources are added to obtain a total ignition probability as a function of the cloud size. This is presented in the table below:

Cloud size (Q9)	< 1 litre	5 litre	10 litre	25 litre	50 litre	125 litre	250 litre
Ignited clouds / year	Jet fire	0.17	0.06	0.035	0.10	0.11	0.16

The next part of a probabilistic risk assessment is the explosion study. Herein, different cloud sizes are ignited. Released gas clouds are also ignited for reference. Equal weighting is given to centre and end ignition. The explosion pressures are calculated as a function of gas cloud sizes. For the scenario under consideration, the results can be summarized in the table below:

Cloud size (Q9)	< 5 litre	10 litre	25 litre	50 litre	125 litre	250 litre*	500 litre*
Pressure mbar (center & end)	0 & 0	43 & 5	47 & 25	60 & 38	70 & 50	55 & 30	57 & 35

* lower pressure for large cloud sizes die to lower turbulence (high turbulence prevented large clouds)

The results are also plotted in Figure 6. The symbols depict the explosion pressures resulting from the actual gas clouds resulting from the release. Reference calculations with homogeneous gas clouds are included. Two different leak locations are used. From these plots it can be seen that the FLACS Q9 equivalent stoichiometric gas cloud size with a quiescent gas cloud seems to give a good indication of expected explosion pressure for the scenarios simulated. It can be seen that some of the non-homogenous explosion pressures are higher than the quiescent and even initial turbulence simulation with a stoichiometric cloud, but the difference is not more than 30-50%. However, since the effect of initial turbulence seems to become very strong with increasing gas cloud size and confining side walls, one should consider whether this has to be taken into account when using such methods with equivalent stoichiometric gas clouds.

Findings like this are important for the development of risk assessment methods, as it is unrealistic to simulate 1000s of ignition locations (variations in time and ignition location for a range of dispersion calculations). If the output from the dispersion calculations can be sorted into different homogenous cloud sizes based on the FLACS Q9 parameter, this can reduce required number of explosion simulations by orders of magnitude. However, this needs to be validated further.

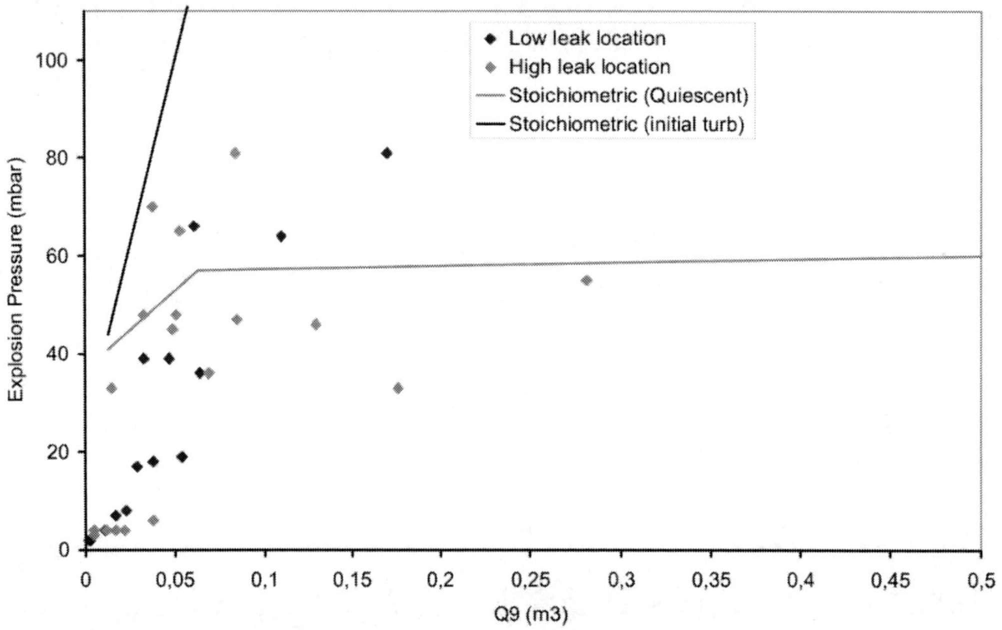

Figure 6. Relation between explosion pressures and cloud size based on FLACS equivalent stoichiometric gas cloud (Q9) for geometry with sidewalls. Reference calculations with homogenous stoichiometric gas clouds are included. Low leak location is shown in blue, high leak location is shown in green, stoichiometric reference calculations are shown in red colour (quiescent) and black colour (pre-ignition turbulence).

In the end, a frequency for explosion load can be estimated based on an accident occurrence of 10 incidents/year and a few transparent assumptions. This is presented in Figure 7. Slightly conservative assumptions have been used, but these may be easily modified with better knowledge available (or more extensive and accurate evaluations). Typically, the impact on person safety and acceptance criteria would decide whether to work to further reduce the estimated risk by modifications or further calculations.

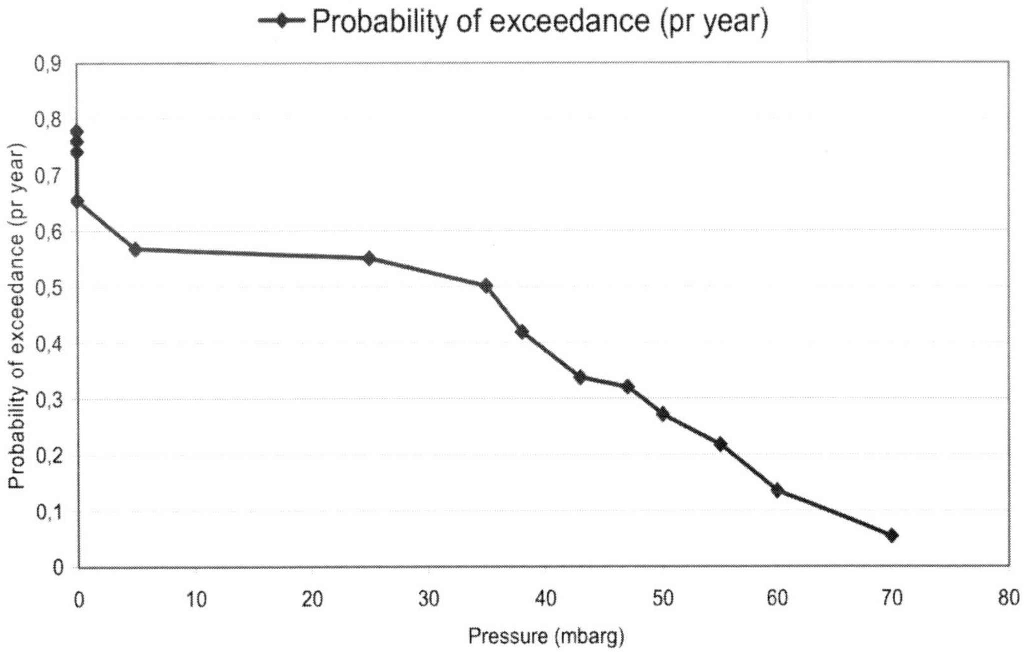

Figure 7. Frequency of a certain explosion load based on an accident occurrence rate of 10 incidents per year.

Thus, this example demonstrates how the complete risk of a given system in terms of the likely explosion pressure load and the frequency with which it is expected to occur can be estimated in a rigorous fashion starting from possible release scenarios.

5. Summary and Conclusions

CFD has become increasingly accepted for risk assessment and explosion safety studies in offshore oil platforms worldwide. Due to the increasing focus on hydrogen safety, significant use of the tool is seen in evaluating the viability of new and existing establishments in terms of probability and consequences of accidental events. It is expected that CFD will be used in carrying out risk assessments for hydrogen risk with the same frequency as that for oil and gas applications. However, the CFD tool needs to be well validated to be able to evaluate risk of hydrogen applications in a truly predictive way.

This article presents a rigorous methodology for carrying out quantitative risk assessment calculations using a validated CFD tool such as FLACS. Three different approaches are presented, with an increasing degree of complexity. This methodology enables the user to estimate the complete risk arising from a hydrogen release in a given geometry in a quantitative fashion. A detailed example of using the methodology is also presented. This example demonstrates how the complete risk of a given system in terms of the likely explosion pressure load and the frequency with which it is expected to occur can be estimated in a rigorous fashion starting from possible release scenarios.

This proposed CFD based risk analysis method for explosion (and fire) risk is advantageous in many respects. These include:

- Physically sound methods will help minimize risk in cost efficient way
- Transparent method will facilitate the identification of knowledge gaps
- Methods and competence developed within oil and gas should be looked to
- More accurate calculation tools and faster computers should be exploited
- Extensive use of simplified methods prevent progress towards more intelligent solutions

It is therefore recommended that such approaches are actively used as a part of the risk "toolbox", and that any RCS being developed will stimulate their use.

6. References

[1] LPG – A study – A comparative analysis of the risks inherent in the storage, transshipment, transport, and use of LPG and motor spirit, TNO, Apeldoorn, the Netherlands, 1983.

[2] Rosyid, O. A., Jablonski, D., and Hauptmanns, U., 2006, Risk Analysis for the Infrastructure of a Hydrogen Economy, Proceedings of the 15th annual conference of the Society for Risk Analysis – Europe, Ljubljana, Slovenia, 11-13 September 2006.

[3] NORSOK Standard Z-013, 2001. Risk and emergency preparedness analysis, Norwegian Technology Center, Oslo, Norway Rev. 2 (September 1, 2001). Available from www.standard.no/pronorm-3/data/f/0/01/50/3_10704_0/Z-013.pdf.

[4] Hansen, O.R., Storvik, I., and van Wingerden, K., 1999. Validation of CFD-models for gas explosions, where FLACS is used as example - model description, experiences and recommendations for model evaluation. Proceedings European Meeting on Chemical Industry and Environment III, Krakow, Poland, available at the Department of Environmental Systems Engineering, Faculty of Process and Environmental Engineering, Technical University of Lodz, Wolczanska 175, 90-924 Lodz, Poland, pp. 365–382.

[5] van Wingerden, K., Hansen, O. R., and Teigland, R., 1995, Prediction of the strength of blast waves is the surroundings of vented offshore modules, ERA-report 95-1173, London, UK.

[6] van Wingerden, K., Hansen, O. R., and Foisselon, P., 1999. Predicting blast over pressures caused by vapor cloud explosions in the vicinity of control rooms. *Proc. Safety Prog.* **18**, 14–17.

[7] Hansen, O. R., Renoult, J., Bakke, J. R., 2001. Explosion risk assessment: how the results vary with the approach chosen. Fall Symposium Proceedings, Mary Kay O' Connor Process Safety Center, Dept. of Chemical Engineering, Texas A&M University, 3122 TAMU, College Station, TX 77843-3122, pp. 395–410.

[8] Dharmavaram, S., Hanna, S. R. and Hansen, O.R., 2005, Consequence Analysis-Using a CFD Model for Industrial Sites. *Proc. Safety Prog.*, **24**, 316-327.

[9] Skjold, T., Arntzen, B. J., Hansen, O. R., Taraldset, O. J, Storvik, I. E. and Eckhoff, R. K., 2005, Simulating Dust Explosions with the First Version of DESC. *Proc. Safety Env. Protec.*, **83**, 151-160.

[10] Hansen, O. R. and van Wingerden, K. 1998, Modeling the Effects of Water Deluge and Inert Gas Dilution on Gas Explosions in Congested Geometries, 9th International Symposium of Loss Prevention and Safety Promotion in the Process Industries, Barcelona, Spain, Addenda to Proceedings, pp. 23-32.

[11] Hansen, O. R., Renoult, J., Sherman, M. P., and Tieszen, S. R. 2005, Validation of FLACS-Hydrogen CFD Consequence Prediction Model Against Large Scale H_2 Explosion Experiments in the FLAME Facility, Proceedings of International Conference on Hydrogen Safety, Pisa, Italy, September 2005.

[12] Wilkins, B. A. and Renoult, J., 2003, Hydrogen Explosion Safety Phase 2 – Small scale dispersion experiments, GexCon-03-F46201-3, GexCon, Bergen, Norway (confidential).

[13] Swain, M. R., Filoso, P. A., and Swain, M. N., 2007, An experimental investigation into the ignition of leaking hydrogen. *Intl. J. Hyd. Ener.*, **32**, 287-295.

[14] Shirvill, L. T. and Roberts, P. T., 2006, Dispersion of hydrogen from high-pressure sources, Proceedings of HAZARDS XIX, Manchester, UK, March 27-30, 2006.

[15] Ruffin, E., Mouilleau, Y., and Chaineaux, J., 1996, Large scale characterization of the concentration field of supercritical jets of hydrogen and methane. *J. Loss Prev. Proc. Ind.*, **9**, 279-284.

[16] Sherman, M. P., Tieszen, S. R., and Benedick, W. B, FLAME Facility, The Effect of Obstacles and Transverse Venting on Flame Acceleration and Transition to Detonation for Hydrogen-Air Mixtures at Large Scale, Sandia National Laboratories, Albuquerque, NM 87185, USA, NUREG/CR-5275, SAND85-1264, R3, April 1989.

[17] Becker T., Ebert F., 1985, Vergleich zwischen experiment und theorie der explosion grober, freier gaswolken. *Chem.-Ing. Tech.*, **57**, 42-45.

[18] Gallego, E., et al., 2005, An Intercomparison Exercise on the Capabilities of CFD Models to Predict Deflagration of a Large-Scale H2-Air Mixture in Open Atmosphere, Proceedings of International Conference on Hydrogen Safety, Pisa, Italy, September 2005.

[19] Sato, Y., Merilo, E., Groethe, M., Colton, J., Chiba, S., and Iwabuchi, H., 2006, Homogeneous Hydrogen Deflagrations in a Sub-Scale Vehicle Tunnel, 15th World Hydrogen Energy Conference, Lyon, France, 12-16 June 2006.

[20] Renoult, J. and Wilkins, B. A., 2003, Hydrogen Explosion Safety Phase 2 – Small scale explosion experiments, GexCon-03-F46201-2, GexCon, Bergen, Norway (confidential).

[21] Peraldi, O., Knystautas, R., and Lee, J. H.S., 1986, Criteria for transition to detonation in tubes. *Proc. Comb. Inst.*, **21**, 1629-1637.

[22] Middha, P, Hansen, O. R., and Storvik, I. E., 2006. Prediction of deflagration to detonation transition in hydrogen explosions. Proceedings of the AIChE Spring National Meeting and 40th Annual Loss Prevention Symposium, Orlando, FL, April 23-27, 2006.

[23] Groethe, M., Colton, J., Chiba, S., 2002, Hydrogen deflagration safety studies in a confined tube. 14th World Hydrogen Energy Conference, Montreal, Québec, Canada, June 9–13, 2002.

[24] Middha, P, Hansen, O. R., Groethe, M., and Arntzen, B. J., Hydrogen Explosion Study in a Confined Tube: FLACS CFD Simulations and Experiments. Paper submitted at 21st International Colloquium of Dynamics of Explosions and Reactive Systems, Poitiers, France, July 23-27, 2007.

[25] Breitung, W., et al., 2005, Integral large scale experiments on hydrogen combustion for severe accident code validation-HYCOM. *Nuc. Engg. Des.* **235**, 253–270.

[26] Schneider, H., 2005, Deflagration and deflagration to detonation transition within a partial confinement similar to a lane Proceedings of International Conference on Hydrogen Safety, Pisa, Italy, September 2005.

[27] Middha, P, Hansen, O. R., and Schneider, H., Deflagration to Detonation Transition (DDT) in Jet Ignited Hydrogen-Air Mixtures: Large Scale Experiments and FLACS

CFD Predictions. To be presented at 12th International Loss Prevention Symposium, Edinburgh, UK, May 22-24, 2007.

[28] Dorofeev, S. B., Bezmelnitsin, A. V., and Sidorov, V. P., 1995, Transition to detonation in vented hydrogen-air explosions. *Comb. Flame*, **103**, 243-246.

[29] Middha, P. and Hansen, O. R., Estimating the consequences of Deflagration to Detonation Transition (DDT) in hydrogen explosions. Paper accepted for presentation at 41st Annual Loss Prevention Symposium, Houston, USA, April 22-26, 2007.

LOSS PREVENTION SYMPOSIUM

Application of CFD Models for the Onshore Process Industry

Gary A. Fitzgerald
Greg R. Knight
Anthony J. Gutierrez

ABS Consulting
14607 San Pedro Ave, Suite 215
San Antonio, Texas 78232
210-495-5195
210-495-5134 (fax)
gfitzgerald@absconsulting.com

Prepared for Presentation at
American Institute of Chemical Engineers
2007 Spring National Meeting
41st Annual Loss Prevention Symposium
Houston, Texas
April 22–26, 2007

UNPUBLISHED

Application of CFD Models for the Onshore Process Industry

Gary A. Fitzgerald
Greg R. Knight
Anthony J. Gutierrez

ABS Consulting
14607 San Pedro Ave, Suite 215
San Antonio, Texas 78232
210-495-5195

ABSTRACT

The evolution of technology has made the application of complex computational fluid dynamics (CFD) models practical for the onshore process industry. Previously, CFD models were extremely complex and required computing capacity in excess of the typical desktop PC and expertise beyond most skilled engineers. Creating CFD models had also been a cumbersome and time intensive task, resulting in a cost exceeding that within reach of most process industries. However, advances in computing capacity together with easier to use programs have allowed skilled engineers the ability to apply these tools to enclosed processes at onshore process plants for a reasonable cost. These enclosed processes previously had few practical means of evaluating explosion hazards. Limitations of commonly used approaches include:

- The use of external vapor cloud explosion blast curves is not a valid application, though often used due to the lack of other available means.
- The Shell Code for Overpressure Predictions in gas Explosions[1] [SCOPE] model is a cost-effective tool, but has limitations in the available venting surface and deluge modeling, limiting its application to onshore facilities as a screening-level analysis.
- NFPA 68 calculations have also been used, but inappropriately in the process industry since congestion levels cannot be accounted for by this method.

Over the last several years, ABS Consulting has performed multiple studies evaluating process buildings for internal explosions to evaluate both dispersion and explosion and their effects on nearby structures using the FLame ACceleration Simulator[2] (FLACS) CFD model. Significant factors that can be accounted for with the CFD models include: fan activation, natural ventilation, hydrocarbon monitoring, leak isolation, sprinkler activation and a significant amount of vent surface area from openings and frangible panels. Most of these studies had been previously evaluated using other methods that were very conservative and resulted in significant damage predictions to nearby buildings. In the majority of cases, the CFD models reduced these damage predictions and provided confidence that accurate and cost-effective modeling methods were now available to the onshore process industry. Examples presented in this paper show how CFD models were used to minimize the blast effects of internal explosions, sometimes by helping to

---

[1] Shell Code for Overpressure Prediction in Gas Explosions (SCOPE), Shell Global Solutions, Chester, United Kingdom.

[2] Flame Acceleration Simulator (FLACS), GexCon, Bergen, Norway.

design fire protection and other automated actions, and examples of actions that can inadvertently increase building damage predictions.

INTRODUCTION

Over the last several years, ABS Consulting has been contracted to perform CFD modeling for potential flammable releases and explosions in process buildings for the onshore process industry. The examples presented in this paper summarize three of these projects. Each of these three examples had a screening study performed using different methods.

Example 1:

1: Description of problem

In this example, an explosion consequence study was performed at a plant for ethylene oxide releases inside a process building. Many buildings were evaluated, but the closest building was a Control Room attached to the process building. The layout is shown in Figure 1 and a photo of the Control Room in Figure 2. The released material was from a ¾ inch diameter hole. The plant did have a water deluge system and it was considered during the evaluation.

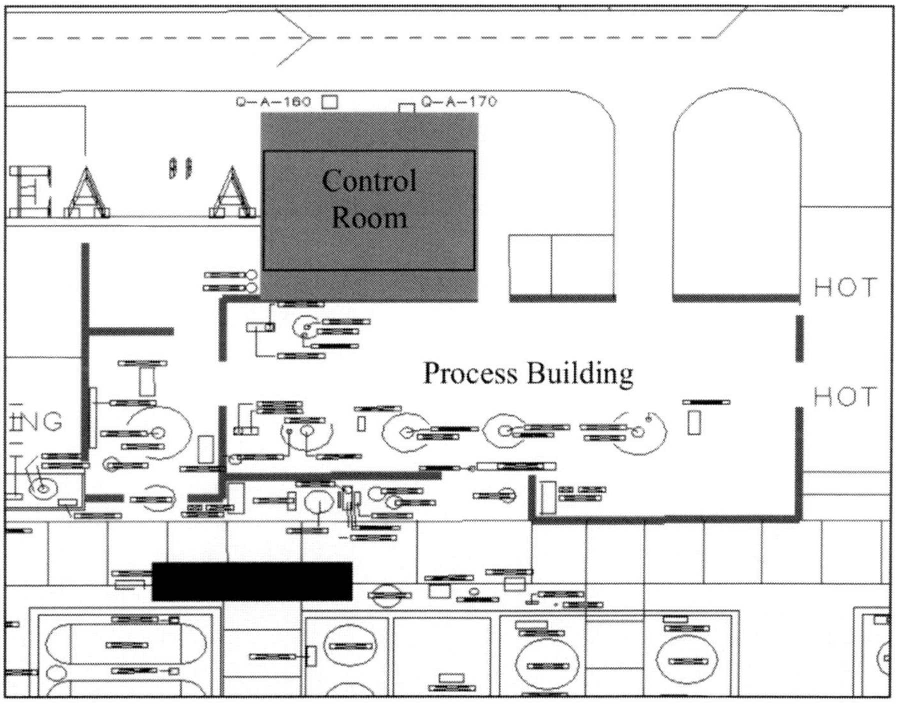

Figure 1 - Example 1 Site Layout

Figure 2 - Example 1 Process Building and Control Room A

1: Review of Previous Analysis

An initial screening level study had been previously performed using PHAST 5.2 for the source/dispersion calculations and the Baker Strehlow Method (BSM, 1999) for the explosion assessment. The dispersion analysis in PHAST was modeled with the best effort to reflect the conditions inside a building, despite the obvious problems with applying an external dispersion code to an internal release. Initial pressure predictions were 5.2 psi on the adjacent control room and the control room would collapse. A ventilation and water spray study was then conducted to determine mitigation necessary to prevent the explosion. Water spray similar to that recommended was installed, but had a delayed activation time of 60 seconds. Ventilation recommendations had not been implemented after the initial study.

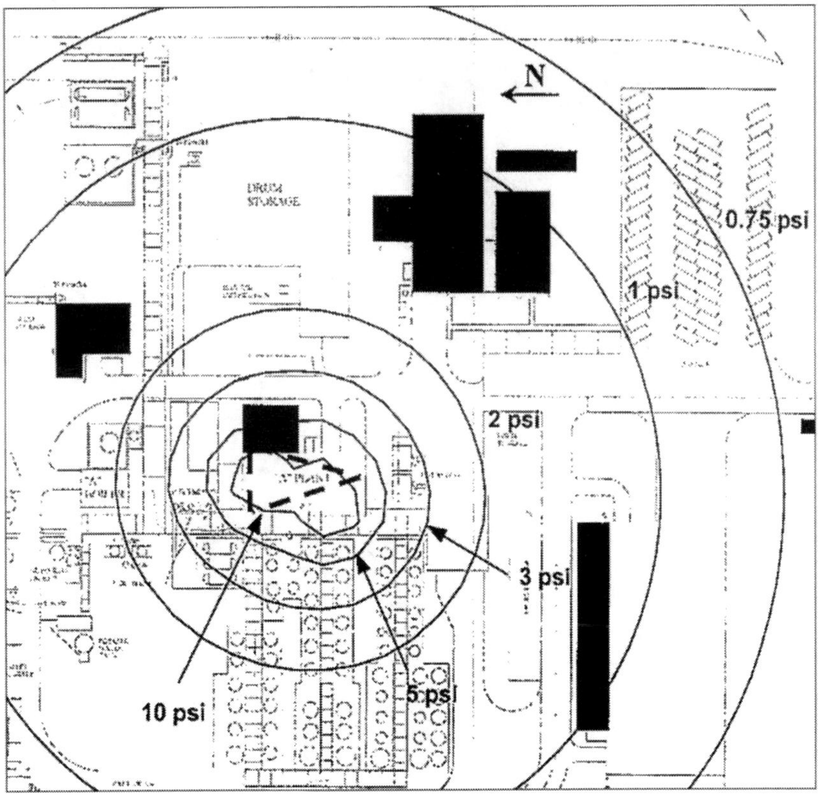

Figure 3. Example 1 Free Field Pressure Contours from the Previous Study Before Mitigation Measures

Subsequently, during the facility siting revalidation, it was determined that advances in modeling could evaluate the implementation of the mitigation measures taken and better evaluate potential consequences.

1: CFD Modeling Performed

CFD modeling was performed to evaluate the installed water spray, existing ventilation and proposed system isolation. A site visit was conducted to document piping in the building. The notes and photographs acquired from the site visit were used to construct a CFD model that represented major piping and equipment in the process building. Smaller diameter piping was represented using equivalent grids of small piping to provide similar blockage ratios as that in the building. This model is shown in Figure 4. A gaseous source term was calculated for a ¾ inch diameter hole and was used in the CFD model. Several releases were considered with different locations and directions evaluated to determine the largest potential flammable vapor cloud. Then several ignition locations were considered for the largest vapor cloud and that ignition location with the highest pressure was taken to represent this scenario. The installed deluge system was included in this evaluation for its effect on the postulated explosion.

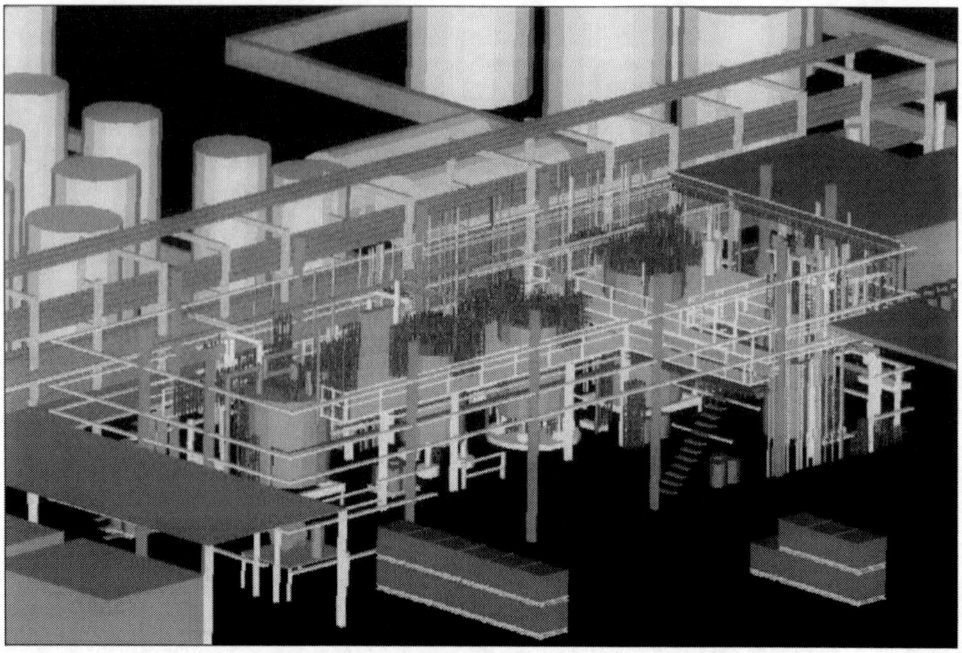

Figure 4 - Example 1 CFD Model Geometry (building walls removed for clarity)

1: CFD Dispersion Modeling Results

As expected, the largest vapor cloud was for an unisolated condition, where the leak was allowed to persist indefinitely. Illustrations of the predicted vapor clouds at different times are shown in Figure 5. Figure 6 shows a plot of flammable vapor cloud volume with time. This figure shows that a steady state volume is reached after about 200 seconds. However, the largest vapor cloud volume does not occur during the steady state, but occurs after the leak is stopped. This is because once the leak is stopped, a significant volume of gas greater then the flammable limit exists, which is then mixed with air to create a large peak of flammable vapors about 38 seconds after the leak is stopped.

The study showed a need to be able to isolate the leak as soon as possible; thus, further study of an isolation system was conducted. The proposed isolation system was assumed to shut the isolation valves in 30 seconds and the isolated volume would deinventory in 30 seconds. The 30 seconds of blowdown was assumed as a linear decrease in flow. Stopping the leak 60 after detection did not produce the large spike in flammable volume as was observed without isolation in the previous figure because there had not been a large build-up of rich vapors above the flammable limit due to reducing leak rate from 30 to 60 seconds. The resulting maximum flammable vapor cloud volume with isolation was found to be approximately 37,000 ft^3, only a moderate increase above the volume at 60 seconds (30,800 ft^3). These can be seen in Figure 7.

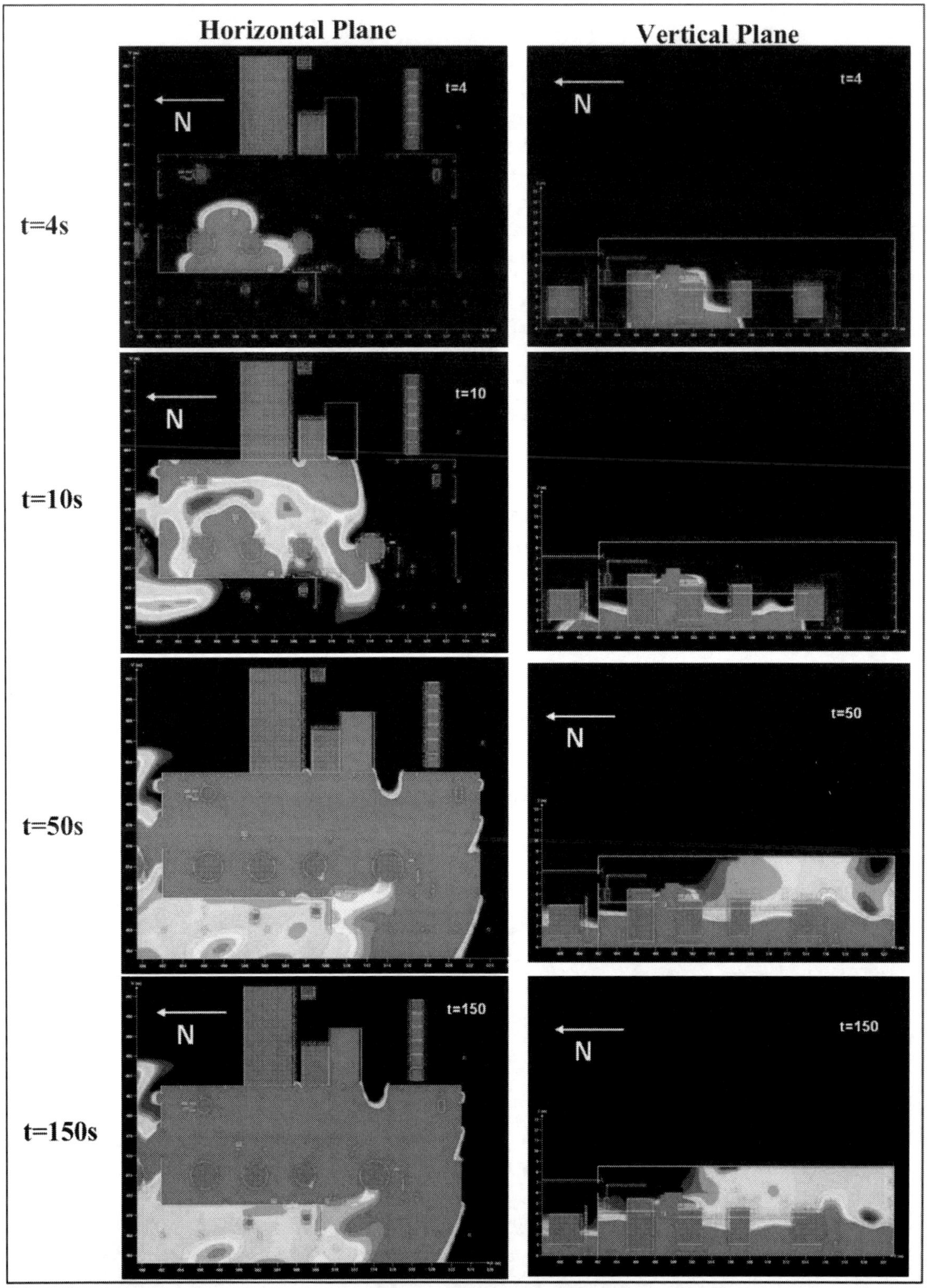

Figure 5 - Example 1 Dispersion

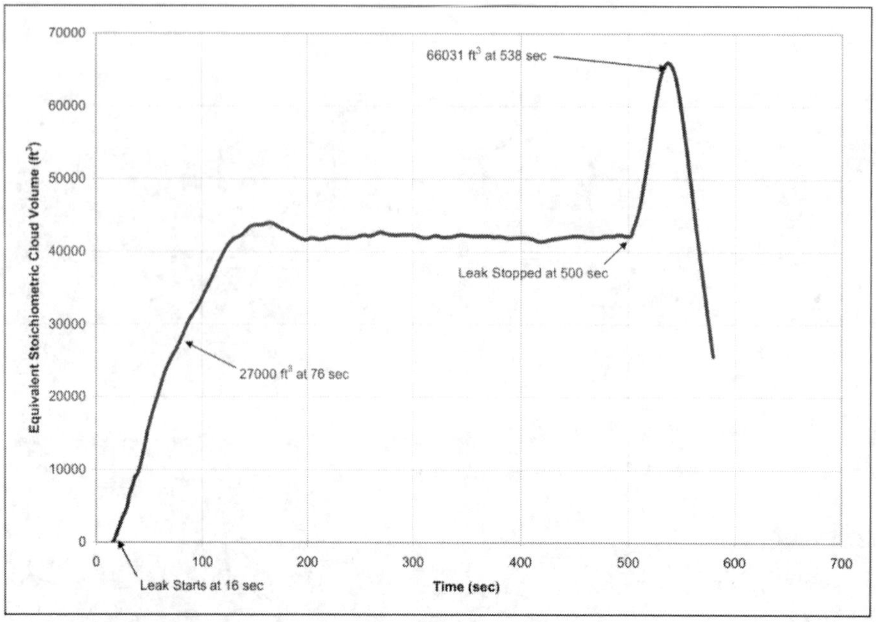

Figure 6 – Example 1 Flammable Cloud Volume Buildup and Dispersion with Time without Isolation

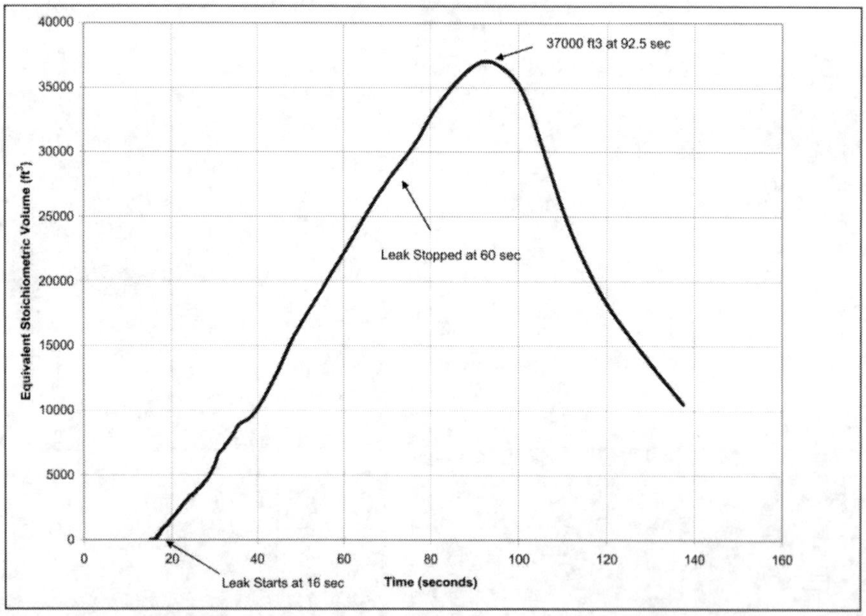

Figure 7 – Example 1 Flammable Cloud Volume Buildup and Dispersion with Time with Isolation

1: CFD Explosion Modeling Results

The explosion scenarios considered in this study were as follows:

1. As-Is Condition (no isolation, with water spray): Ignition at the maximum flammable mass while water spray is running (46% building volume, 66,031 ft^3)
2. Isolation and Pre-deluge (system isolation, before deluge initiates): Ignition occurs before deluge activation at 60 seconds (30,800 ft^3)

3. Isolation and Deluge (system isolation and ignition occurs after deluge activation): Ignition occurs after deluge activation at 93 seconds (37,000 ft^3)

The potential for a deflagration to detonation transition was first evaluated since the material was a high reactivity fuel that has been known to detonate. It was found that all flame speeds were in the range of Mach 1.0 or less for the case without deluge or isolation. Thus, it was determined that a detonation was not likely and this study would be representative of the postulated event.

Multiple ignition locations were then evaluated to determine the location resulting in the greatest blast loads. A summary of predicted blast loads on Control Room A is given in Table 1. Site contours for Scenario 3 are shown in Figure 8.

Table 1 - Example 1 CFD Blast Loads Summary

Evaluations	Control Room A Applied Loads	
	P (psi)	I (psi-ms)
1. FLACS study (as-is), no deluge, no isolation	3.3	59
2. FLACS study, pre-deluge, with isolation	1.6	61
3. FLACS study, post-deluge, with isolation	1.4	45

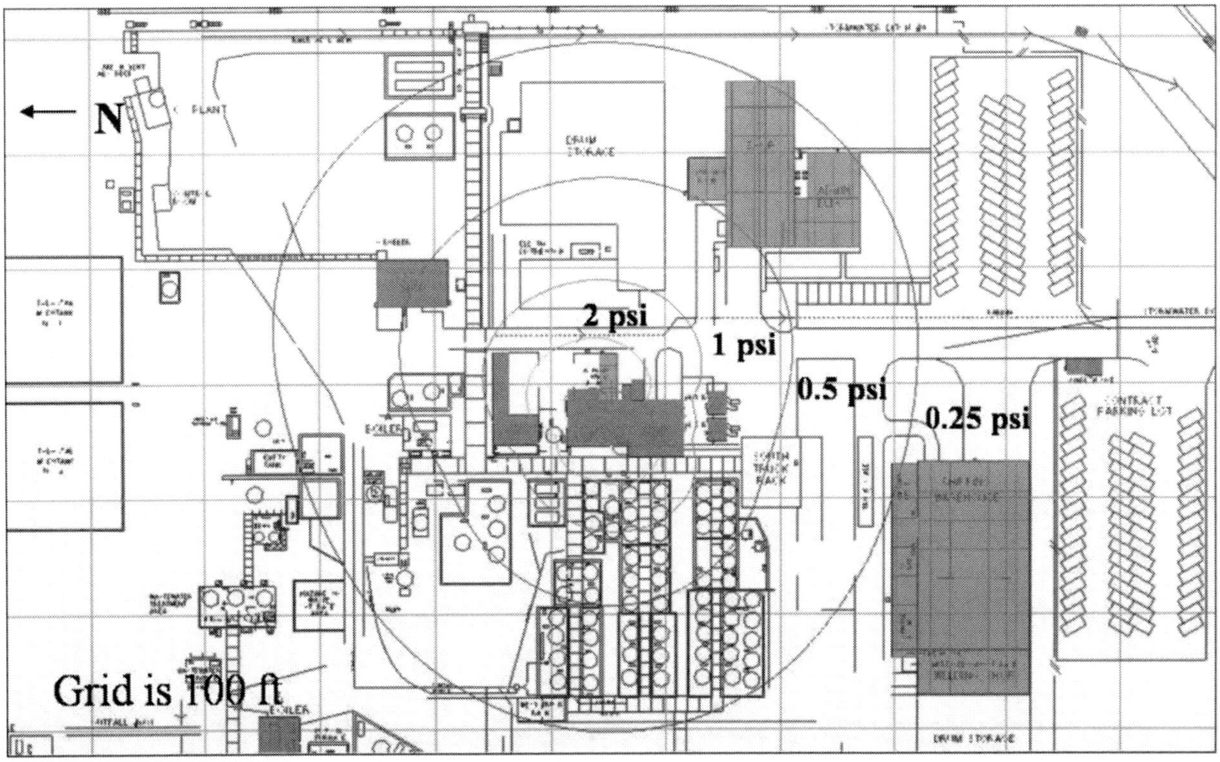

Figure 8 – Example 1 Pressure Contours for FLACS Study with Deluge and Isolation (Scenario 3)

The study concluded that the water spray would significantly reduce the explosion consequences, but not mitigate the explosion entirely. Control Room building damage was estimated to be high, but not collapse. It was also noted that allowing immediate deluge activation upon detection (removing the 60-second delay) would be highly beneficial.

Example 2:

In this example, a PVC processing plant requested that its polymerization and additive building (Poly Building) be analyzed for explosion threats originating inside the building. The Poly Building was a large partially clad metal building approximately 155' by 51' by 69' high. It contained six reactors and two feed tanks, which processed vinyl chloride monomer (VCM) at 175 psig at ambient temperature. A 1 inch leak scenario was considered for this study. The Poly Building did have a water deluge system activated by gas detectors, as well as passive ventilation in the form of openings in the building walls as well as metal panels for the walls which were expected to act as relief panels at their failure pressure. Nearby occupied buildings included a blast resistant control room 270 ft west of the Poly Building, an engineer's office 213 ft west of the Poly Building, a PVC Maintenance Building 350 ft south-west of the Poly Building, and a Lab/I&E Building 150 ft south of the Poly Building. The site layout can be seen below in Figure 9.

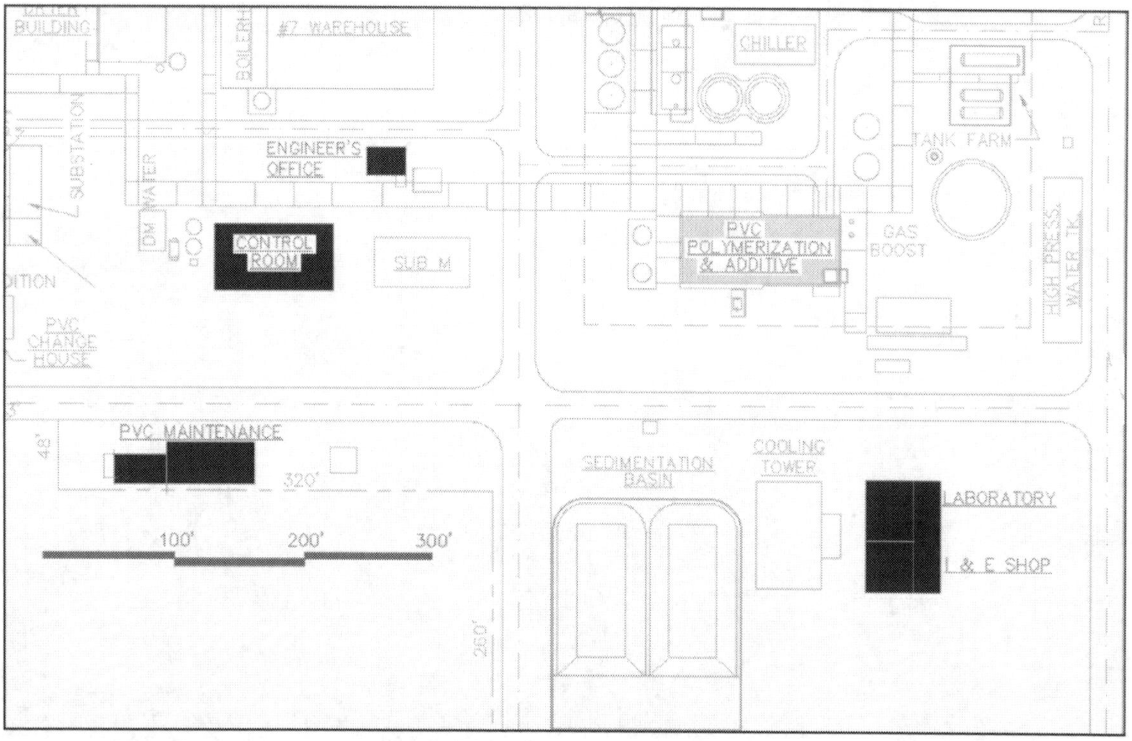

Figure 9. Example 2 Site Layout

2: Review of Previous Analysis

An initial screening level study was done using Trace for the source/dispersion calculations and the TNO multi-energy method (curve 7) for the explosion assessment. The dispersion analysis in Trace was modeled to best reflect the conditions inside a building, despite the obvious problems with applying an external dispersion code to an internal release. The results indicated that flammable vapors would reach the ground floor from an elevated release. The explosion assessment using the TNO curve 7 for overpressure calculations assumed a congested region 26' by 20' by 10' high. The effects of the building containment, frangible wall panels and the deluge system were not considered. Overpressure contours can be seen below in Figure 10.

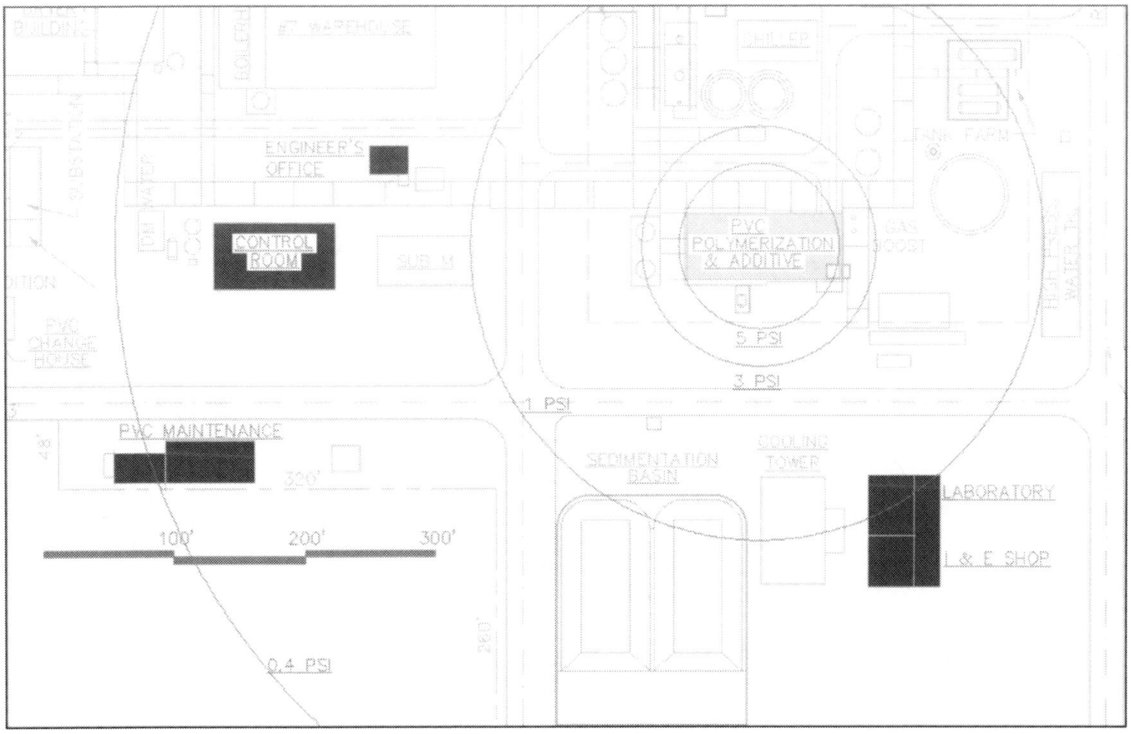

Figure 10. Example 2 Free Field Pressure Contours from the Previous Study

2: CFD Modeling Performed

A CFD model of the Poly Building was built using information gathered during a site visit, including photographs, measurements, and drawings. Pipes down to a few inches were included in the model explicitly, while other areas of high congestion were modeled using arrays of pipes that approximated the area and volume blockages seen during the site visit. In addition to pipes, equipment was included in the model as well as building framing and wall panels. Wall panels were included as frangible panels in FLACS that release (at 2 psi) during the explosion event. The 3D cad model can be seen below in Figure 11.

The source term input into the CFD model for the release was calculated using PHAST. Multiple release locations were evaluated; all on the third floor where the reactor feed lines were located. Gas monitor regions encompassing the Poly Building were used to track the flammable volume inside the building during the dispersion. In this way, the effects of release location, wind velocities and wind directions were evaluated in order to determine a worst case.

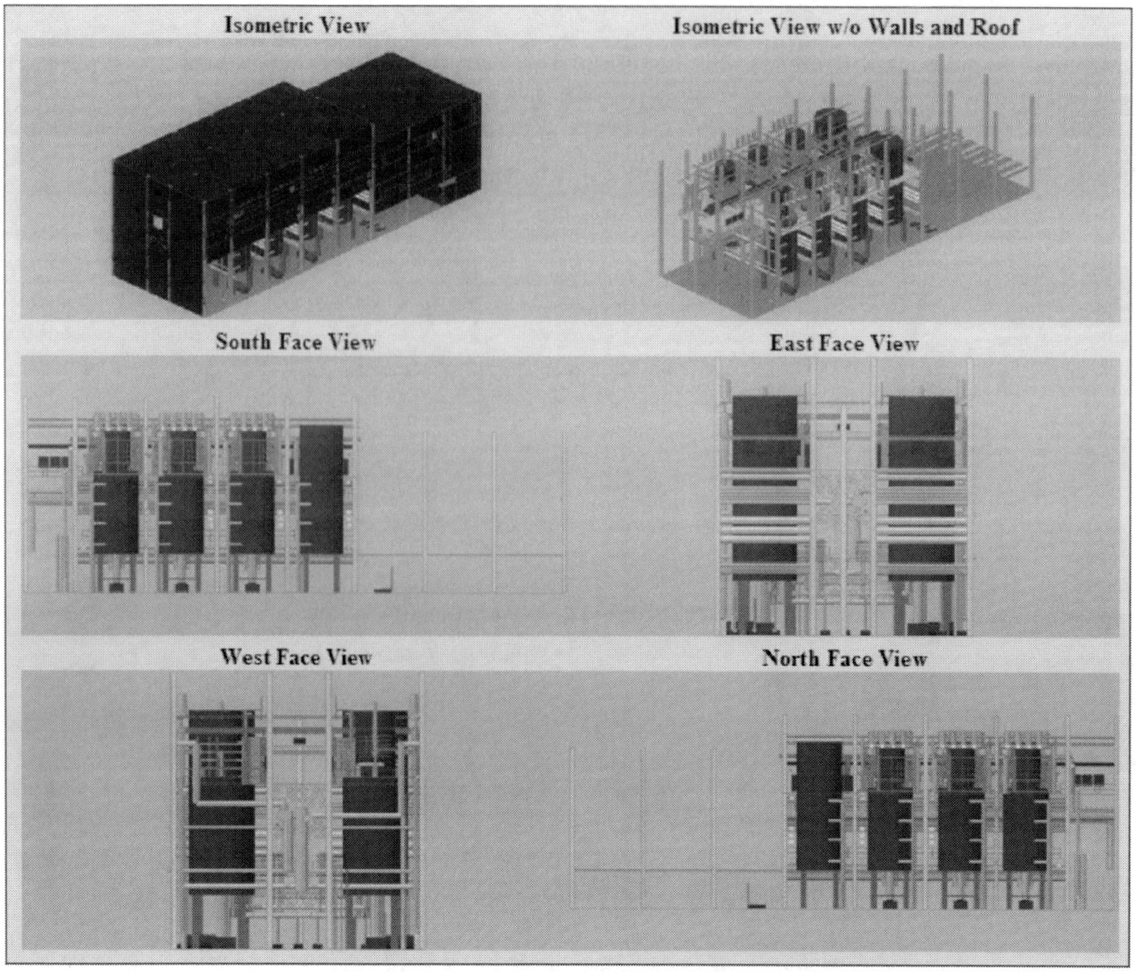

Figure 11. Example 2 CFD Model Geometry

2: CFD Dispersion Modeling Results

Dispersion modeling investigated multiple wind velocities and directions in order to determine the worst case. Wind direction will influence the dispersion inside the building when penetrations in the building exterior allow ventilation and subsequent mixing of flammables with air. The effects of air exchange fans were also included in the model. Hydrocarbon monitors on all three levels were included. Since there was no automatic isolation based on flammable detection, but only water deluge activation, the hydrocarbon detectors were primarily used to determine at what times the largest flammable cloud could exist both before deluge activated and after deluge activated. Peak flammable mass would typically occur at one or two minutes, after which the flammable mass would begin to decrease as the vapors in the building became too rich to support combustion. Flammable contours inside the building can be seen below (Figure 12) as well as a flammable volume plot (Figure 13) for the worst case dispersions both before and after deluge activation.

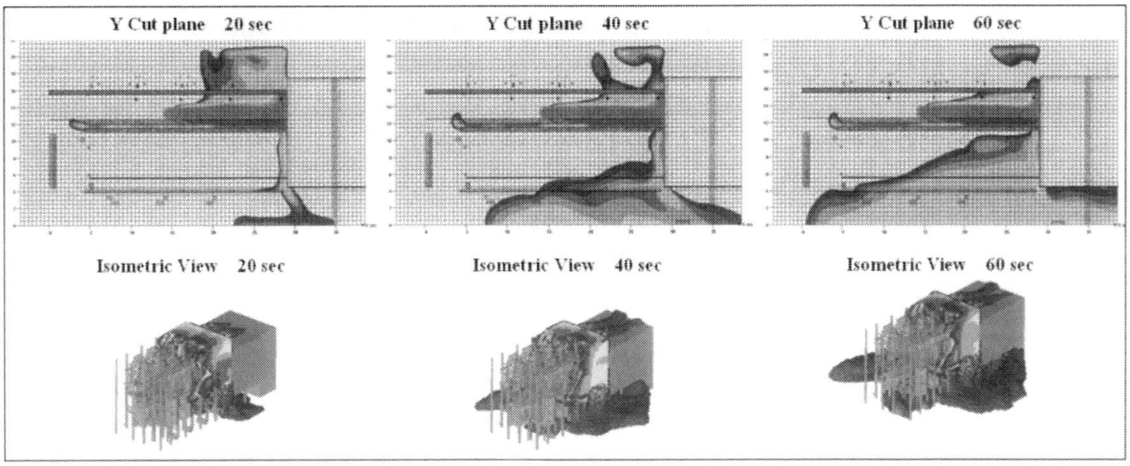

Figure 12. Example 2 Dispersion Contours for D04 (Flammable Limits Shown)

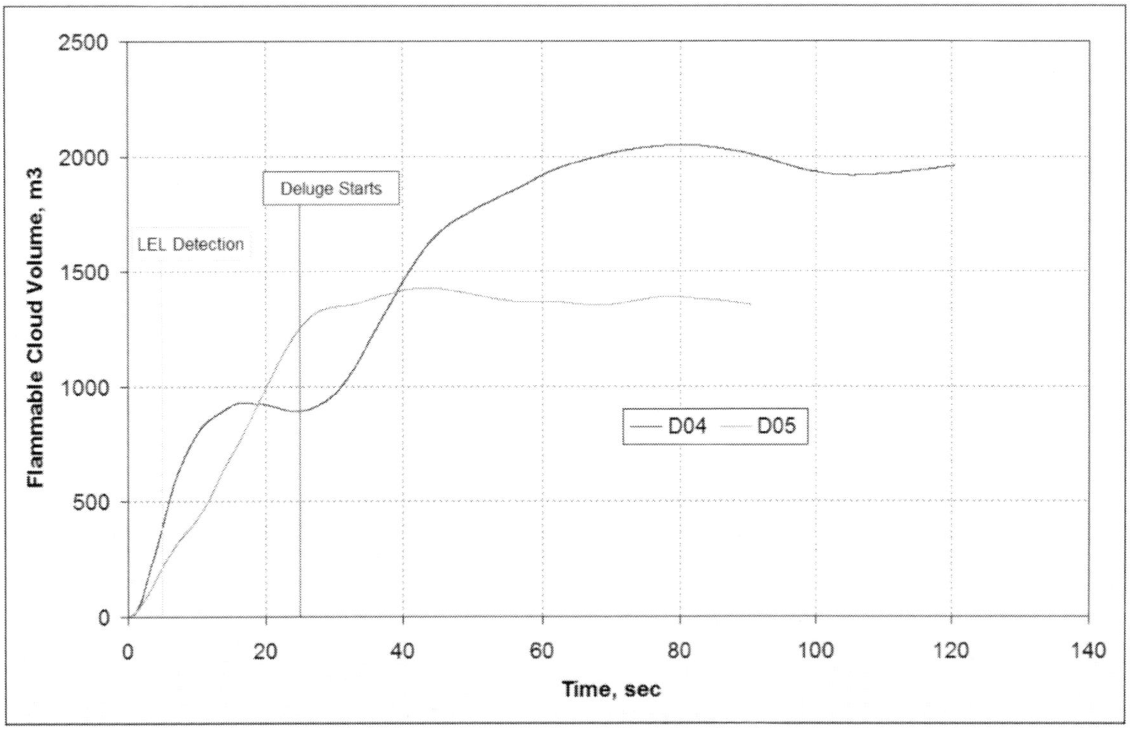

Figure 13. Example 2 Flammable Volume Plots

2: CFD Explosion Modeling Results

Explosion modeling examined the effects of cloud location and ignition location on peak pressures. Cloud size was based on the maximum flammable volumes calculated in the dispersion study both before and after deluge. Cloud location was based on observation of the dispersion on where the flammables were centered. A rectangular stoichiometric cloud was used in place of the actual inhomogeneous cloud due to a requirement in FLACS. FLACS automatically calculates the volume of this stoichiometric cloud in the dispersion model assuming a certain level of congestion inside the gas monitor region. Multiple ignition locations were chosen including locations inside highly congested regions, in corners formed by external walls, and between reactors. The worst case location was found to be inside a congestion region rather than directly adjacent to one, due to the rapid initial flame acceleration.

Maximum overpressures were generated by clouds before the activation of deluge, despite the fact that the post deluge clouds were approximately 40 percent larger than the pre deluge clouds. As seen below in Figure 14, these overpressures were much lower than those calculated in the previous screening study using simplified methods.

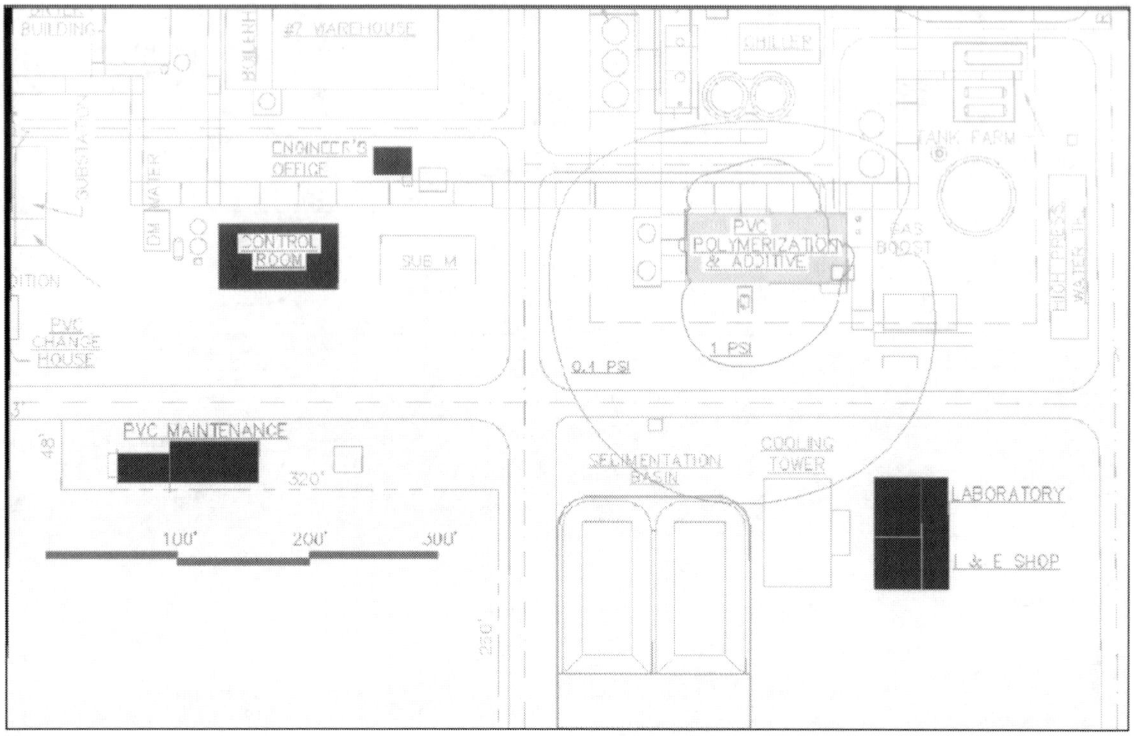

Figure 14. Example 2 CFD Free Field Pressure Contours Before Deluge Activation

A comparison of the previous study contours to the CFD study contours shows a significant reduction in blast contours. Whereas the previous study had the north end of the Laboratory at 1 psi or greater, the CFD study shows the north end of the Laboratory at less than 0.4 psi.

Building damage assessments using the CFD loads resulted in damage to the west and north wall panels as in the screening study. The revised damage levels did result in the frame sway being classified as acceptable and the overall possibilities of retrofit to the wall panels as good based on the level of response. The remaining buildings were all classified as acceptable response, with the directional nature of the blast loads as revealed by the CFD analysis being a large factor in the response of the buildings west of the Poly building.

Example 3:

3: Description of problem

In this example, a process plant needed an explosion consequence study performed for releases inside their process building with reactors. The released material was a flammable liquid from a 2 inch diameter hole. The plant did not have, but was willing to consider an area water deluge explosion mitigation system (in addition to the already installed process equipment deluge system) and explosion vent panels. Many plant buildings were evaluated, but the closest buildings were a Control Room attached to the process building and a Laboratory next to the Control Room. This layout can be seen in Figure 15. The Control Room was designed for blast-resistance, but the Laboratory next door was not.

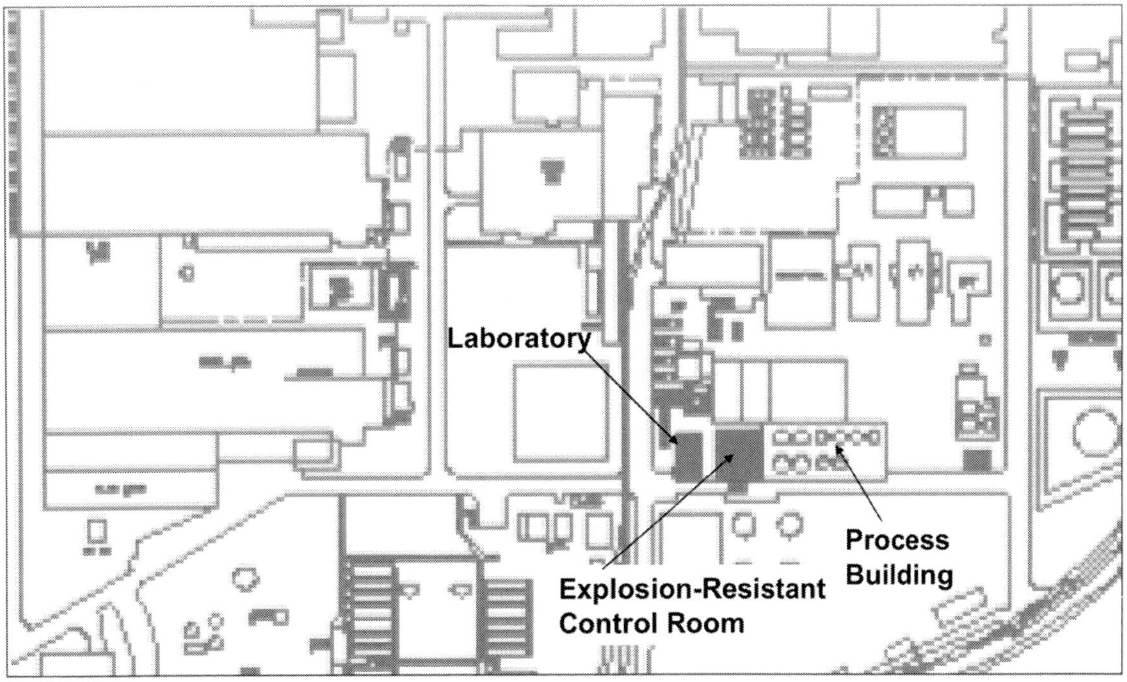

Figure 15 – Example 3 Site Layout

3: Review of Previous Analysis

A screening-level study was first performed using SCOPE. Two scenarios were considered; one with the building assumed to be one-half filled with stoichiometric vapors and another assuming the building was one-third filled with stoichiometric vapors. The resulting pressure contours are shown in Figure 16 for without area deluge and with area deluge conditions.

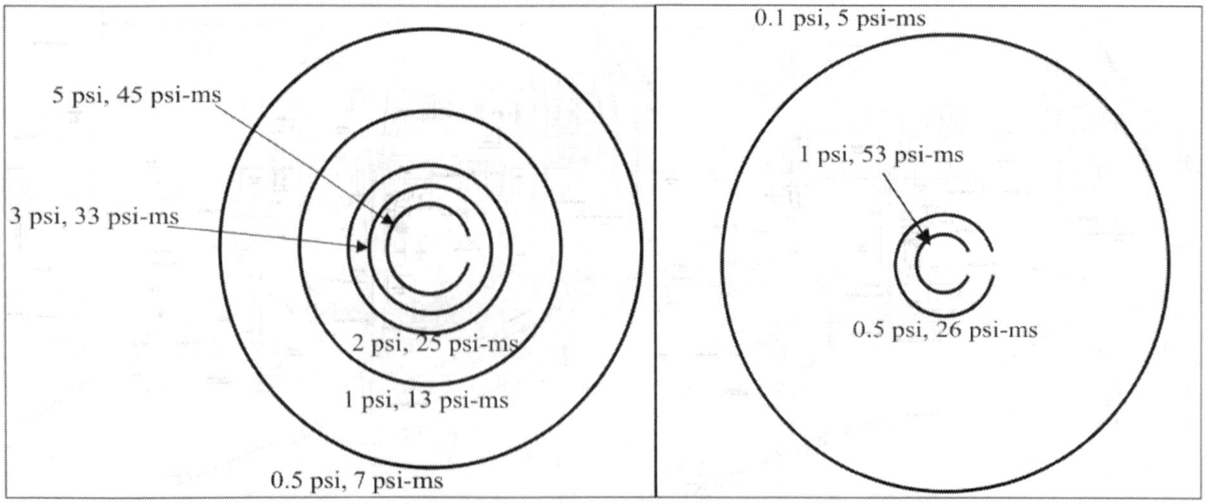

Figure 16 – Example 3 Free Field Pressure Contours from the Previous Study (without deluge on the left and with deluge on the right)

That study concluded that with deluge, the only buildings with excessive damage were the Control Room and Laboratory. However, the deluge capabilities in SCOPE were limited and it was not practical for the client to install an area deluge system similar to the deluge assumptions used by SCOPE. Thus, a better method of evaluating an area deluge system was needed.

Additionally, a small space existed between the Control Room and the Laboratory. Since Scope is not able to evaluate potential shielding, the blast loads were developed without considering the geometry between the Laboratory and the Control Room. The Laboratory wall facing the process building was predicted to have high damage and it was also the wall facing this gap with the Control Room. Thus, it was proposed to install a roof over this gap, spanning the Control Room and Laboratory, to prevent the blast wave from coming over the Control Room and damaging that wall of the Laboratory. Installing walls and a roof to enclose this area was not feasible for the client, so only the roof was considered to mitigate the damage to the Laboratory. It was assumed this would resolve the problem with the Laboratory, but this needed to be confirmed.

3: CFD Modeling Performed

A CFD model was built from notes and photographs to represent major piping greater than 4 inch diameter. Smaller diameter piping was represented by using equivalent grids of small piping to provide similar blockage ratios as that in the building. This model is shown in Figure 17. Significant portions of the walls and roof were frangible and would be expected to vent internal blast loads. A structural analysis from the SCOPE study determined these panels would vent at 1.5 psi internal loading. A gaseous source term was calculated for a 2 inch diameter hole and used in the CFD model. Several release locations and spray directions were evaluated to determine the largest flammable vapor cloud. Then several ignition locations were considered for the largest vapor cloud and that ignition location with the highest pressure was taken to represent this scenario. The area deluge system being considered for installation was then evaluated for its effect on the postulated explosion.

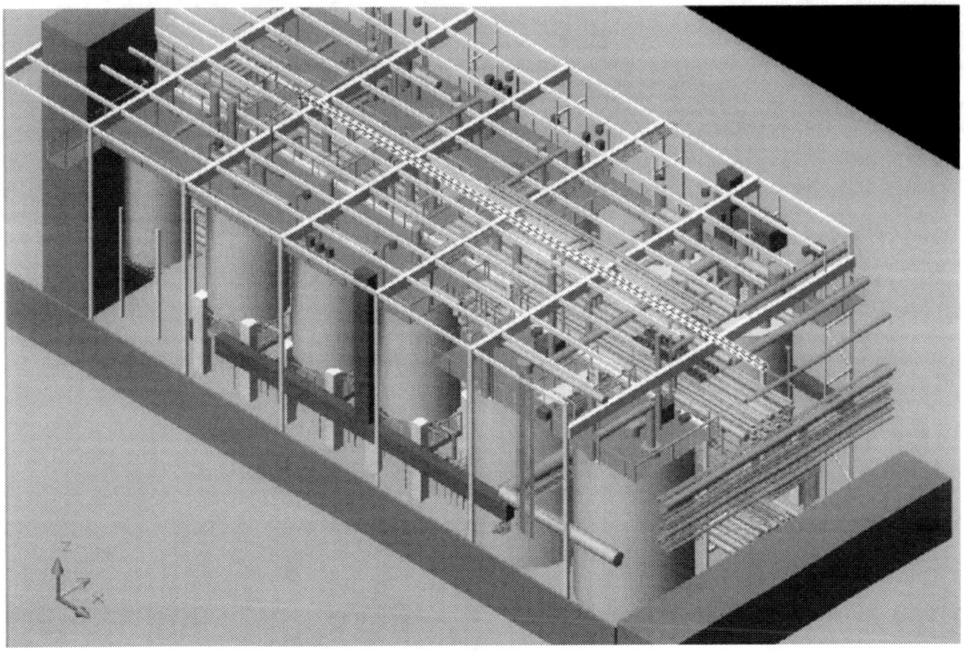

Figure 17 – Example 3 CFD Model Geometry (building walls removed for clarity)

3: CFD Dispersion Modeling Results

As expected, the largest vapor cloud was for an unisolated condition, where the leak was allowed to persist indefinitely. It was found that for this condition, the building could quickly fill with vapors. Existing hydrocarbon monitor locations were then used along with additional monitor locations to evaluate the effect of leak isolation upon flammable vapor detection. With automated leak detection and isolation (allowing time for leak detection, isolation and deinventory), it was found that approximately 60% of the building could be filled with an equivalent stoichiometric vapor cloud. This vapor cloud was then used in the explosion model to evaluate as-built conditions, proposed vent panels and a sprinkler system being considered for this building. Illustrations of the predicted vapor clouds at different times are shown in Figure 18.

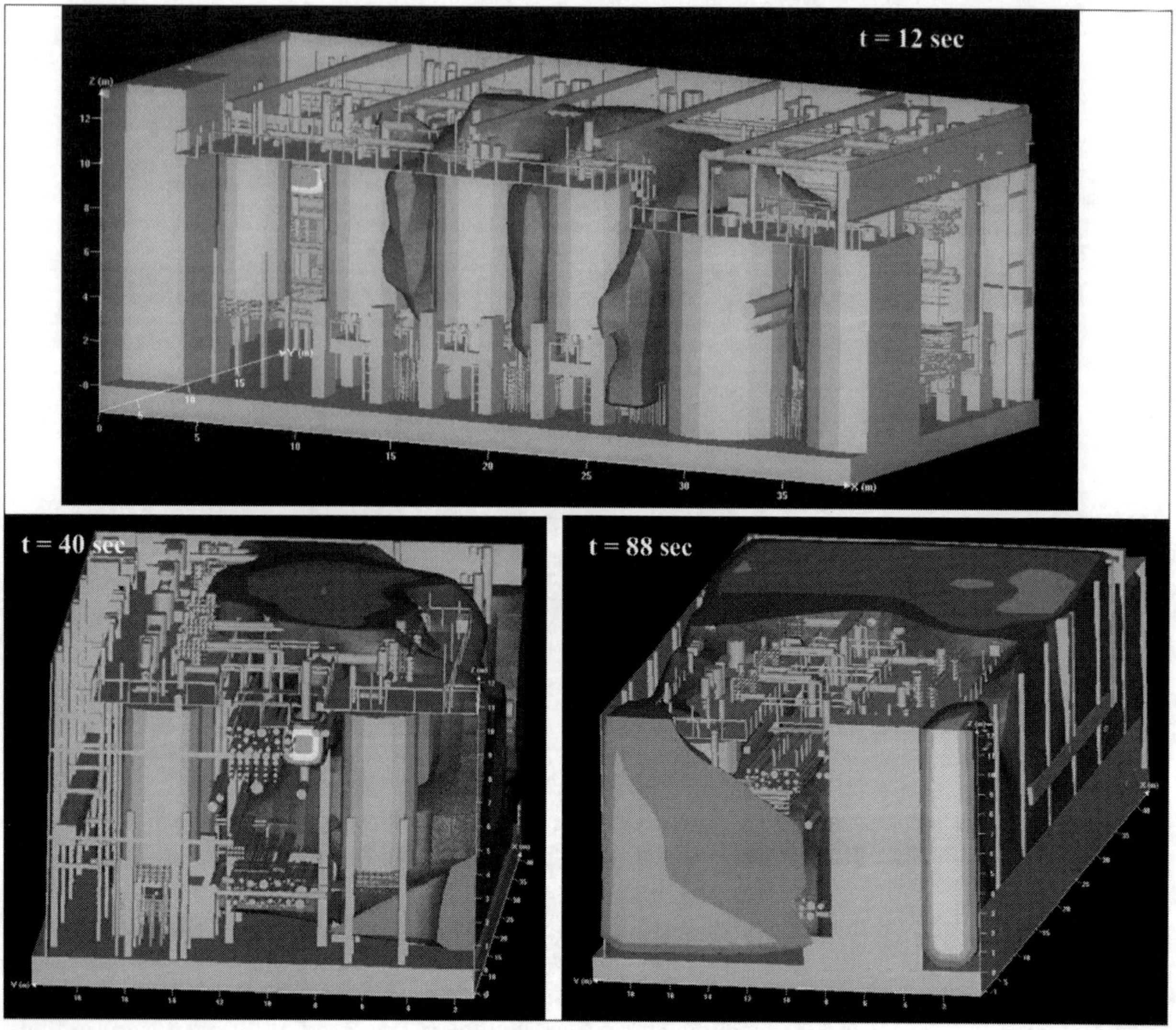

Figure 18 – Example 3 Dispersion Results

3: CFD Explosion Modeling Results

Isolation of the leak was found to be fundamental to effective blast load reduction. It was found that extremely high blast loads inside the process building resulted without leak isolation, indicating a potential deflagration to detonation transition.

Installation of explosion vent panels on the process building (to replace the installed metal panels) to relieve at 0.14 psi was found to be of no benefit. Blast loads were reduced slightly, but not enough to make any difference in building damage predictions. If a significant portion of the process building were not covered with lightweight metal panels, this result may have been different.

The proposed shield roof spanning the gap between the Laboratory and Control Room was shown to shield a blast wave traveling over the roof, but it also trapped a blast wave entering the end of the corridor (ends of the corridor could not be walled off), increasing the applied blast loads on the Laboratory. Thus, the recommendation to place a roof over this corridor was withdrawn.

The originally proposed explosion mitigation deluge system design was shown to highly effective, but the system still resulted in damaging blast loads. A parameter study was conducted to optimize the deluge system such that blast loads were reduced to the point of acceptable damage to the Laboratory. This optimization involved increasing the following: water droplet diameter, number of water nozzles and fire water system pressure (installing an additional fire water pump and modifying sections of the underground fire water main). A summary of the predicted blast loads on the Laboratory is given in Table 2.

Table 2 - Example 3 Blast Loads Summary

Evaluations	Laboratory Applied Loads	
	P (psi)	I (psi-ms)
SCOPE study, no deluge, with isolation	23	155
SCOPE study, with deluge, with isolation	2.4	124
FLACS study, no deluge, no isolation	50	508
FLACS study, no deluge, with isolation	27	400
FLACS study, with as-designed deluge, with isolation	1.5	58
FLACS study, with optimized deluge, with isolation	0.7	54

CONCLUSION

This paper shows that screening methods can conservatively estimate blast loads from releases inside buildings, but limitations in these methods produced very high blast loads compared to the CFD evaluations. However, the screening methods were successful in narrowing the focus of problem scenario/building combinations to minimize the efforts in the more detailed analysis. Accurate evaluations of releases inside buildings required CFD dispersion and explosion evaluations. Water spray and leak isolation was necessary in all cases, but could not be adequately evaluated using screening methods. The FLACS CFD model was able to evaluate water spray and assist in designing optimal water spray parameters. Additionally, mitigation measures such as blast panels, ventilation (natural and forced) and blast shielding can be evaluated using CFD, sometimes showing adverse results that were not intuitively obvious.

The Relationship between Flash Point and LFL with Application to Hybrid Mixtures

Richard W. Prugh
Chilworth Technology, Inc.
250 Plainsboro Road
Plainsboro, New Jersey 08536

Prepared for Presentation at
American Institute of Chemical Engineers
2007 Spring National Meeting
41st Annual Loss Prevention Symposium
Houston, Texas
April 22–26, 2007

UNPUBLISHED

The Relationship between Flash Point and LFL with Application to Hybrid Mixtures

Richard W. Prugh
Chilworth Technology, Inc.
250 Plainsboro Road
Plainsboro, New Jersey 08536

ABSTRACT

There usually is a significant difference between the Flash Point and the equilibrium liquid/vapor ["saturation"] temperature that corresponds to the Lower Flammable Limit [LFL]. This occurs because (1) the LFL is frequently not measured at the Flash Point temperature, and (2) the LFL concentrations are typically for *upward* propagation of flame, whereas the Flash Point is obtained for *downward* propagation of flame, and upward propagation occurs at lower ["conservative"] concentrations. In this paper, guidance is given for "correcting" this difference. Since many "hybrid" mixtures of flammable vapors and combustible dusts occur at elevated temperatures (as in dryers), it may be important that the effect of temperature on the LFL be taken into account.

For several hybrid mixtures, there is a linear relationship between the dust concentration and the vapor concentration that are required to make the mixture explosible. This relationship is more-readily visualized if vapor concentrations are expressed in terms of grams per cubic meter [g./cu.m.], like dust concentrations. A linear relationship is observed when the deflagration indices [K_{st} and K_g] are about equal, with strong deviations if the indices differ significantly. Testing is advisable to establish the relationship with confidence.

<u>Lower Flammable Limit versus Flash Point</u>

There is a sound basis for the observation that the vapor concentration at the Flash Point of a flammable liquid is near the Lower Flammable Limit [LFL] concentration: the Flash Point occurs at a temperature at which the lowest concentration of vapor in a mixture with air – thus, the Lower Flammable Limit – can propagate combustion. Therefore, one might expect that the Flash-Point and LFL concentrations would be (or should be) identical. However, there usually is a significant difference between the Flash Point and the equilibrium liquid/vapor ["saturation"] temperature that corresponds to the Lower Flammable Limit [LFL]. There are two reasons why they should not be expected to be identical:

1. In most cases, the LFL is measured at a temperature that is not the Flash Point temperature. Since the LFL's are typically measured at "room temperature" (20°C to 25°C), only if the Flash Point was also near 20°C to 25°C would the temperatures be the same. In fact, for some flammable liquids and many combustible liquids, the

LFL is measured at temperatures that are well above the Flash Point [Reference 1]. Somewhat contrarily, for flammable liquids having Flash Points below "room temperature", the LFL typically would be measured at temperatures well above the Flash Point.

2. The reported LFL concentrations are typically for *upward* propagation of flame in cylindrical or spherical apparatus. Historical LFL data [Reference 2] clearly show that upward propagation of flame requires a lower ["conservative"] concentration of flammable vapor, as compared to downward propagation. The Flash Point determinations are determined in an apparatus where *downward* propagation in a cup produces a "flash". Thus, the concentration of flammable gas or vapor that corresponds to the Flash Point is greater than the LFL concentration.

Figure 1 shows the vapor pressure curve for Cyclohexanone, the Flash Point, and the reported Lower Flammable Limit. There is a significant difference between (1) the LFL concentration [1.1 volume-percent] and (2) the equilibrium concentration of flammable vapor over a liquid at the Flash Point [1.7 volume-percent]. For this material, the difference in concentrations is about 0.6 volume-percent (12.8 mmHg minus 8.35 mmHg partial pressures, or 1.68 volume-percent minus 1.10 volume-percent), or difference of about 50 relative-percent. Also shown on this graph is the essentially straight-line relationship between the decrease in LFL concentration as the initial temperature increases toward the flame temperature of the LFL mixture [Reference 3], as based on the heat of combustion and the specific heats of the products of combustion. [On semi-log graph paper, the straight line becomes a curve.] If this relationship is extended to the vapor pressure curve, the difference in LFL and Flash-Point concentrations is about 0.38 volume-percent, or about 27 relative-percent.

Table I presents the LFL and Flash Point data for all of the flammable materials that are listed in the National Fire Protection Association tabulation [Reference 1], where the temperature is recorded at which the LFL was measured. Figure 2 shows that the difference between the LFL concentration at the Flash Point temperature and the liquid/vapor equilibrium concentration at the Flash-Point temperature averages about 0.35%, with some considerable "scatter". However, about 75% of the data show that the Flash-Point concentration is above the LFL concentration at the Flash Point temperature, and this represents the difference between downward and upward propagation.

Thus, as described above, there are two components to the difference in LFL and Flash Point concentrations. Although Flash Points and LFL concentrations are best determined by test, the flame-temperature extrapolation and the downward/upward propagation difference can be used to estimate (or confirm a test result for) one of these properties if the other is experimentally determined.

<u>Upper Flammable Limit versus UFL Temperature</u>

Recent investigations have indicated that the UFL also is strongly affected by the test temperature [Reference 4], but the temperature at which combustion occurs for a 100% concentration of the flammable substance [thus, 0% air] has not been determined. It is likely that this occurs only at "infinite" temperature, thus giving an "exponential" relationship between UFL and test temperature, of the form $UFL = 100 - (100 - UFL_{25})(e^{-k(T-25)})$. For example, the value of $\underline{k}$ for UFL mixtures of ethane and air at atmospheric pressure [Reference 4] would be about 0.00024 per degree Celsius.

Lower Flammable Limits in Mass-Concentration Terms

The Lower Flammable Limits [LFL] for gases and vapors are usually expressed in terms of volume-percent [v%]. The values of LFL range from a high of about 2.5 v% [excluding halogen-containing compounds] to a low of about 0.3 v%, with the LFL generally decreasing as the molecular weight increases. However, it has been observed that the values of LFL for hydrocarbons – when expressed in terms of mass per volume (such as grams per cubic meter [g./cu.m.]) – are nearly constant at about 48 grams per cubic meter [Reference 5].

As shown in Table I, the range of LFLs is relatively narrow, when they are expressed in terms of grams per cubic meter. The LFLs range from 27 g./cu.m. to 77 g./cu.m. [excluding halogenated compounds], with an average that is near 52 grams per cubic meter, and with a range of about a factor of 3.0 (as compared to the range of about 8 for volume-percentages). Also, the values of LFL are relatively independent of molecular weight, when expressed in terms of grams per cubic meter. Thus, a rough approximation (+/- 50%, relative) of the LFL can be obtained from:

$$LFL = 100 / [\{ 12{,}200\ M / [52 (T + 273)] \} + 1] \qquad \text{volume-percent}$$

Where M is the molecular weight of the gas or vapor;
T is the temperature of interest (in °C); and
12,200 is 14.7/0.001206 gmol-°K/cu.m.

Flammability of "Hybrid" Mixtures

Mixtures of dust suspensions, gases and/or vapors, and air are termed "hybrid" mixtures [Reference 6]. This term could also apply to mixtures of mists, gases and/or vapors, and air. It is known that such mixtures can explode even when the concentration of the dust is less than the Minimum Explosible Concentration [MEC] and the concentration of the gas or vapor is less than the Lower Flammable Limit [LFL].

The explosibility data for hybrid mixtures are extremely limited, and the data that are available in the open literature [References 7 and 8] are presented in Table I. These data (when vapor concentrations are expressed in terms of grams per cubic meter [g./cu.m.]) can be summarized as follows [References 9 and 10]:

DUST	DUST MEC (g./cu.m.)	DUST CONC. (g./cu.m.)	CONC. /MEC	GAS/VAPOR	G/V LFL (g./cu.m.)	G/V CONC. (g./cu.m.)	CONC. /LFL	TOTAL CONC. (g./cu.m.)
PVC 3,950 cal./g. Kst=168	200	102	0.51	Methane 12,000 cal./g. Kg=55	33	7.2	0.22	135
		17	0.09			22	0.67	39
		140	0.70	Propane 11,150 cal./g. Kg=100	23	8	0.36	148
		60	0.30			16	0.72	76
Pea Flour Kst=110	60	15	0.25			16	0.72	76
Cellulose 3,860 cal./g. Kst=220	30	6	0.20			16	0.72	22
Cornstarch	60	32	0.53	Ethanol	62	32	0.52	64

DUST	DUST MEC (g./cu.m.)	DUST CONC. (g./cu.m.)	CONC. /MEC	GAS/VAPOR	G/V LFL (g./cu.m.)	G/V CONC. (g./cu.m.)	CONC. /LFL	TOTAL CONC. (g./cu.m.)
4,220 cal./g. Kst=128		13	0.22	6,430 cal./g. Kg=78		45	0.73	58
		8	0.13			53	0.85	61
	70	64	0.91	Hydrogen 33,000 cal./g. Kg=550	3.6	1.7	0.46	66
		60	0.86			2.1	0.58	62
		50	0.71			2.6	0.74	53
		30	0.43			3.1	0.88	33
		20	0.29			3.5	0.93	24

The above data can be conveniently plotted in terms of the ratios of (1) the mass concentrations of interest to (2) the lower-limit mass concentrations, as shown in Figure 3.

It might be expected that – for each combination of dust and gas or vapor – the sum of the dust concentrations and the vapor concentrations would be a constant. However, as shown in the above Table, the sum of the concentrations is not constant (except for cornstarch and ethanol). From Figure 3, it appears that the relationship between dust and gas/vapor concentrations depends on stoichiometry (as indicated by the values of the MEC and LFL) and on the ratios of the Deflagration Indices (K_g / K_{st}). The diagonal line on this Figure can be described as:

$$C_{DUST} / MEC = 1 - (C_{VAPOR} / LFL)$$

Alternatively, similarly to the Le Chatelier equation [References 3 and 6]:

$$(C_{DUST} / MEC) + (C_{VAPOR} / LFL) = 1$$

There is a significant departure from linear behavior for hydrogen (which has a very high K_g, at 550 bar-meters per second, but a very high specific heat of about 3.5 calories per gram-°C), and for methane (which has a relatively low K_g, at 55 bar-meters per second, and a moderate specific heat of about 0.52 calorie per gram-°C). Thus, it appears that the straight-line relationship applies only to mixtures where the ratio of the heat capacities of the dust and gas (or vapor) are similar and where the Deflagration Indices [K_g and K_{st}] are roughly equivalent.

The behavior that is shown in Figure 3 is "counter-intuitive", since it might be expected that the presence of hydrogen would serve to reduce the total concentration of flammable materials that would be required to propagate combustion in a hybrid mixture. However, the high specific heat of hydrogen apparently increases the requirement for a high concentration of flammables. That is, at concentrations below the "Lower Flammable Limit" for the hybrid mixture, the hydrogen "soaks-up" the incipient combustion and prevents propagation. Confirming this behavior is the very low flame temperature for LFL-concentration mixtures of hydrogen and air (below 350 °C [Reference 11]).

It is likely that the curves on Figure 3 connect points of equal flame temperature. Calculations of flame temperature for the mixtures of methane and polyvinyl chloride dust indicate that the flame temperatures of the mixtures increase with dust concentration – ranging from about 1,000 °C for LFL methane/air mixtures to about 1,900 °C for MEC mixtures of polyvinyl chloride dust and air. However, these calculations are based on calculated heats of combustion that were obtained from temperatures that were attained by burning polyvinyl chloride powder in oxygen [Reference 9], presumably with back-corrections for dissociation of carbon dioxide, water vapor [Reference 12], and hydrogen chloride. Also, since polyvinyl chloride dust is not a vapor, the heat of sublimation may need to be taken into account in flame-temperature calculations, and heats of sublimation are not readily available for dusts.

Effects of Temperature on Hybrid Flammability

Since many "hybrid" mixtures of flammable vapors and combustible dusts occur at elevated temperatures, it is important that the effect of temperature on the LFL and MEC be taken into account. This is of particular concern if the Flash Point was used to estimate the LFL and/or if the LFL and MEC were obtained at room temperature.

Conclusion

This analysis of the data for "hybrid" explosion hazards may provide some insight into explosion phenomena and a method for estimating the concentrations at which mixtures of combustible dusts and flammable vapors are explosible. However, testing of hybrid mixtures is advisable, to establish the relationship with confidence. From the results of this analysis, it appears that one or two tests may be sufficient to establish the relationship between dust and vapor concentrations that could result in explosion.

References

1. National Fire Protection Association, "Fire Protection Guide to Hazardous Materials", 13th Ed., NFPA 49 and NFPA 325 (2002).

2. Coward, H. F., and G. W. Jones [U. S. Bureau of Mines], "Limits of Flammability of Gases and Vapors", Bulletin 503, pages 16 to 100, 110, and 112 (1952).

3. Zabetakis, M. G. [U. S. Bureau of Mines], "Flammability Characteristics of Combustible Gases and Vapors", Bulletin 627, pages 21 and 31, and Figures 20, 21, 22, and 23 (1965).

4. Van den Schoor, "Journal of Hazardous Materials, 128 (1): 3 (Jan. 2006)])

5. Zabetakis, M. G. [U. S. Bureau of Mines], "Flammability Characteristics of Combustible Gases and Vapors", Bulletin 627, equation (27), page 21 (1965).

6. Mannan, S., "Lees' Loss Prevention in the Process Industries", 3rd Ed., page 17/259 (2005).

7. Bartknecht, W., "Dust Explosions – Course, Prevention, Protection", pages 87 and 88 (1981).

8. Eckhoff, R. K., "Dust Explosions in the Process Industries", 2nd Ed., page 51 (1999).

9. Society of Fire Protection Engineers, "The SFPE Handbook of Fire Protection Engineering", pages 1-148 to 1-152 and 1-186 (1988).

10. National Fire Protection Association, "Fire Protection Handbook", 18th Ed., page A-6 (1997).

11. Coward, H. F., and G. W. Jones [U. S. Bureau of Mines], "Limits of Flammability of Gases and Vapors", Bulletin 503, pages 11 and 15 (1952).

12. Perry, J. H., "Chemical Engineers' Handbook", 4th Ed., page 9-40 (1963).

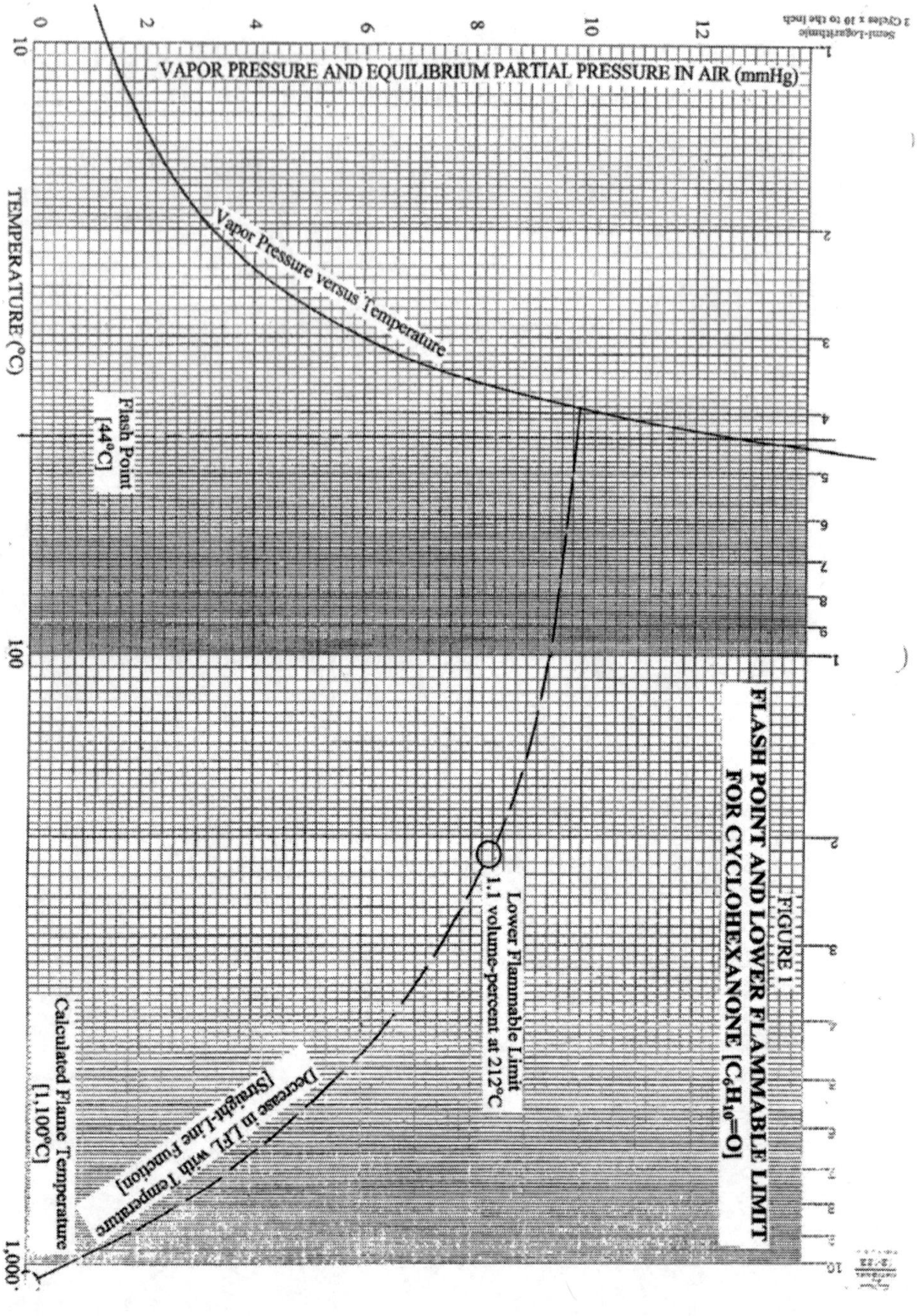

FIGURE 1
FLASH POINT AND LOWER FLAMMABLE LIMIT FOR CYCLOHEXANONE [$C_6H_{10}=O$]

LPS 2007 Paper 4a

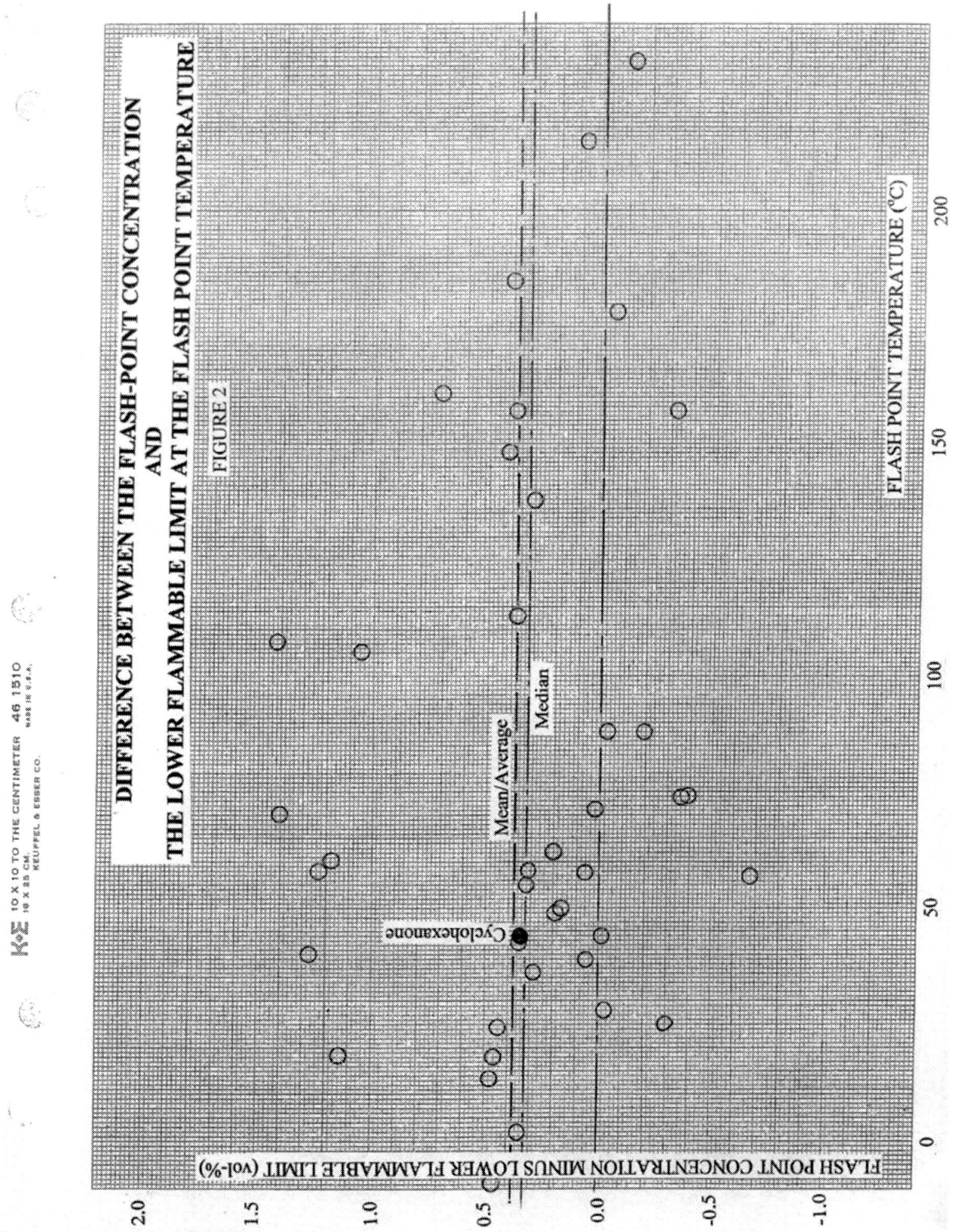
DIFFERENCE BETWEEN THE FLASH-POINT CONCENTRATION
AND
THE LOWER FLAMMABLE LIMIT AT THE FLASH POINT TEMPERATURE
FIGURE 2
Mean/Average
Median
Cyclohexanone
FLASH POINT CONCENTRATION MINUS LOWER FLAMMABLE LIMIT (vol-%)
FLASH POINT TEMPERATURE (°C)
2.0
1.5
1.0
0.5
0.0
-0.5
-1.0
0
50
100
150
200

FIGURE 3

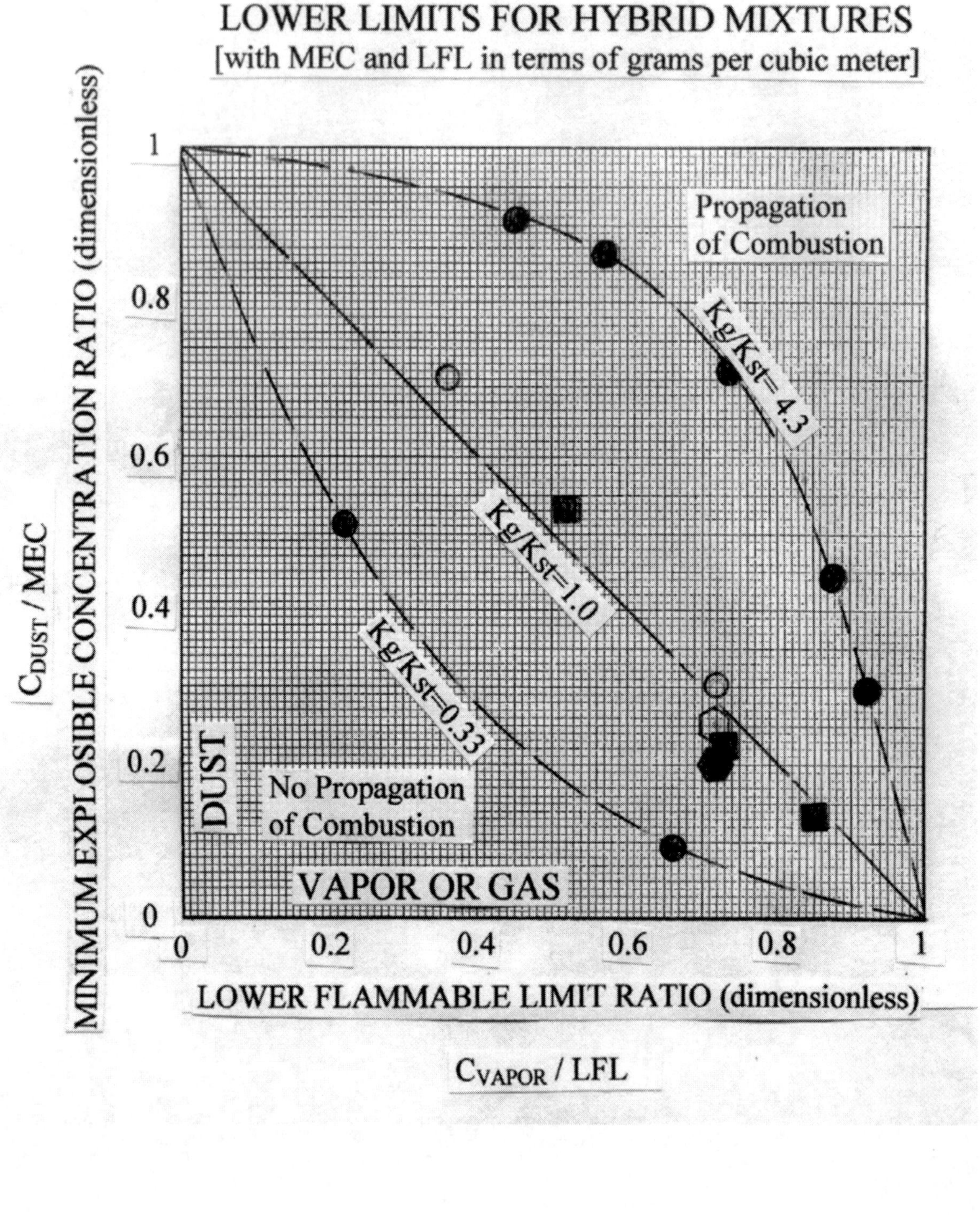
LOWER LIMITS FOR HYBRID MIXTURES
[with MEC and LFL in terms of grams per cubic meter]
MINIMUM EXPLOSIBLE CONCENTRATION RATIO (dimensionless)
C_{DUST} / MEC
1
0.8
0.6
0.4
0.2
0
Propagation of Combustion
Kg/Kst= 4.3
Kg/Kst=1.0
Kg/Kst=0.33
DUST
No Propagation of Combustion
VAPOR OR GAS
0
0.2
0.4
0.6
0.8
1
LOWER FLAMMABLE LIMIT RATIO (dimensionless)
C_{VAPOR} / LFL

COMPARISON OF FLASH POINT WITH LOWER FLAMMABLE LIMIT

MATERIAL	Flash Point (°C)	Lower Limit (vol%)	LFL Temp. (°C)	Mole Wgt.	Heat of Form. (cal/mol)	Heat of Combust. (cal./g.)	Flame Temp. (°C)	Concentrations (vol-%) Vapor [down]	LFL [upward]	Difference [up-down]	Conc. @LFL (g/cu.m)	Stoich. Conc.. (v-%)	LFL /Cst (%)
BiCycloHexyl	74	0.70	100	166.3	-272.0	10,136	1,468	0.35	0.71	-0.36	38.3	1.18	59
BiPhenyl	113	0.60	111	154.2	182.1	9,348	1,121	0.97	0.60	0.37	29.5	1.42	42
sec-Butyl Alcohol	24	1.70	100	74.1	-292.3	7,869	1,242	2.26	1.81	0.44	41.9	3.37	50
t-ButylAmine	-9	1.70	100	73.1	-119.9	8,998	1,391	10.18	1.84	??	41.3	3.01	56
Butyl Bromide	18	2.60	100	137.0	-107.3	4,361	1,865	3.86	2.72	1.14	119.5	3.24	80
N,p-Cresol	86	1.10	150	108.1	-125.4	7,785	1,184	1.14	1.17	-0.03	34.7	2.41	46
CycloHexanone	44	1.10	212	98.1	-230.1	8,033	1,109	1.66	1.31	0.35	27.4	2.55	43
p-Cymene	50	0.70	100	134.2	-29.0	9,882	1,189	0.86	0.73	0.13	30.9	1.53	46
DecaHydronapthalene	58	0.70	100	138.3	-169.0	10,186	1,250	0.78	0.73	0.06	31.9	1.42	49
N,n-DiSecButyl	132	0.60	165	220.4							37.0	1.04	58
DiButylPhthalate	157	0.50	235	278.3	-750.9	6,961	1,215	0.21	0.54	-0.33	33.6	1.06	47
DiButylSecabate	178	0.44	243	314.3				0.27	0.32	-0.05	32.8	0.85	52
DiEthylCycloHexane	49	0.80	60	140.3							41.4	1.38	58
DiEthyleneGEE	91	1.20	182	134.2							43.7	2.55	47
DiEthGlyMEA	107	1.00	135	176.2	-776.0	5,751	1,244	2.44	1.03	1.42	53.2	2.05	49
DiEthylPhthalate	161	0.70	186	222.2	-688.3	6,061	1,183	1.43	0.72	0.71	41.6	1.53	46
DiIsoButylCarbinol	74	0.80	100	144.3	-398.0	9,311	1,340	0.42	0.82	-0.39	38.0	1.53	52
DiIsoButylKetone	49	0.80	93	142.2	-357.6	8,922	1,274	1.01	0.83	0.18	38.2	1.59	50
DiIsoButylPhthalate	185	0.40	231	278.3	-750.9	6,961	993	0.82	0.42	0.40	27.0	1.06	38
DiDecylPhthalate	232	0.30	264	446.7	-1060.1	8,314	1,382	0.18	0.31	-0.13	30.5	0.56	54
DiMethoxyEthylPhth	210	0.70	227	250.3							43.0	1.25	56
DiMethylAcetAmide	70	1.80	100	87.1	-225.0	6,529	1,273	3.25	1.85	1.40	52.2	3.52	51
DiMethylDecalin	84	0.70	95	164.3							38.4	1.22	58
DiMethylFormamide	58	2.20	100	73.1	-191.7	5,849	1,177	3.52	2.29	1.23	53.7	4.70	47
DiMethylPhthalate	148	0.90	180	194.2	-663.0	5,426	1,187	1.34	0.93	0.41	47.5	1.96	46
DiOctylAdipate	206	0.40	242	374.6							35.6	0.66	61
DiOctylAz	227	0.30	266	412.7							28.1	0.60	50
DiOctylPhthalate	215	0.30	245	390.6	-966.7	7,928	1,172	0.39	0.31	0.08	27.7	0.66	45
DiPentene	45	0.70	150	136.2							27.7	1.47	47
DiPropyleneGME	86	1.10	200	148.2							42.5	2.16	51
EthylAceto	57	1.40	93	130.1	-560.0	5,436	1,225	0.77	1.44	-0.68	61.5	2.91	48
EthyleneDiAmine	40	2.50	100	60.1	-17.3	6,725	1,255	3.90	2.63	1.27	50.4	4.98	50
EthGlyMBE	62	1.10	93	118.2	-441.0	7,180	1,168	1.33	1.13	0.20	43.8	2.41	46
EthGlyMBEA	71	0.88	93	160.2							47.4	1.96	45
EthGlyEE	43	1.70	93	90.1	-400.0	6,206	1,182	2.12	1.78	0.35	51.9	3.67	46
EthGlyMBE	58	1.20	93	118.2	-441.0	7,180		1.05	0.75	0.31	47.8	2.41	50
EthGlyMEA	49	1.50	93	118.1							59.9	3.37	44
EthylHexanal	44	0.85	93	128.2	-299.6	8,825	1,214	0.87	0.89	-0.02	36.6	1.79	48
EthylLactate	50	1.50	100	118.1	-635.0	5,021	1,104	1.90	1.57	0.32	58.8	3.37	44
GlycerolTriAcetate	138	1.00	189	218.2	-1252.7	4,275	1,138	1.36	1.05	0.30	58.2	2.16	46
GlycerolTriButyrate	180	0.50	208	302.4							38.5	1.12	45

GlycerolTriPropionate	167	0.80	186	260.3							55.7	1.47	54

COMPARISON OF FLASH POINT WITH LOWER FLAMMABLE LIMIT

	Flash	Lower	LFL		Heat of	Heat of	Flame	Concentrations (vol-%)			Conc.	Stoich.	LFL
	Point	Limit	Temp.	Mole	Form.	Combust.	Temp.	Vapor	LFL	Difference	@LFL	Conc.	/Cst
MATERIAL	(°C)	(vol%)	(°C)	Wgt.	(cal/mol)	(cal./g.)	(°C)	[down]	[upward]	[up-down]	(g/cu.m)	(v-%)	(%)
HydroxyEthylAcrylate	101	1.80	100	116.1							69.6	3.67	49
IsoAmylAcetate	25	1.00	100	130.2	-511.9	7,141	1,167	0.74	1.07	-0.33	43.0	2.16	46
IsoButylAlcohol	28	1.70	51	74.1	-426.7	7,896	1,246	1.70	1.73	-0.03	48.2	3.37	50
IsoPropylAcetate	2	1.80	100	102.1	-481.7	6,220	1,398	2.29	1.94	0.35	61.2	3.12	58
IsoBiCycloHexane	124	0.50	150	208.4							30.2	0.94	53
IsoBiPhen	141	0.50	175	196.3							26.8	1.09	46
MethylAmylKetone	39	1.10	66	114.2	-300.5	8,580	1,345	1.18	1.12	0.06	45.7	2.05	54
MethCycloPentadiene	49	1.30	100	80.1	100.0	10,141	1,350	43.62	1.35	??	34.5	2.55	51
MethylEthylKetone	-9	1.40	93	72.1	-238.4	7,496	976	2.03	1.56	0.46	34.1	3.67	38
MethylHeptylKetone	60	0.90	82	142.2		8,922		1.84	0.66	1.18	44.4	1.59	57
MethylIsoAmylKetone	36	1.00	93	114.2	-304.0	8,581	1,235	1.33	1.05	0.28	38.4	2.05	49
MethylIsoButylKetone	18	1.20	93	100.2	-288.5	8,321	1,258	1.74	1.28	0.46	40.5	2.41	50
MethylLactate	49	2.20	100	104.1							76.5	4.45	49
MethylPentylAcetate	45	0.90	100	144.2							42.8	1.87	48
NitroBenzene	88	1.80	93	123.1	67.6	5,782	1,582	1.61	1.81	-0.20	75.2	3.24	56
Propyl Acetate	13	1.70	38	102.1	-464.8	6,253	1,336	2.22	1.73	0.48	69.2	3.12	54
PropylGlycolAcrylate	97	1.40	100	130.1							60.4	2.91	48
PropylGlycolMEA	42	1.50	200	132.2							51.9	2.72	55
Resorcinol	127	1.40	200	110.1							40.3	3.12	45
Salicylic Acid	157	1.10	200	138.1	-466.4	5,009	974	2.55	1.16	1.39	39.6	2.91	38
TetraHydroNaphthylene	71	0.80	100	132.2	26.6	9,685	1,305	0.84	0.82	0.02	34.8	1.59	50
TriChloroBenzene	105	2.50	150	181.4	-11.8	3,499	1,905	3.61	2.56	1.05	134.1	3.01	83
TriChloroEthylene	10	7.80	100	131.4	-280.8	1,572	1,851	4.63	8.20	??	363.3	8.52	92
TriChloroPropane	71	3.20	120	147.4	-185.8	2,519	1,423	5.34	3.32	2.02	151.2	4.70	68
TriMethylPentene	120	0.60	149	216.3							37.7	1.25	48
								Avg.:		0.39			
													49.4
	Avg:	1.07							Avg:		43.6		366
	High*:	2.50	233	%					Highest*:		76.5	176	%(+)
	Low:	0.30	28	%					Lowest:		26.8	62	%(-)

* Excluding halogenated hydrocarbons

* Excluding halogenated hydrocarbons

LOSS PREVENTION SYMPOSIUM

Fire Hazards at Power Generation Facilities

John A. LeBlanc and Christopher J. Wieczorek, Ph.D.
FM Global
1151 Boston Providence Turnpike
Norwood, MA 02062 USA
Phone: 781-255-4776
john.leblanc@fmglobal.com

Prepared for Presentation at
American Institute of Chemical Engineers
2007 Spring National Meeting
41st Annual Loss Prevention Symposium
Houston, Texas
April 22–26, 2007

UNPUBLISHED

Fire Hazards at Power Generation Facilities

John A. LeBlanc
FM Global
1151 Boston Providence Turnpike
Norwood, MA 02062 USA
Phone: 781-255-4776
John.leblanc@fmglobal.com

Abstract

Power generation plants present several challenging hazards. Some are well understood (e.g., rotating equipment) while others remain poorly defined or understood (e.g., lubricating oil systems, control oil systems, and hydraulic oil systems). In an effort to better define the hazards created by pumped oil systems for our clients and to help us develop better protection options, FM Global initiated a research project to investigate oil fire hazards expected at a power generation plant without an operating floor. The objectives of the project were to 1) provide detailed visual documentation of the potential fire scenarios associated with turbine halls within power generation facilities without an operating floor and 2) examine new and existing protection schemes to mitigate losses due to fires. A final project was aimed at defining a functional spray shield that could be used to mitigate spray fires at flanges.

BACKGROUND

Power generation plants present several challenging hazards. Some are well understood (e.g., rotating equipment) while others remain poorly defined or understood (e.g., lubricating oil systems, control oil systems, and hydraulic oil systems). In an effort to better define the hazards created by pumped oil systems for our clients and to help us develop better protection options, FM Global initiated a research project to investigate oil fire hazards expected at a power generation plant without an operating floor.

FM Global loss statistics have shown the majority of fire loses within power generation facilities are due to mechanical failures and human error. These failures have resulted in numerous oil fires including spray fires, three-dimensional spill fires, and pool fires, and have involved lubricating oil, control oil and seal oil systems.

A fire generated by flowing oil, either as a spray or three-dimensional spill, presents a very challenging situation. Water-based automatic sprinkler systems have limited effectiveness against these types of fires. The only way to extinguish spray or three-dimensional spill fires is to stop the oil flow; however, suddenly shutting down any of the oil systems on a turbine can result in severe damage to the equipment. Turbine operators generally are resistant to shutting down oil systems under any circumstances and turbine coast down times can be as high as 45 minutes. Therefore, a multi-level solution is required. Fire protection measures must be taken to minimize the impact of the fire to the facility while the oil is flowing, and operators must look for opportunities to shut down oil systems or at least reduce pressures and flow rates.

TESTING OBJECTIVES

The objectives of the project were to 1) provide detailed visual documentation of the potential fire scenarios associated with turbine halls within power generation facilities without an operating floor and 2) examine new and existing protection schemes to mitigate losses due to fires. A final project was aimed at defining a functional spray shield that could be used to mitigate spray fires at flanges.

Test Arrangement and Conditions

To meet these objectives a large-scale mock-up of a turbine unit was constructed under an 18.3 m (60 ft) high ceiling within the large burn laboratory at the FM Global Research Campus. To simulate a power generation facility, a turbine generator mock-up was created. The mock-up consisted of the following elements (see Figure 1):

- Lube Oil Tank
  3.05 m x 2.13 m x 3.66 m (10 ft x 7 ft x 12 ft)

- Containment Area Around Lube Oil Tanks
  7.32 m x 3.66 m x 0.91 m (24 ft x 12 ft x 3 ft)

- Turbine and Pedestal
  6.1 m x 4.57 m x 5.49 m (20 ft x 15 ft x 18 ft)

- Grated Walkway

In addition to creating a visually credible array, the test conditions needed a fuel that was credible to our clients. Unheated mineral oil, closed cup flash point of 140.6°C (285°F), was used for all of the tests using a number of different sprinkler protection schemes. Many lube, control and hydraulic oil systems use this type of oil and it is generally at temperatures above ambient.

One recognized limitation of the testing was the size fire that could be produced. Loss history has shown that oil releases within a power generation facility can be very large and well beyond what the FM Global Research Campus could safely control. Test conditions were chosen to ensure a visually stunning fire while also allowing the development of protection concepts that would be expected to be applicable with significantly larger fire events.

The test array allowed the testing of a spray fire condition initiated at the pumps located over the lube oil tank, a three dimensional spill fire initiated at the turbine bearing and spilling below the operating floor, a pool fire below the operating floor, and a pool fire in the containment area around the lube oil tanks. A total of 22 large-scale tests were conducted. The testing broke down to:

Table 1. Fire Test Summary

Number of Tests	Fire Type / Test Location
10	Spray Fires (top of lube tank)
2	Three-Dimensional Spill Fires (spill at bearing flow below pedestal)
11	Pool Fires (7 – lube oil containment area) (4 – under turbine pedestal)

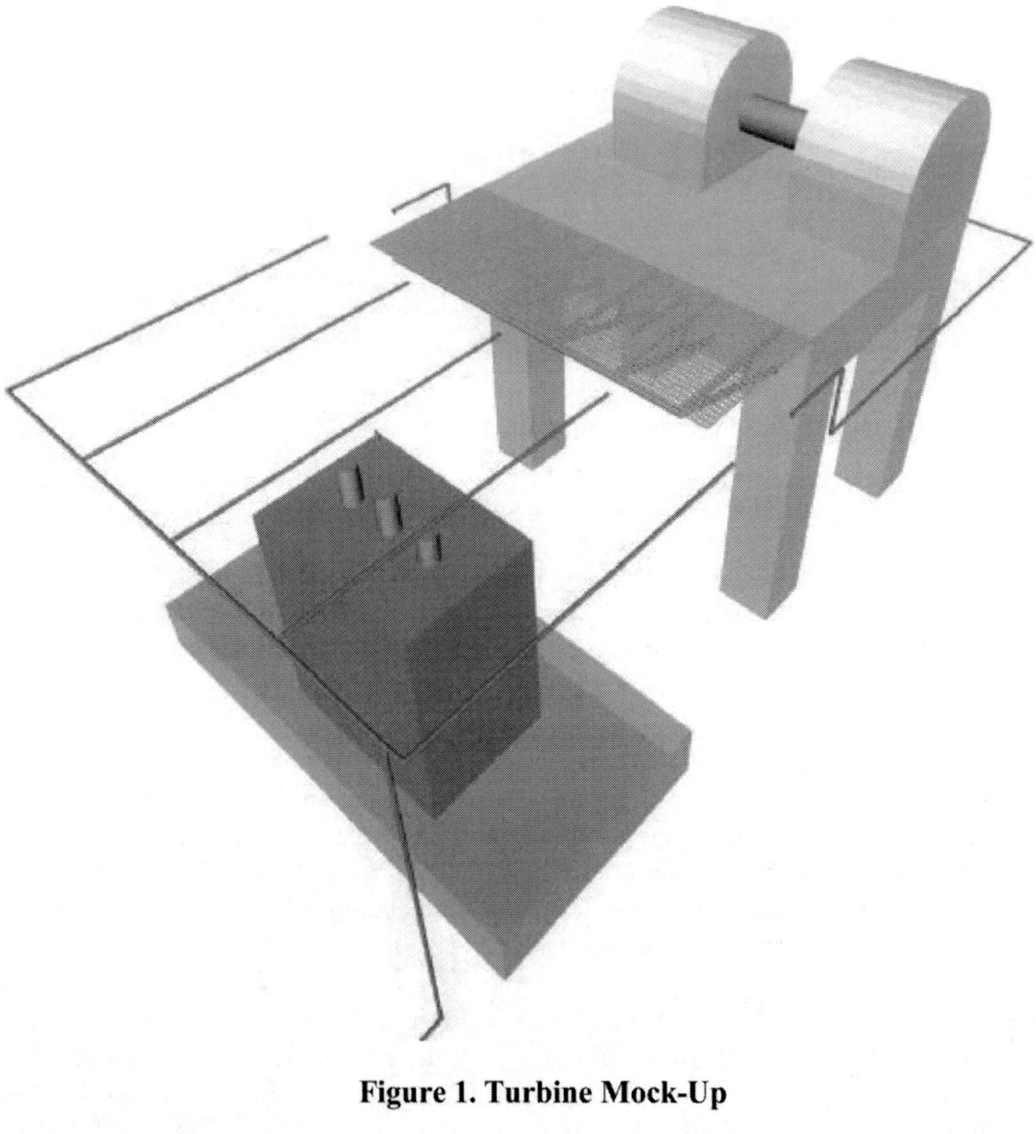

Figure 1. Turbine Mock-Up

Spray Fire Testing

The spray fire testing was designed to simulate a failure at a fitting in an oil system located near the highest system pressure. In the case of a lube oil system, the lube oil pumps tend to be located above the lube oil tanks and the discharge of the pump represents the highest system pressure. To limit the overall fire size a leak flow rate of 20 gpm (76 lpm) at a pressure of 90 psig (6.2 bar) was chosen. The pressure is representative of the operating pressure for a lube oil system; however, the flow rate is well below common flow rates which can be more than 100 gpm (380 lpm). The testing began with a freeburn to develop a set of baseline conditions that could be used to evaluate various protection approaches. The protection options started with a ceiling only sprinkler system and then went to a local area protection system.

The ceiling only protection scheme consisted of closed, standard response sprinklers spaced on a 10 ft x 10 ft (3 m x 3 m) grid at the roof deck. The local area protection system consisted of a grid of sprinklers located just above the spray fire source. Two different grid spacing were evaluated, a 5 ft x 5 ft (1.5 m x 1.5 m) and 8 ft x 10 ft (24 m x 3 m). The variables that were considered for the testing were:

- Protection type (i.e. local or ceiling)
- Sprinkler layout
- Sprinkler type (i.e. water nozzle or sprinkler)
- Sprinkler/nozzle K-Factor
- Water discharge pressure
- Water density

The freeburn produced a heat release rate of 40 MW and a flame height of almost 70 ft (21 m) (this is the vertical height plus the spread across the ceiling). Gas temperatures of 820°C (1500°F) were measured consistently during the fire. Since a water based sprinkler system cannot extinguish a spray fire, the key parameters that were measured to determine how effective a protection scheme was were ceiling gas temperatures and flame height. A summary of the test results are in Table 2.

Table 2. Spray Fire Test Summary

Protection Type		Maximum Ceiling Gas Temperature	
		°C	°F
None (baseline)		806	1483
Ceiling Only	Density #1	759	1398
	Density #2	792	1458
Local Scenario 1		541	1006
Local Scenario 2		553	1027

Protection Type		Maximum Ceiling Gas Temperature	
		°C	°F
Local Scenario 3		296	565
Local Scenario 4	Density #1	220	428
	Density #2	159	318
Local Scenario 5	Density #1	271	520
	Density #2	281	538

The testing clearly demonstrated that sprinkler protection, even a very high density delivered locally over the leak point cannot extinguish a spray fire. The testing also showed:

- That a standard ceiling sprinkler system has limited capability to provide adequate cooling for ceiling steel,
- A locally installed deluge water spray system can reduce the fire severity and provide cooling at the roof deck,
- Only sprinklers directly impacted by the spray fire in a local area system will operate, so a closed sprinkler system will have limited impact on fire severity.

Three-Dimensional Spill Fires

The three-dimensional spill fire testing was designed to simulate a leak at the turbine bearing. To limit the overall fire size leak flow rates of 6 gpm (23 lpm) to 10 gpm (38 lpm) were used. The leak produced a local pool fire near the bearing and excess burning oil spilled over the pedestal and through the grated walkway to the containment area below the pedestal. The first test utilized closed sprinklers. The second test utilized open sprinklers. Past testing on spills through process structures has shown that sprinklers have limited effectiveness against this type of fire. They can provide cooling to areas outside the spilling/burning liquid but equipment directly impacted by the spill and the ceiling above will only see very limited cooling. The goal of this testing was to see if the mineral oil created a more easily controlled 3-D spill fire than lower flash point liquids like heptane.

The sprinkler protection for both tests was located below the turbine pedestal and below the grated walkway. The sprinklers were installed on a 8 ft x 8 ft (2.4 m x 2.4 m) grid. The variables that were considered included:

- Local Protection (closed vs. deluge)
- Fixed Sprinkler Layout, Type, and K-Factor
- Water Discharge Pressure
- Water Density

The parameters measured during the testing included:

- Temperatures on the Underside of the Pedestal
- Ceiling Gas Temperatures
- Flame Height

The results of the two tests showed that water based fire protection would not extinguish or control this type of fire even when the fuel consisted of a high flash point liquid. In both cases, the test duration was limited due to the lack of cooling provided by the installed sprinklers. One test outcome reinforced the importance of a solid ceiling above closed sprinklers to ensure prompt operation. In the first test, only the sprinklers directly in the spilling/burning liquid below the grated walkway operated. Adjacent sprinklers never operated.

<u>Pool Fires at Lube Oil Tank</u>

The pool fire tests in the containment area around the lube oil tank were designed to simulate a large oil release at the tank. The oil in the containment area was floated on water to protect the metal pan from heat damage. Each test used 136 gal (515 l) of oil in the pan. Since the ceiling clearance above the lube tank tends to be in excess of 30 ft (9 m), only local area protection was tested. This type of fire protection has been provided at existing power generation facilities where a lube oil containment area is located adjacent to a turbine pedestal.

The sprinklers/open nozzles in this testing were provided over the containment area in an approximate 8 ft x 10 ft (2.4 m x 3 m) grid. The variables considered by the testing included:

- Local Protection
- Sprinkler Layout
- Sprinkler Type (i.e. water nozzle or sprinkler)
- Water Discharge Pressure
- Water Density

The parameters measured during the tests included:

- Ceiling Gas Temperatures
- Flame Height
- Time to Extinguishment

The goal of this testing was to extinguish the pool fire. The fires were ignited with a small amount of heptane poured in one corner of the containment pan. This ignition source was used to provide reliable and consistent ignition of the oil. The first test was run without any protection to provide an understanding of how big the fire was and what type of ceiling temperatures could be expected. The freeburn test last approximately 6 minutes. This pool fire produced a heat release rate in excess of 20 MW and gas temperatures at the roof deck of 250°C (482°F). Ceiling steel temperatures never stabilized and were continuing to rise at test termination.

A summary of the time to achieve extinguishment of the pool fires are provided in Table 3.

Table 3. Pool Fire Test Summary - Lube Oil Tank Containment Area

Protection Type	Time to Extinguish (min)
Freeburn (baseline)	-
Local Scenario 1	6
Local Scenario 2	12.5
Local Scenario 3	8.5
Local Scenario 4	10
Local Scenario 5	-
Local Scenario 6	9

The key result of this testing was showing that higher than expected sprinkler discharge rates are needed to extinguish a high flash point liquid pool fire even when using a locally installed sprinkler system.

Pool Fires Under the Turbine Pedestal

This series of pool fire tests were done in a containment pan located below the turbine pedestal and were meant to simulate a large release of oil below the operating level. The containment area was 10 ft x 10 ft (3 m x 3 m) and 62 gal (236 l) of oil was floated on a layer of water in the pan. Only local protection directly below the pedestal was installed. The sprinklers were arranged on a 8 ft by 8 ft (2.4 m by 2.4 m) grid. The tests used a fixed sprinkler layout, sprinkler type and K-factor. The variables considered for the testing included:

- Local Protection (closed vs. deluge)
- Water Discharge Pressure
- Water Density

The parameters measured during the testing included:

- Temperatures on the Underside of the Pedestal
- Time to Extinguishment

The goal of this testing was to determine what density was needed to extinguish an oil pool fire with 15 ft (4.6 m) of clearance and to see if a lesser density that cannot extinguish the fire could at least provide some cooling. Table 4 provides a summary of extinguishment times observed in the tests.

Table 4. Pool Fire Summary – Below Turbine Pedestal

Protection Type	Time to Extinguish (min)
Closed Sprinklers	
Scenario 1	-
Scenario 2	20
Deluge	
Scenario 1	9
Scenario 2	7.5

This testing showed the following:

- The density needed to extinguish a pool using closed sprinklers with a 4.6 m (15 ft) clearance.
- The density needed to extinguish a pool fire using open sprinklers with a 4.6 m (15 ft) clearance is less than required for a closed sprinkler system.
- Dramatically showed that providing less water than needed to extinguish the fire can significantly increase the fire severity by producing large fireballs.

APPROVED SPRAY SHIELDS

The fire testing demonstrated the challenges in using water-based fire protection to control or limit the severity of spray fires. Recognizing the difficulty in providing a tightly spaced water spray system over any potential leak point, FM Global also undertook an effort to understand how effective commonly used spray shields might be in eliminating spray fires.

The first step in this process was to define the performance characteristics of a spray shield. The first goal of a spray shield is to eliminate the momentum of a high velocity liquid release which will result in a limited area being affected by the liquid discharge. This would not stop the leak but simply change its characteristics from a strong directional flame to a weaker 3-D spill fire. Secondly, since the base assumption of the power generation project is the oil cannot be shutdown until the turbine winds down which could take 45 minutes, the shield would need to be able to resist exposure to a hydrocarbon fire for that amount of time.

Many of the commercially available shields may have had some testing completed; however, none have been fire tested and the pass/fail criteria for the spray testing are not well defined. A test program was initiated to determine the spray performance criteria, define a test to measure spray performance and to develop a fire test for spray shields. The initial selection of shields was limited to those that could potential survive a severe fire exposure, i.e., metal shields. Most, if not all, of the bag type spray shields are not expected to withstand the spray fire.

The final spray test performance measure required that a spray shield limited a pressurized leak from a flange to within 1.5 m (5 ft) of the point of liquid release. This prescriptive cut-off was chosen to ensure that a pressurized leak did not create a spray fire that would impact horizontally adjacent equipment.

The fire exposure needed to simulate a hydrocarbon fire was developed based on a literature review. ASTM E1529, Standard Test Method for Determining Effects of Large Hydrocarbon Pool Fires on Structural Members and Assemblies, indicated that a heat flux of 158 kW/m^2 was needed. A test method that can provide an effective heat flux in the range needed was developed using a combination of a propane sand burner and two parallel panels (See Figure 2).

Figure 2. Spray Shield Fire Exposure Test

To date, the testing has shown that developing a spray shield that will actually limit the distance a pressurized release will travel is very difficult. All of the commercially sold units that were tested reduced the spray distance perpendicular to the release point. However, they also redirected the spray parallel to the release point producing sprays that traveled up to 6 m (20 ft) away. From a fire resistance standpoint, all of the units tested were stainless steel which resisted the fire exposure. However, on units that needed some type of mesh or other lighter weight obstruction inside the outer shell to break up the fluid flow, the meshes were consumed by the fire exposure. All of the meshes were metallic.

CONCLUSIONS

This full-scale test series provided a clear confirmation of the severity of oil spray and three-dimensional spill fires and the challenge of providing effective fixed protection for the tested scenarios. In addition, the testing also demonstrated:

- The best protection is to shut off the flow of oil!

- Pool fires can be extinguished.
  - Extinguishment is affected by:
    - Ceiling height
    - Type of system
    - Water density
  - A deluge system is more effective than closed sprinklers
  - Inadequate amounts of water may change the dynamics of a fire and produce larger event.

- Closed sprinklers or heat detectors will not activate under grated walkways or without the presence of a ceiling unless they are in the fire plume.

- Ceiling sprinkler protection can not extinguish a spray or 3-D spill fire.

- Local sprinkler protection can not extinguish a spray or 3-D spill fire.

- A properly designed local sprinkler system will control the exposure created by a spray or 3-D spill fire until the flow of oil is terminated.
  - Close spaced sprinklers are needed to impact spray fire intensity.

- Currently available spray shields may not remove the momentum of a spray leak; instead they may redirect the spray.

- To determine the effectiveness of a spray shield, it needs to be tested against the conditions it will need to perform under.

Fire and Explosion Framework Assessment on Oil and Gas Floating Production Storage Offloading (FPSO) : an Effective Screening and Comparison Tool

Jaffee Suardin[a]
Jeff McPhate[b]
Anthony Sipkema[b]
Matt Childs[c]
M Sam Mannan[a]

[a] **Mary Kay O'Connor Process Safety Center, Artie McFerrin Department of Chemical Engineering, Texas A&M University, College Station, 77843-3122, USA**

[b] **Shell International Exploration Production Inc, 200 N Dairy Ashford, Houston, Texas, USA**
Correspondence: Phone: (979) 739-2251, Fax: (979) 458-1493

[c]**Riskbytes, Inc, PO Box 2005, Houston, Texas 77252.**
Correspondence: Phone: (713) 224-5740, Fax (713) 224-5030

jaffee.suardin@chemail.tamu.edu

Prepared for Presentation at
American Institute of Chemical Engineers
2007 Spring National Meeting
41st Annual Loss Prevention Symposium
Houston, Texas
April 22–26, 2007

UNPUBLISHED

Fire and Explosion Framework Assessment on Oil and Gas Floating Production Storage Offloading (FPSO) : an Effective Screening and Comparison Tool

Jaffee Suardin[a], Jeff McPhate[b], Anthony Sipkema[b], Matt Childs[c], M Sam Mannan[a]

[a] Mary Kay O'Connor Process Safety Center, Artie McFerrin Department of Chemical Engineering, Texas A&M University, College Station, 77843-3122, USA

[b] Shell International Exploration Production Inc, 200 N Dairy Ashford, Houston, Texas, USA
Correspondence: Phone: (979) 739-2251, Fax: (979) 458-1493

[c]Riskbytes, Inc, PO Box 2005, Houston, Texas 77252.
Correspondence: Phone: (713) 224-5740, Fax (713) 224-5030

jaffee.suardin@chemail.tamu.edu

Abstract

Fires and explosions have been identified as major potential hazards for Oil and Gas Floating Production Storage Offloading (FPSO) installations and pose risk to personnel, assets, and the environment. Current fire and explosion assessment (FEA) tools require physical effect modeling software and follows standards from API, ISO, and engineering practices. However, the tools are not specific to any particular system such as an FPSO, and do not provide comprehensive guidance for safety engineers to perform FEA.

This paper discusses the development of a screening and comparison tool for FEA on FPSOs and the incorporation of an expert system into the tool. The results are computerized using MS Excel/VBA to provide a structured and comprehensive assessment on each equipment and module handling natural gas, crude oil, methanol and diesel on FPSO topsides.

This tool features built-in calculations for jet and pool fire size estimation for gas/liquid releases, and the ability to perform Quantitative Risk Analysis (QRA) to specify the personnel and equipment risk for varying leak sizes and process conditions. Control and recovery measures are incorporated as an expert system based on report findings, engineering practices, and relevant standards. Bowtie analysis is applied in the tool to define detailed control and recovery measures for the FPSO based on the incident scenarios. An explosion assessment is performed by incorporating physical effect modeling software results.

Unique features provided in the tool include fire and radiation contour mapping on an FPSO layout to help determine personnel and equipment risk more accurately and fire pump sizing that can be used to verify the amount of water deluge system required to mitigate fires and explosions. In addition, flexibility of data input (process data, failure rate data, etc.) and user interfaces assist safety engineers to screen and compare process alternatives, check design quality, and evaluate design options at any design stage.

1 Introduction

1.1 Oil and Gas Floating Production Storage and Offloading (FPSO)

An FPSO is one concept that can lower the economic field size and make possible the development of the small or remote fields in deeper water, as the offshore industry goes towards deeper water [1]. The FPSO is typically based on a ship vessel with equipment built in modules which then positioned in vessel deck, as shown in Figure 1. The FPSO receives production fluids from one or a number of undersea oil reservoirs via risers. The fluid is separated at the topsides (vessel deck) into oil, gas, and water then contaminants in those is reduced by a separation system. Oil is stored in the vessel hull then offloaded to an oil tanker. Depending on the amount of gas produced, gas is either exported, reinjected, used as fuel, or a combination of the above. Since it is based on a ship, the FPSO features several advantages [1]. An FPSO is faster to other floating structures such as TLP, spars, semisubs, etc. thus saves production and project cycle times. The FPSO has the ability to adapt with different water depth, movable and easy to relocate, and adaptable to work with or without other facilities such as platform [1]. Since FPSO acts as storage, it does not need expensive oil pipelines to onshore thus projects in remote area is more viable to develop.

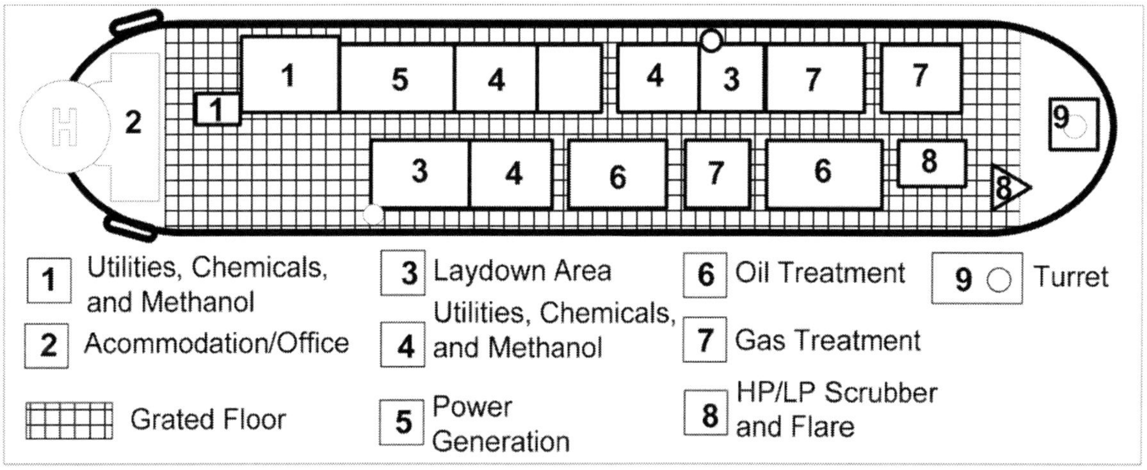

Figure 1. Typical FPSO modules layout (adapted from [2])

1.2 Captivation of Expertise to Help Performing Fire and Explosion Assessment and Providing Control and Recovery Measures (CRM) Solutions

Fire and Explosion CRM used in industry has been developed through a number of researches and experiments with contributions from different backgrounds such as engineers, safety technologists, scientists, and process designer expert. With the arising of complexity from the combination of FPSO design, hazards presented, CRM engineering, and regulations, it is almost impossible for safety professionals to manage all aspects of fire protection into practice even though it is necessary for them to support engineering solutions or compliance with legislation [3]. Hence, it is important to find way(s) to reduce the burden of expertise by compiling all pertinent expertise into one easy-to-use and guided system called an expert system. This system provides proven solutions required for CRM practice in industries without violating both technical/engineering constraint and regulations. With the increasing of computer ability, it is therefore important to combine the expert system with risk analysis thus CRM solutions for any incident scenario can be found.

1.3 Proposed Framework

The proposed tools called Fire and Explosion Assessment Framework (FEAF) is a combination of Quantitative Risk Analysis (QRA), Bowtie Analysis, and Control and recovery measures (CRM) which include fire water pump sizing, as shown in Figure 2. The comparison between proposed framework and current consequence analysis is also shown in Figure 2. FEAF is intended to be used as a screening tool, options evaluation, and assessment quality check, easy to build with MS Excel/VBA, and ready to apply for current industrial project.

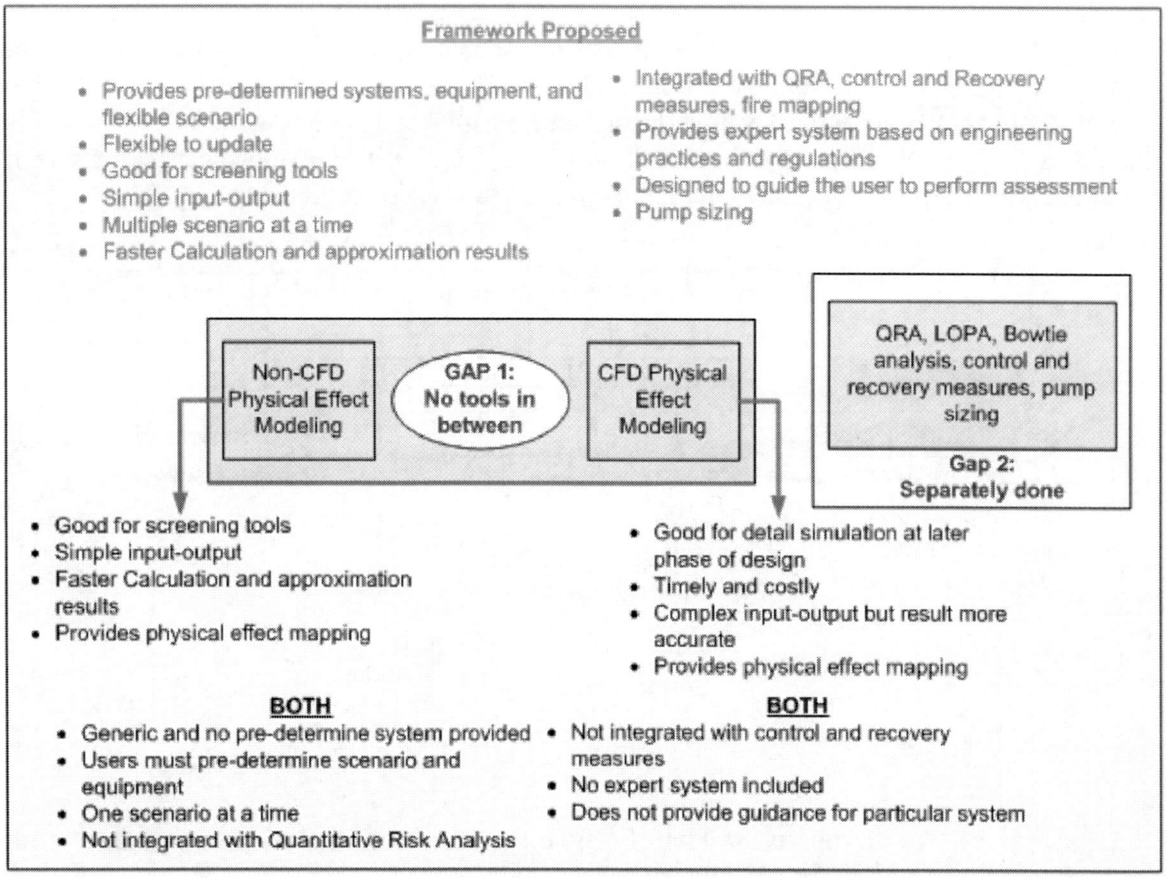

Figure 2. Proposed Framework

2 Relevant Methodologies

The logical thinking behind this FEAF is to utilize not-so-detail process conditions for each modules and equipment on an FPSO, quantify fire and explosion hazards with physical effect modeling, map fire hazards on FPSO layout, provide the control and recovery measures expert system, and calculate water deluge requirement for each scenario to help fire water pump sizing. The applicable methodologies are described in the following.

2.1 Physical Effect Modeling (PEM)

This section describes all the Fire PEM (FPEM) methodologies in the assessment included in the tool. Gas and liquid release rate of a particular leak size, jet fire length (at t=0, 5 min, 10 min, and 30 min), pool diameter, pool fire length, release duration, burning duration, and release rate are the main FPEM output required. Explosion PEM is performed separately and is discussed in detail in results and discussion section later on.
The initial gas release rate for a leak gas under pressure is calculated by [4]:

$$Q_0 = C_D\ A\ P_0 \sqrt{\frac{M\ \gamma}{R\ T_0}\left(\frac{2}{\gamma+1}\right)^{\frac{\gamma+1}{\gamma-1}}} \quad \text{If } P_0 > P_a \left(\frac{2}{\gamma+1}\right)^{\frac{\gamma-1}{\gamma}} \qquad (1)$$

With Q_0 =Initial release rate (kg/s), C_D =Discharge coefficient, A =Whole area (m^2), P_0 =Gas operating pressure (Pa), R =Universal gas constant = 8314 j/kgmol K, M =Molecular weight of gas, γ =Ratio of specific heats, and T_0 =Operating temperature (K).

Jet fire flame length is approximated by Chamberlain equation for offshore conditions [4]:

$$Jet\ Flame\ Length\,(m) = 11.14 \times Q_0^{0.447} \qquad (2)$$

Jet fire length at the time of 0 minute, 5 minute, 10 minute, and 30 minute are estimated from the exponential release rate at a particular time then use the Chamberlain equation to determine jet fire length. The equation is given by [5]:

$$Q_t = Q_0 \exp\left(\frac{-Q_0}{m_G} t\right) \qquad (3)$$

Where m_G is given by [5]:

$$m_G = \frac{PM}{0.08314T} \Pi\, r^2\ L \qquad (4)$$

With Q_t =Gas release rate at time t, m_G =Mass of gas (kg), and t =Time of releases (sec).

Crude oil produced, methanol for well injection, and diesel fuel provide the main liquid pool releases hazards and the FEAF is designed to simulate those. Since all equipment containing liquid on the FPSO are provided with bund to prevent liquid spreading, pool diameter is derived from bund area (as the worst case) and the relationship is given by:

$$Pool\ Diameter\,(m) \approx equivalent\ diameter\ of\ bund = \sqrt{\frac{4\,A}{\Pi}} \qquad (5)$$

Then, the pool fire length is estimated by [5]:

$$Pool\ Fire\ Length = 42\,D\left(\frac{burning\ rate}{\rho_{air}\sqrt{(9.81\,D)}}\right)^{0.61} \quad (6)$$

2.2 Risk Estimation

Risk is defined as the product of probability or frequency (P/F) and consequence hence the P/F data should be provided. For this FEAF, built in P/F is taken from E&P Forum [6] and used for both personnel and equipment risk estimation, which is the frequency of fire or explosion. The framework provides four different leak sizes for each of the equipment thus the outcome frequency for each of the equipment would be [4]:

$$F_{scenario} = \sum_{outcome} F_{outcome\ of\ each\ leak\ size} \quad (7)$$

Personnel Risk is determined by calculating the number of personnel affected by the fire, which depends on the module types, and types of fires. Personnel are assumed distributed evenly in the area of module. Personnel risk estimation includes LSIR – Location Specific Individual Risk, ISIR – Individual Specific Individual Risk, and Fatality Annual Rate (FAR), which are:

- Location-specific individual risk (LSIR) for a hypothetical continuously present in a particular area of the FPSO, which is given by [4]:

$$LSIR_{equipment} = \sum_{scenario} F_s\ P_F \quad (8)$$

$$LSIR_{module} = \sum_{equipment} LSIR_{each\ module} \quad (9)$$

- Individual-specific individual risk (ISIR) for the specific group of workers from events on the FPSO, which is given by [4]:

$$ISIR_{equipment} = \sum_{scenario} LSIR \times P_L \quad (10)$$

$$ISIR_{module} = \sum_{equipment} ISIR_{each\ module} \quad (11)$$

- Fatality Accident Rate (FAR) [4]:

$$FAR_{equipment} = \frac{ISIR_{equipment} \times 10^8}{H} \quad (12)$$

$$FAR_{module} = \sum_{equipment} FAR_{equipment} \quad (13)$$

- Annual Damaged Costs (equipment risk) [4]:
  It is based on the damage fraction for each scenario, which is the fraction of each module's volume damaged.

$$D_F = \sum_{modules} \frac{D_m \times C_m}{C_p} \quad (14)$$

Annual damage fraction (ADF) is given by [4]:

$$ADF = \sum_{scenario} F_s D_F \quad (15)$$

2.3 Water Deluge System Requirement

This feature consists of three steps of calculation, which are built in the framework and calculate the results automatically, as the following:

- Determine the module water requirement for each based on equipment and independent to fire and explosion scenario.

$$water_{equipment} = area \times water\ required_{each\ \ equipment} \quad (16)$$

- Based on fire mapping, approximate the number of modules affected by assuming that once a part of the module is in fire, water will be provided for the total module to prevent escalation. The total water is:

$$Total\ Water = \sum_n \text{Total water for each module covered by fire} \quad (17)$$

Since the water application rate is different from one type of equipment to another, the total water application rate depends on the footprint of each area on the FPSO. The area applied in the framework is:

$$\text{Area} = \frac{Module\ area}{number\ of\ equipment} \quad (18)$$

2.4 Fire Mapping

The tool provides the features to map both jet and pool fires onto the scaled FPSO layout with the ability to automatically size the fire according to the physical effect modeling for the worst case scenario and most likely scenario. This helps the user to see the fire engulfment of modules and equipment.

2.5 Control and Recovery Measures

This feature provides an indication of generally used control and recovery measures for both fires and explosions. By following Bowtie analysis methodology as shown in Figure 3, the framework provides the barriers for several threats and recovery measures for consequences.

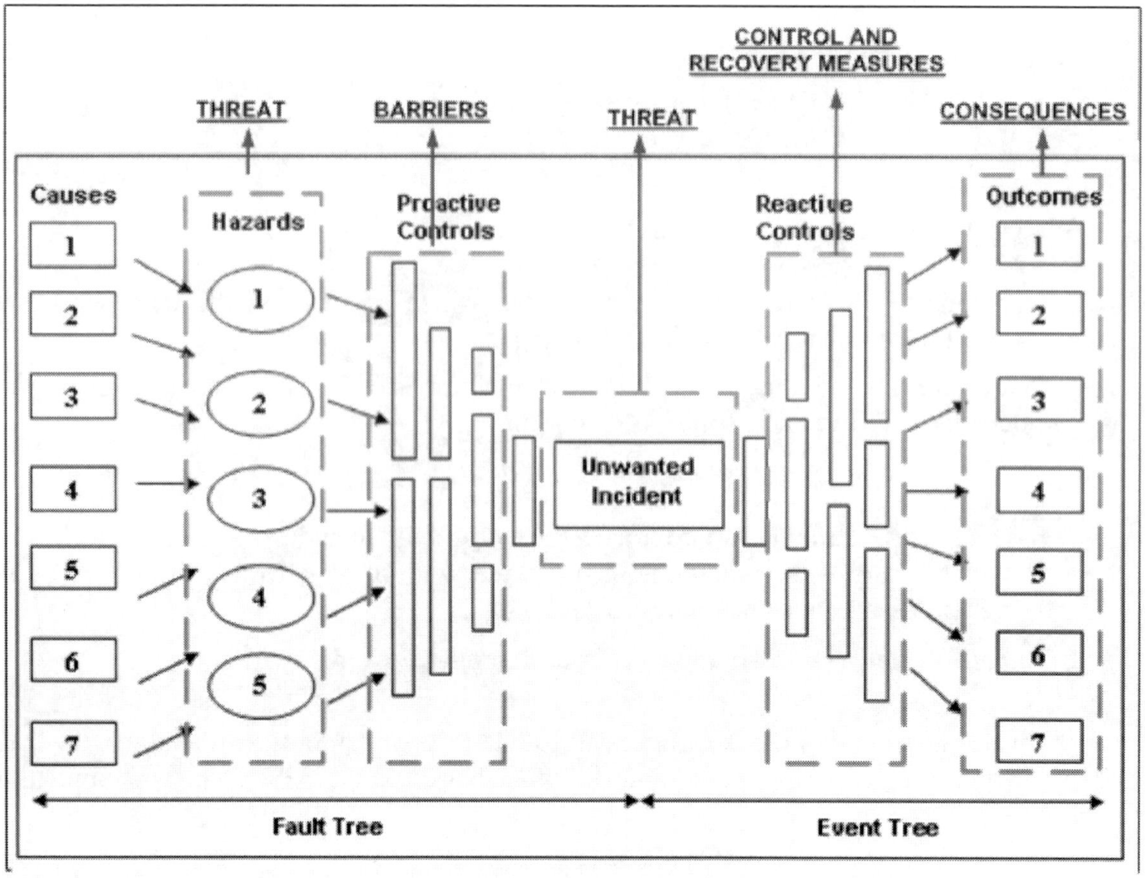

Figure 3. Bowtie analysis [7]

In addition, the FEAF provides another important aspect in control and recovery measures, which is to define the amount of water deluge system as control and recovery measures for fire and/or explosion and to determine the required pump size.

The water application rate depends on the type of equipment covered by the deluge water and whether personnel and structure passive fire protection are included. The framework provides a table to specify the number and type of equipment for each module, footprint area, and the fire mapping results to show fire coverage. This helps to determine the water deluge requirement for different equipment/module arrangements and scenarios.

3 Results and Discussion

The logic diagrams and snapshots of the tool developed are presented and discussed in this section. Input is divided into two parts: process data as shown in Figure 4 and probability or failure rate data. Since the QRA is prepared for all modules and equipment on the FPSO with four different leak sizes, general FPSO modules and equipment and four flexible leak sizes are provided. These can be changed or updated according to the user's particular scenario. Process data input includes pressure, temperature, and types of equipment. User interface is managed to help the user to go appropriate section of the tool.

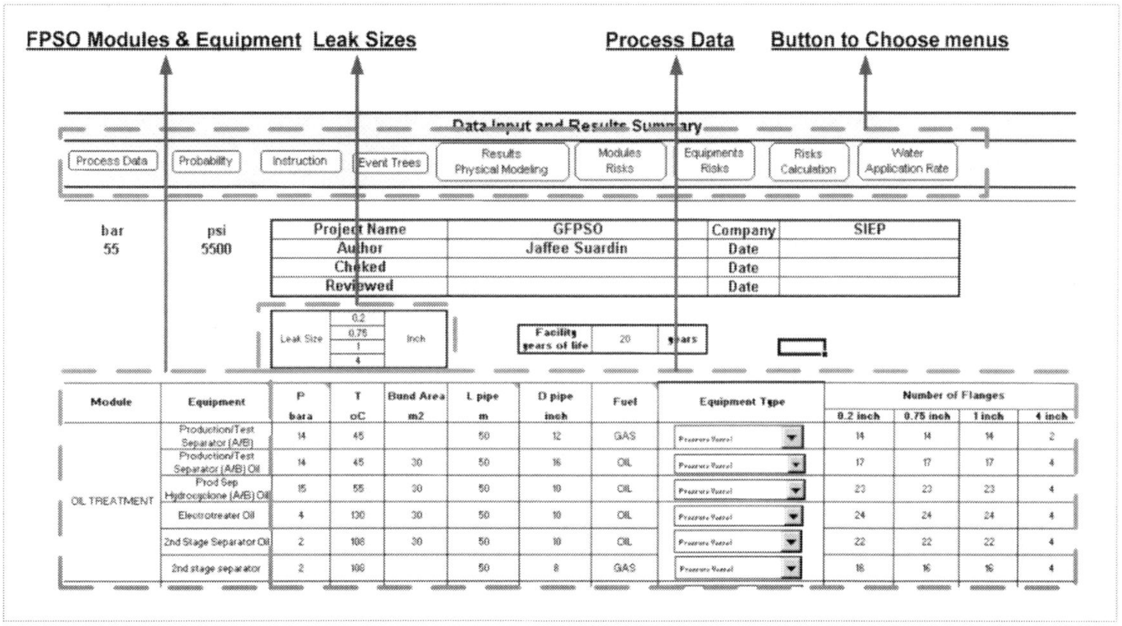

Figure 4. **Process data input form**

This framework provides physical effect modeling for fires and explosions separately due to different complexity of phenomena and physical effects. As the fire physical effect modeling is performed using the builtin equation entirely, explosion physical effect modeling requires the user to use an explosion simulation model.

3.1 Fire

The fire physical effect modeling (FPEM) flow diagram and the usage of its output for calculating risk are shown in Figure 5 and Figure 6. All pertinent modules and equipments are provided in detail and guide the user on which equipment should be considered and the types of data needed before starting the simulation. A snapshot of the fire PEM is illustrated in Figure 7 which shows that jet fire lengths at 0, 5, 10, and 30 minutes after release and the pool fire size are calculated for each of equipments with four different leak sizes simulation scenarios. Input required for the fire PEM are leak sizes, facility years of life, general FPSO modules and equipment names/types, operating pressures and temperatures, bund areas, pipe diameters and lengths to define inventory volume, types of fuel, equipment types, and number of flanges (optional).

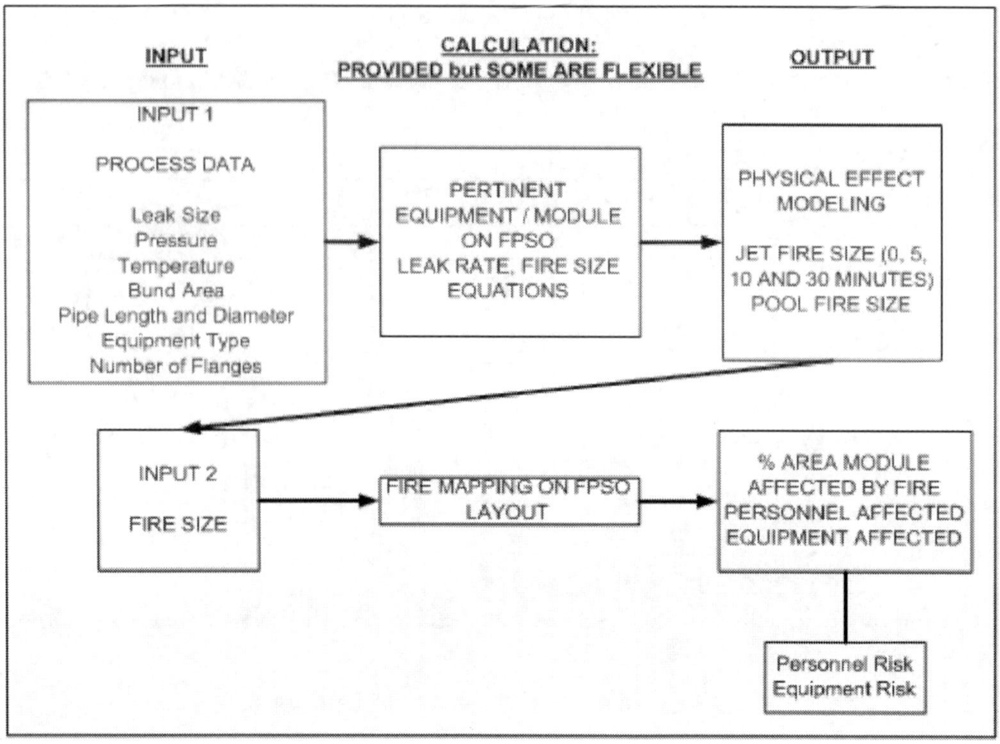

Figure 5. Fire physical effect modeling

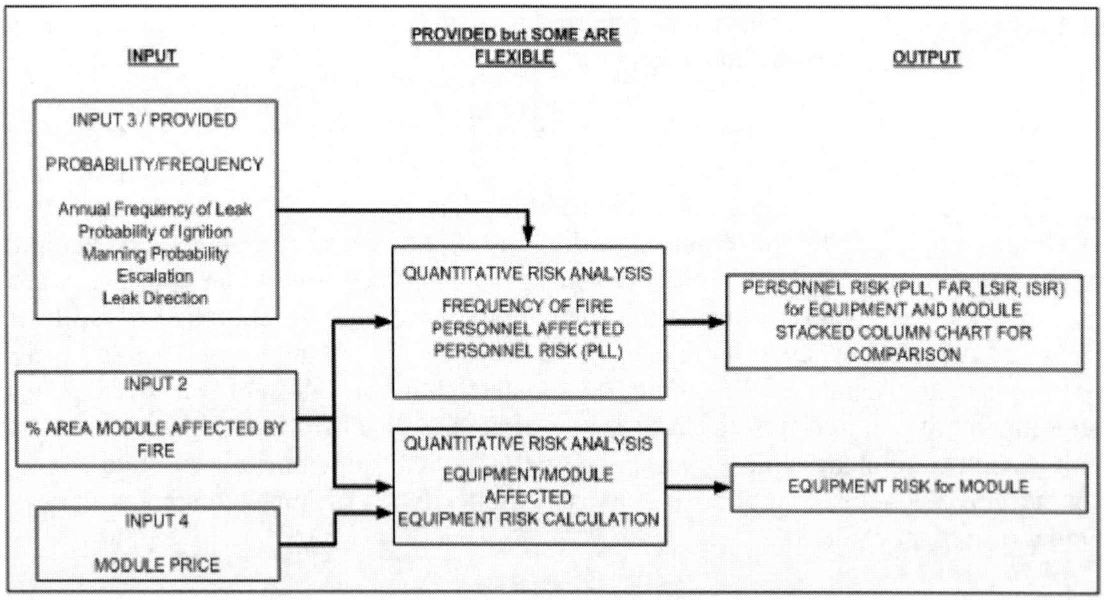

Figure 6. Fire risk analysis

The most commonly used leak sizes are 0.2 in, 0.75 in, 1 in, and full bore (pipe diameter). The pool fire size is simulated conservatively by assuming that each module has a bund collecting the liquid spilled, liquid is prevented from spreading and liquid will cover the entire bund area. Thus, the pool diameter is the diameter that represents the bund area. The types of fuel used in this tool are natural gas, liquid, diesel and methanol. The equipment type is to define the type of the equipment used; the leak frequency will be automatically calculated based on historical experience. Number of flanges as optional input provides significant contribution to the frequency of the release. This could be obtained from equipment data specification or approximation by engineering judgment.

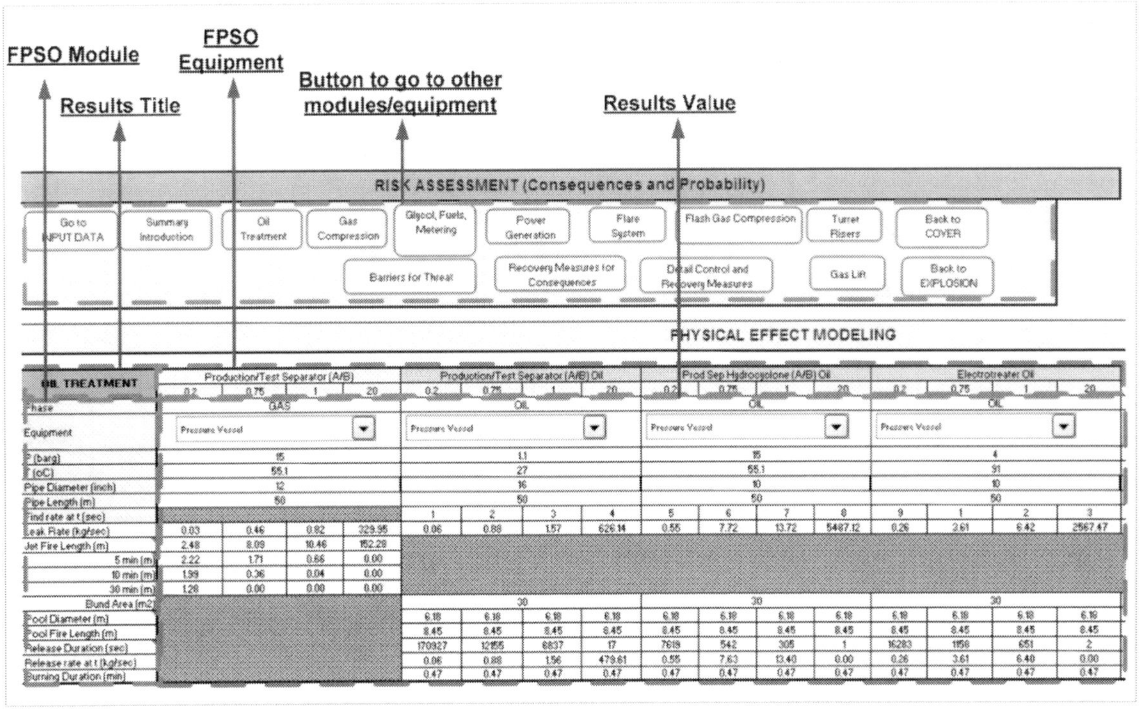

Figure 7. Fire physical effect modelling tool snapshot

The fire mapping feature plots the fire size onto a diagram of the FPSO layout. The red circle represents the jet fire and pool fire dimensions. The jet fire is taken as a circle to account for the 360 degrees of leak direction. The pool fire width is taken as the pool diameter. The red circle is sized automatically. As shown in Figure 8, the table provides the jet fire and pool diameter as the worst case (the longest fire) and most likely fire (highest probability) based on the process conditions and equipment type. Fire mapping is used to determine the personnel and equipment risk and the water application rate needed. The red circle will help the user to graphically visualize the area of module covered by the fire and based on that, the user must percentage input area of module affected into the table.

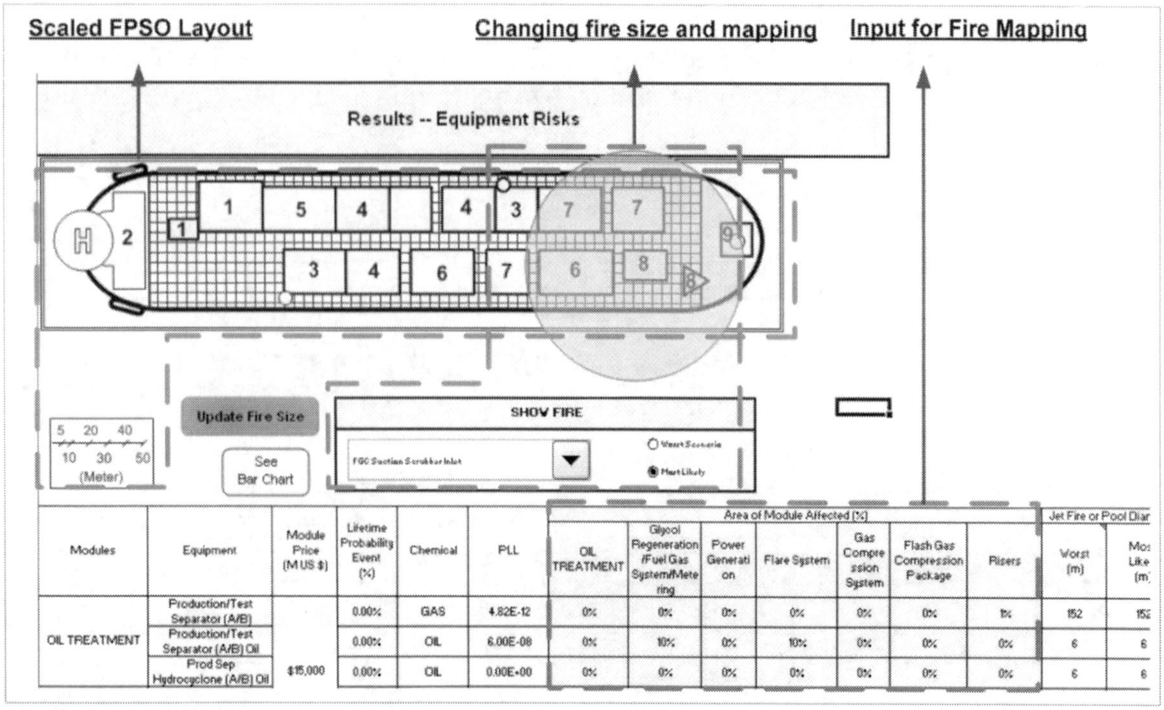

Figure 8. Fire Mapping

3.2 Explosion

The explosion physical Effect Modeling is shown in Figure 9.

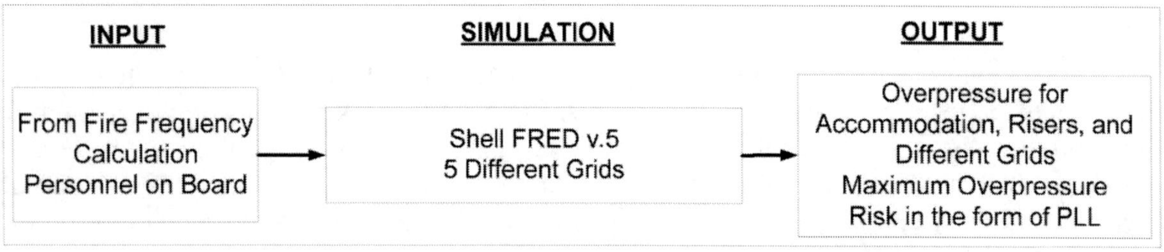

Figure 9. Explosion physical effect modelling

This tool applies Shell FRED v.5.0 software to simulate explosion physical effects. Thus, the results are only for a particular FPSO with specific process conditions and layout, which is Shell FPSO Base Case A. It is used only as an approximation for other FPSOs. Look at the evaluated-FPSO process data and layout, find the closest condition of grids in this framework then use the results as an approximation. The simulation is divided into 5 different grids of simulations, as shown in Figure 10.

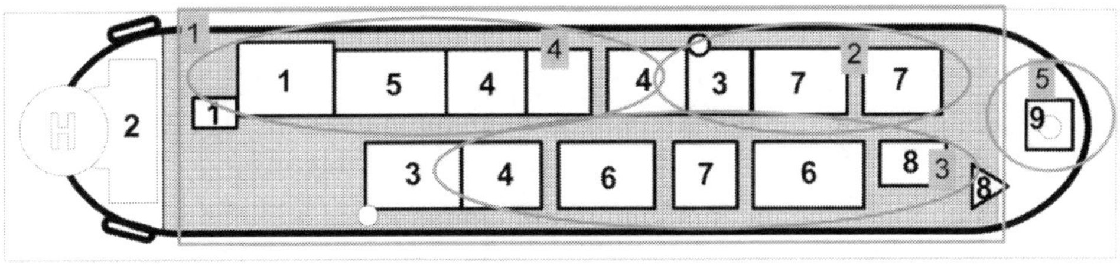

Figure 10. FPSO explosion simulation grids development

The division into five grids is based on open spaces in between each grid, which is considered as 10 m of open space. There is a lay down area in between grid 2 and grid 4. The open space between module 3 and 2 and 4 are considered having enough open space for explosion simulation. Risers are located far from the other modules. Grid 1 consists of all modules thus explosion overpressure is an effect of congested area of all modules. Leak sizes chosen are 0.75 inch, 1 inch, 2 inch, and 4 inch as the representative leak scenario.

3.3 Control and Recovery Measures and Water Deluge System and Pump sizing

Control and recovery measures are provided based on standard practices and regulations in the oil and gas industries as well as the past experiences CRM strategies [8][9]. Figure 11 shows the logic diagram of the control and recovery measures framework expert system.

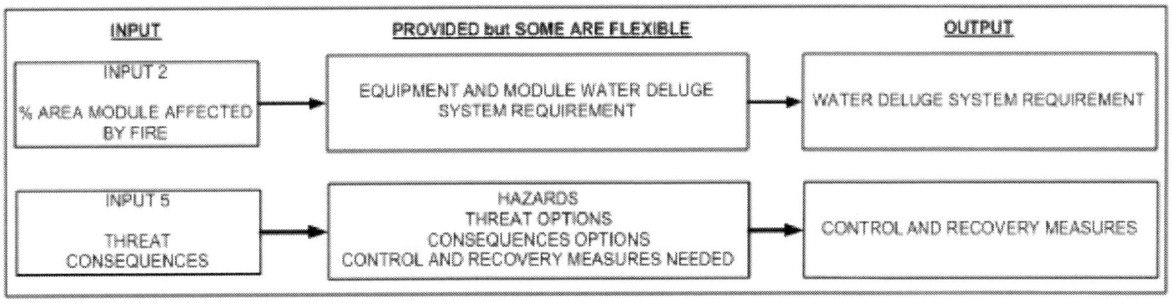

Figure 11. Control and Recovery Measures

Based on the Bowtie analysis methodology, the framework presents a list of options of threats and consequences. By choosing from the list according to the FPSO under evaluation, the framework will give types of CRM as potential engineering solutions. Snapshots of the framework are shown in Figure 12 and Figure 13. The left side shows the list of options of threats for hazards of hydrocarbon and flammable gas/liquid under pressure. By clicking on the options button, the framework shows the potential barriers for a particular threat. The framework provides barrier solutions for ten different threats (pressure, high temperature, seal, failure or degradation, corrosion, erosion, material fatigue, mechanical failure, topsides, containment, human error, and leak location) and recovery measures engineering solutions for five consequences (fire, explosion, smoke, hydrocarbon release, and personnel injury).

In addition, the framework provides detailed control and recovery measures obtained from several standards and previous reports such as Shell EP, Shell DEP, API, NFPA, SINTEF, BONGA reports, as shown in Figure 14. Control and recovery measures engineering solutions are provided based on general regulation requirements, type of fuels, equipment, and types of hazards. Examples of criteria or guidance used in the expert system are shown in Table 1 and Table 2.

The suggested water deluge system for each module and equipment are determined based on the fire mapping, module footprint area, and standardized water deluge requirement complied from references such as NFPA and ISO [8][9], assuming that equipment in a particular module triggers the deluge system for module. Thus, results do not differ from regulations as well as good engineering practices. For example, pump or compressors require 20 l/min/m^2 of water deluge system. By knowing compressor footprint and the number of them affected by fire, water deluge required can be calculated. The same is applied to any equipment on the FPSO topsides. The water requirement is determined based on the scenario by using fire mapping as input. Results are presented in stack bar as shown in Figure 15.

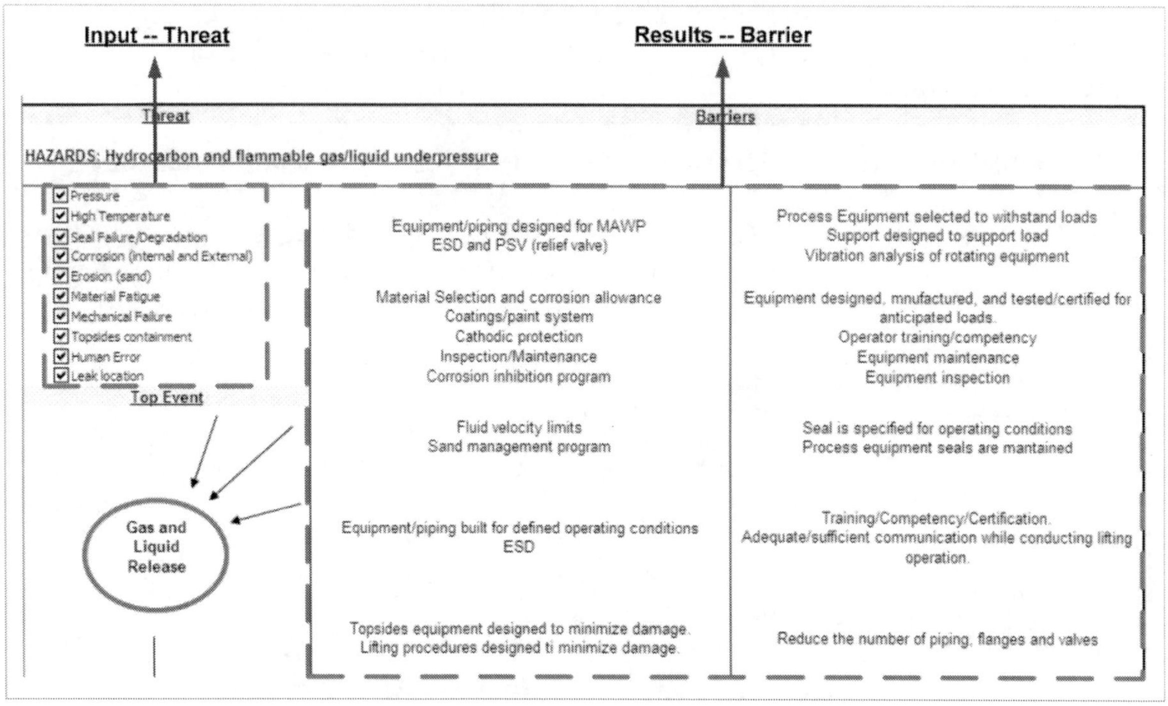

Figure 12. Bowtie analysis results for general CRM – Barrier prevention

Table 1. Examples of criteria and guidance for Water deluge system [10,13]

References	Function	Criteria / Guidance
API 13702: 1999(E)	Process Area deluge	10 l/min/m^2
		Pumps or Compressors 20 l/min/m^2
		Gas Treatment Area 10 l/min/m^2
		Methanol area 10 l/min/m2
	Turret area deluge	10 l/min/m^2
	Vertical structures	10 l/min/m^2
	Horizontal Structures	4 l/min/m^2
	Escape and Evacuation	15 - 45 l/min/m^2
NFPA 15 Year	Controls of burning	pumps, compressors and related eq. > 20.4 l/min/m^2

2001 edition, Water Spray Fixed Systems for Fire Protection		Flammable and combustible liquid pool fires requires not less than 12.2 l/min/m^2
	Exposure Protection	Vessels surface requires not less than 10.2 l/min/m^2
		Horizontal structures requires not less than 4.1 l/min/m^2
		Vertical structures require not less than 10.2 l/min/m^2
		Vertically stacked piping require < 6.1 l/min/m^2

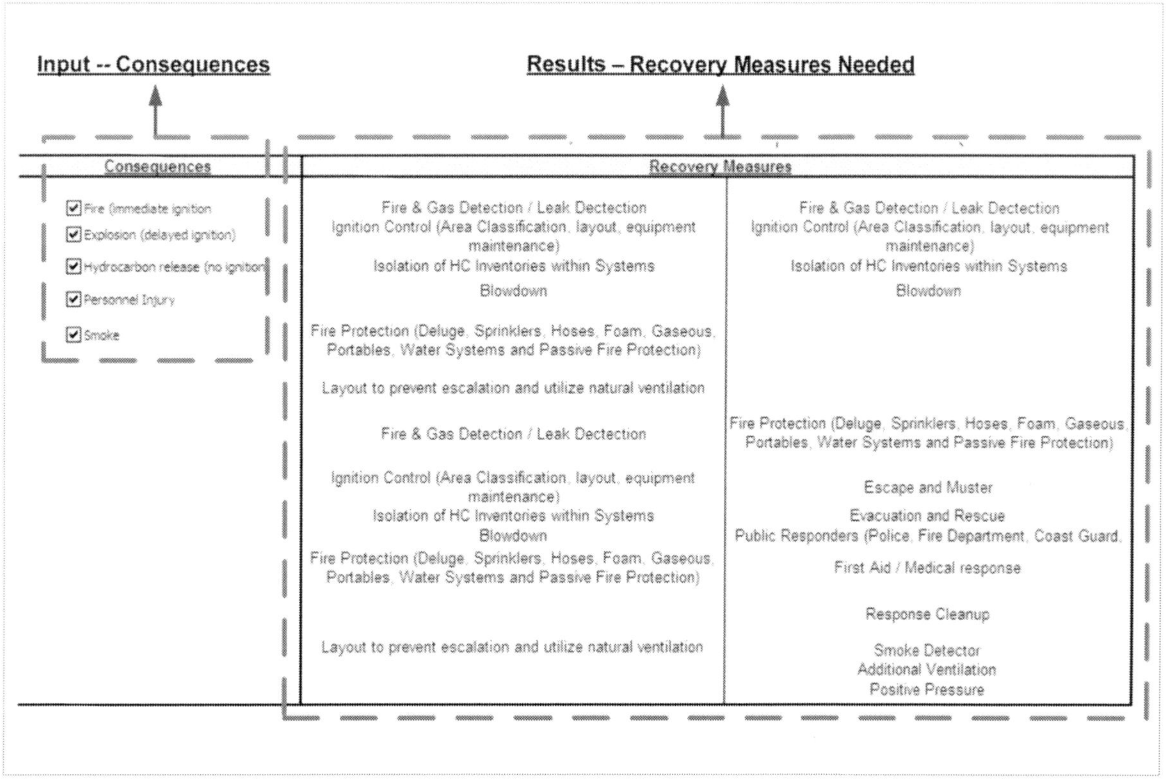

Figure 13. Bowtie analysis results for general CRM – Recovery measures

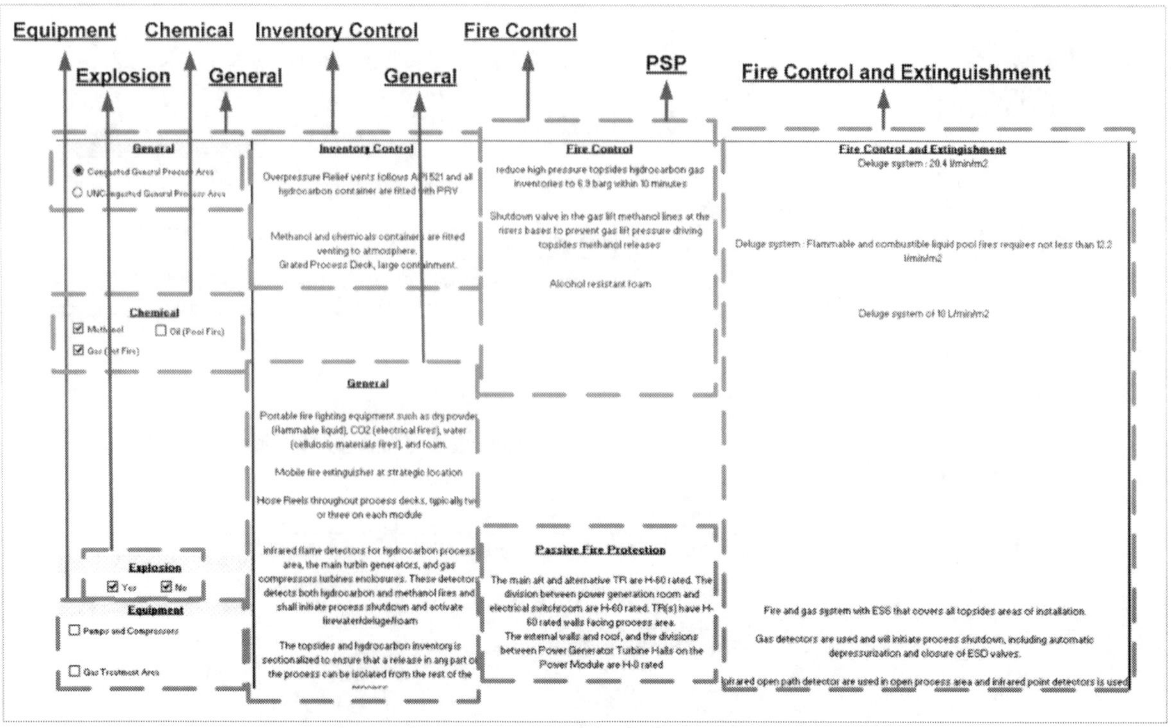

Figure 14. Detail CRM based on chemical, general conditions, explosion, and equipment

Table 2. Examples of explosion control and recovery measures for expert system [10]

Prevention of Hydrocarbon Releases

Design equipment to handle potential mechanical damage

Reduction in release frequency by reducing number of flanges, connections, etc

Blowdown to reduce inventories

Prevention and Limitations of Explosion

Ventilation by orienting the equipment to permit through-ventilation under most wind condition

Grated decks

Control of Explosion

Gas Detection that initiated process shutdown, including automatic isolation and depressurization

Mitigation and Recovery Measures

Water Deluge with firewater pumps (1360m3/hr at 12 bara)

Water deluge for process area of 10 l/min/m2

Overpressure Protection

The list below must be survive during explosion

The Hull is built to resists an internal overpressure of 2.3 barg

the TR's and accommodation should able to handle 0.5 barg

Protect escape and evacuation route

3.4 Result Presentations

Final results are communicated in the form of a stacked bar chart, as shown in Figure 15. Total FPSO risks, equipment risks, module risks, and equipment and water deluge requirements are presented in a stacked bar chart, as shown in Figure 15. This way, the most contributing modules and equipments to the risks and water deluge requirement are viewed clearly. The maximum water deluge system suggested by the system is based on the maximum calculated water requirement among the modules.

4 Conclusions

Modeling software is available to aid in oil and gas fire and explosion assessments. This paper presents an Excel/VBA based tool to apply these techniques to a specific FPSO and identify the associated risks and suggested control and recovery measures.

The framework provides a fire and explosion assessment with risk analysis and expert system that can be used as a screening tool will help the user make better decisions when dealing with process alternatives.

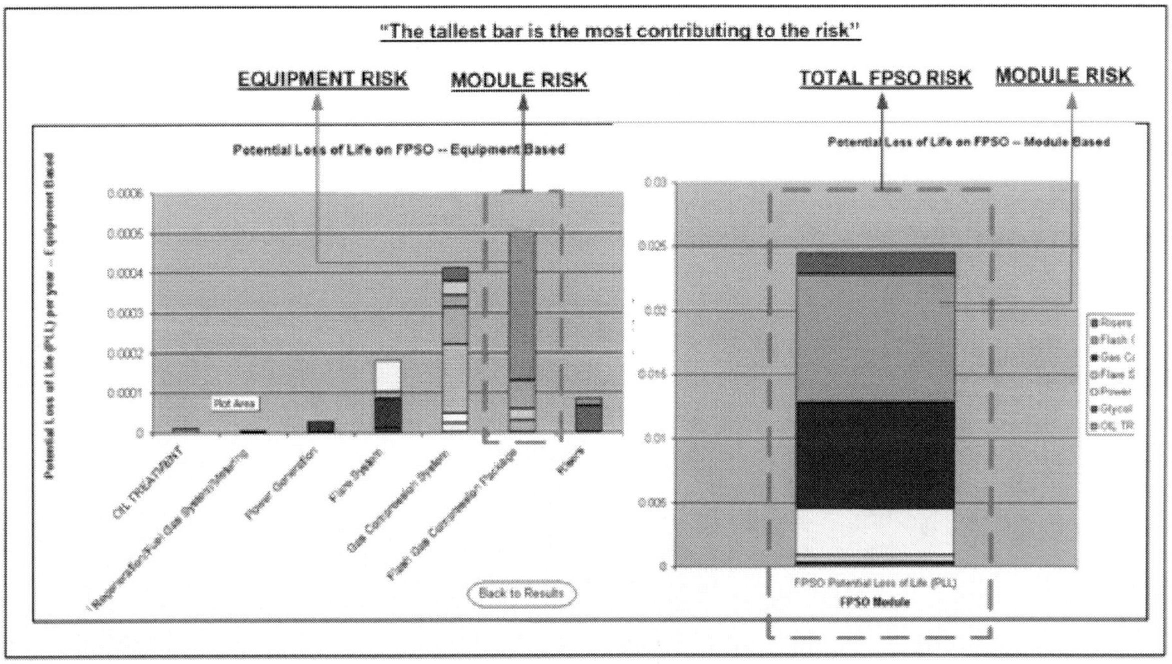

Figure 15. Final results stacked bar

5 Acknowledgment

This project was completed as part of 3-month summer internship program at Shell International Exploration and Production Inc (SIEP), a division of Shell Oil Company. Opportunities, facilities, guidance, and advices provided by Mr. Jeff Edward, Mr. Jeff McPhate, Mr. Anthony Sipkema, Mr. Matt Childs, other SIEP personnel and Dr. Sam Mannan of Mary Kay O'Connor Process Safety Center are greatly appreciated.

6 Disclaimer

This paper does not reflect any data or design used by Shell International Exploration and Production Inc (SIEP). The data were used only for case study whereas this paper concentrates on the development of the framework. SIEP does not assume any responsibility or liability for use of or reliance upon any of the data, information, or concepts presented in this paper.

7 References

[1] Y. Shimamura, FPSO/FSO: State of the Art, *J. Marine Science and Technology 7* (2002), 59-70.

[2] www.naodefencevfm.org/downloads/pdf/bg_gpp_to_nao2.pdf.

[3] H. A. Donegan and I. R. Taylor, "The Potential of expert system in fire safety evaluation", *Computer Applications in Fire Protection Engineering Volume III – Applied Fire Science in Transition Series,* P. R. DeCicco (Series Editor), Baywood Publishing Company Inc, New York (2002), pp. 35-52.

[4] J. Spouge, *A guide to quantitative risk assessment for offshore installations*, The centre for Marine and Petroleum Technology (CMPT) (1999).

[5] SINTEF

[6] E&P Forum., *Quantitative Risk Assessment Datasheet Directory*. E&P Forum Report No. 11.8/250 (1996), London.

[7] http://www.mishc.uq.edu.au/NMIRAG/images/Figure4.19_Bow_Tie_diagram.gif

[8] International Standards Organization (ISO), *ISO 13702 – Petroleum and natural gas industries: Control and mitigation of fires and explosions on offshore production installations – Requirements and guidelines*, 1st Ed. (1999).

[9] National Fire Protection Association (NFPA), NFPA 15 - Water Spray Fixed Systems for Fire Protection (2001).

LOSS PREVENTION SYMPOSIUM

Oleum and Chlorosulfonic Acid Spill Tests – Field Data for Model Validation

S. Dharmavaram

E.I. du Pont de Nemours and Co.
DuPont Engineering Research and Technology
Wilmington, DE 19898 U.S.A.
Phone: 302-774-8046
seshu.dharmavaram@usa.dupont.com

Prepared for Presentation at
American Institute of Chemical Engineers
2007 Spring National Meeting
41st Annual Loss Prevention Symposium
Houston, Texas
April 22–26, 2007

UNPUBLISHED

Oleum and Chlorosulfonic Acid Spill Tests – Field Data for Model Validation

S. Dharmavaram

E.I. du Pont de Nemours and Co.
DuPont Engineering Research and Technology
Wilmington, DE 19898 U.S.A.
Phone: 302-774-8046
seshu.dharmavaram@usa.dupont.com

Abstract

Oleums are mixtures of sulfur trioxide in sulfuric acid and are produced in several strengths. Chlorosulfonic acid is formed by reacting sulfur trioxide with hydrochloric acid. These chemicals are used as sulfating and sulfonating agents in many applications. When accidentally released from vessels or pipes they react instantaneously with water available from all sources like atmosphere, concrete, soil, etc., to form a fine acid mist that disperses downwind, based on atmospheric conditions. Unlike most other chemicals, the vaporization of sulfur trioxide from oleum spills depends not just on its partial pressure but a variety of conditions. Complex chemical reaction and heat generation that occur in the liquid phase determine the amount of sulfur trioxide released above a pool of liquid. The sulfur trioxide or chlorosulfonic acid vapors then react instantaneously with the moisture in the atmosphere generating sulfuric acid and/or hydrochloric acid and heat. Water and/or foam are used effectively in mitigating oleum and chlorosulfonic acid spills. However, very limited laboratory or field data are currently available that describe the complex behavior of these chemicals. Several theoretical models have been developed to predict the vaporization and dispersion of the chemical upon loss of containment. These models and methods have not been validated. In this paper, details are provided on field tests conducted in Nevada in 2003 & 2006 for spills of 65% oleum and chlorosulfonic acid. Included is a description of the spill mitigation, field measurement methods, and some preliminary results.

1. Introduction

"Oleum" is a name/term used for describing mixtures of sulfur trioxide (SO_3) in 100% sulfuric acid (H_2SO_4) and is produced in several strengths. Some of the typical concentrations are 20%, 30% and 65% oleums, and/or pure 100% SO_3 which are used as sulfating agents in many applications such as production of plastics, detergents and dyes. Upon release from vessels or pipes, because of accidents, oleums react instantaneously with water available from all sources including the ground (i.e., asphalt, concrete, soil, etc.) and atmosphere. Oleums are very corrosive and tend to extract water and char surfaces.

Chlorosulfonic acid (CSA) is produced by reacting SO_3 and hydrochloric acid (HCl), and is used as a sulfonating agent. The behavior of CSA upon release from loss of containment is similar to that of oleums.

The reaction with water available results in an increase in initial temperature of a pool that affects the vaporization rate. The SO_3 released from the liquid surface, for an oleum spill, then reacts with the moisture available in the atmosphere to form a fine mist of H_2SO_4 that then disperses downwind. Figure 1 is a picture of a train derailment, which occurred a few years ago, that shows the dispersion behavior of the acid mist plume from a spill of 65% oleum. It illustrates the plume buoyancy that has been observed in many such accidents.

Figure 1: Oleum rail car derailment

Most of the models available for pool vaporization and dispersion calculations do not account for the complex behavior of oleum and CSA spills and the acid mist dispersion. Some efforts have been made in the past [1, 2] to understand the impact of oleum releases. However, the approach used in these earlier works is rather simplistic and contains significant gaps. A major research project was conducted at the University of Manchester Institute of Science and Technology (UMIST), in Manchester, U.K., that led to development of the REACTPOOL model by Kapias and Griffiths [3, 4, 5]. In many of the publications these authors have reiterated that the modeling methodology is theoretical and needs to be validated. In DuPont, a similar spill model has been developed based partly on the UMIST work described further in U.K. Health and Safety Executive publications [6, 7]. The Aspen Custom Modeling (ACM) tool kit [8] was used to develop a custom model [ACM-Spill] for determining the vaporization from oleum spills.

In this paper details are provided on attempts to gather data for validation of spill models (i.e., REACTPOOL and ACM-Spill) and acid mist dispersion models. The focus was on oleum spills, however data was collected for spills of CSA and 65% oleum.

2. Properties of Oleums and CSA

Oleums are extremely aggressive materials that are usually stored and transported in a liquid form because of the relatively high boiling points as shown in Table 1. Table 1 also shows some other physical properties of SO_3 and oleums [9, 10, 11, 12]. The melting points are also high and variable depending on the concentration which could lead to solidification under certain ambient conditions.

Table 1: Physical properties of Oleums and SO_3

Properties	100% H_2SO_4	20% Oleum	30% Oleum	65% Oleum	100% SO_3
Boiling Point, °C	337	142	121	58	45
Melting Point, °C	10	5	19	2	17
Density at 16 °C, kg/m^3	1837	1915	1952	1992	1947
Specific Heat at 20 °C, cal/g °C	0.34	0.32	0.32	0.41	0.77
Viscosity at 20 °C, cP	28	39	48	55	2.2
Vapor pressure at 20 °C, atm	4E-08	0.0008	0.002	0.141	0.254

The vapor pressure of pure 100% H_2SO_4 is significantly lower than that of oleums and SO_3. Oleums are non-ideal liquid mixtures of SO_3 and H_2SO_4. The higher vapor pressures for the oleums are because of the SO_3 present in much larger amounts than H_2SO_4 in the vapor at equilibrium with the liquid.

SO_3 is a highly reactive chemical. When SO_3 or oleums are released into the environment the SO_3 in the liquid mixture reacts violently with water in an exothermic manner as follows.

$$SO_3\,(\ell) + H_2O\,(\ell) \rightarrow H_2SO_4\,(\ell)\,(\Delta H_{rxn} = -\,21\ \text{kcal/mol}) \qquad \text{[Eq. 1]}$$

Like oleums, CSA is also a very aggressive material. Table 2 shows some of the basic physical properties of CSA.

Table 2: Physical properties of CSA

Properties	CSA
Boiling Point, °C	154
Melting Point, °C	-80
Density at 20 °C, kg/m^3	1747
Specific Heat at 20 °C, cal/g °C	0.27
Viscosity at 16 °C, cP	3
Vapor pressure at 20 °C, atm	0.003

Similar to SO_3, CSA is also a highly reactive chemical. When CSA is released into the environment it reacts violently with water in an exothermic manner as follows.

$$ClSO_2OH\ (\ell) + H_2O\ (\ell) \rightarrow H_2SO_4\ (\ell) + HCl\ (\ell)\ (\Delta H_{rxn} = -\ 14\ \text{kcal/mol}) \quad \text{[Eq. 2]}$$

When oleum or CSA are released into the environment, water is the limiting reactant and the total heat generated depends on the amount of water available from sources such as the ground and atmosphere. Presence of standing water on the ground can have a large effect on the initial spill temperature increases; however humidity of the atmosphere usually is the predominant reactive force in an ongoing release. In both cases the heat generated creates a fine visible mist and leads to buoyant plumes.

Handling, production, and transport of Oleums, SO_3, H_2SO_4, and CSA are covered by many regulations. SO_3, H_2SO_4, HCl, and CSA, are toxic and have specific Emergency Response Planning Guideline concentrations [13] and Acute Exposure Guideline Level concentrations [14] as shown in Table 2. The ERPG concentrations are based on 1 hour exposure. The corresponding AEGL-2 and AEGL-3 concentrations (for 1 hour exposure) shown are defined as follows:

> ***AEGL-2*** *is the airborne concentration (expressed as ppm or mg/m3) of a substance above which it is predicted that the general population, including susceptible individuals, could experience irreversible or other serious, long-lasting adverse health effects or an impaired ability to escape.*

> ***AEGL-3*** *is the airborne concentration (expressed as ppm or mg/m3) of a substance above which it is predicted that the general population, including susceptible individuals, could experience life-threatening health effects or death.*

Table 3: Toxicological endpoints of concern for Oleums, SO3 and H2SO4

<table>
<tr><th>Name (CAS No.)</th><th>ERPG-2</th><th>ERPG-3</th><th>AEGL-2</th><th>AEGL-3</th></tr>
<tr><td>Sulfur Trioxide (7446-11-9)</td><td rowspan="4">10 mg/m³</td><td rowspan="4">30 mg/m³</td><td rowspan="3">8.7 mg/m³</td><td rowspan="3">160 mg/m³</td></tr>
<tr><td>Sulfuric Acid (7664-93-9)</td></tr>
<tr><td>Oleum (8014-95-7)</td></tr>
<tr><td>Chlorosulfonic Acid (7790-94-5)</td><td>n/a</td><td>n/a</td></tr>
<tr><td>Hydrochloric Acid (7647-01-0)</td><td>20 ppm</td><td>150 ppm</td><td>22 ppm</td><td>100 ppm</td></tr>
</table>

The toxicological endpoints, shown in Table 3, are used in modeling the acid mist dispersion to determine the extent (i.e., footprint) of a dispersing plume or puff where potential injuries could occur, under a range of weather conditions.

3. Spill Protocol

The U.S. Department of Energy established the Nonproliferation Test and Evaluation Complex (NPTEC) at the Nevada Test Site (NTS), in 1982, where large-scale field tests could be performed on hazardous chemicals and liquefied gaseous fuels. The DOE was required by U.S. law to site and manage construction of the facility and to develop agreements allowing use by other Federal agencies and industry on a user fee basis. The DOE selected the NTS, where the original atom bomb tests were conducted in 1930s to 1960's, after a survey of several potential sites.

The location of NPTEC is along the eastern edge of the NTS within the general western side of Frenchman Lake (elevation of 3100 ft above sea level), which is dry during most of the testing season. The area is extremely flat for several miles both upwind and downwind of the spill point. Good meteorological data exist covering a period of at least 15 years and detailed, high frequency meteorological data have been taken and analyzed covering a recent three-year period. These data have shown that there is a preferred testing period between April and November and that during this period the wind direction is predictably and regularly from the southwest (225°). This wind pattern prevails with a variety of speeds and atmospheric stability conditions and with very little deviation about 225° from about 1200 hours until 2-3 hours past sundown. Tests are performed when the wind direction is 225° ± 45° with a wind velocity of between 2 m/sec. and 10 m/sec.

The NPTEC facility is equipped with television cameras and radio networks. The Command, Control, and Data Acquisition (CCDAS) area is equipped with a public address system. Closed circuit television is used to maintain visual contact between the spill site and the CCDAS Building to enhance safety. For the purposes of this test series, several components of NPTEC used included the data acquisition system, water, nitrogen, and power supplies and the videotaping system.

The main objective of the DuPont Spill tests was to provide hands-on training in the mitigation and clean-up of small spills of sulfur-based fuming acids using water fog, mid-expansion foam or "Spill-X-A®" agent. This training is useful to manufacturers and users of these chemicals and to those who are called upon to respond to transportation spills. The testing allows a qualitative comparison between the different mitigation techniques. Manufacturers of these chemicals are also informed on effective ways of dealing with accidental spills or releases and to recommend these procedures to their facilities and to their customers.

DuPont and NPTEC Staff were present to provide mitigation for any hazards that might develop, including as a result of trainee error. Instructor oversight during this training also minimizes the potential for trainee error. The chemicals on which spill control training was conducted were CSA and 65% Oleum. In each test, up to a maximum of 300 lbs. of the chemical is spilled into a 4.65 square meter steel pan from a ½" stainless tube using nitrogen to transfer the material from containers. The pan is weighed by means of load cells to provide one method of controlling the amount of the material to be spilled, as shown in Figure 2. The rim of the pan is at a height of 0.8 m above the ground.

During each spill the following information was collected at the CCDAS:

- Material spilled, whether CSA or 65% Oleum
- Amount spilled, which varied from 100 to 250 lb
- Amount of water added before spill, which was up to 120 lb.
- Load cell readings before, during & after fuming, which provide a measure of weight of the pan.
- Temperature readings, collected by thermocouples in and above the pan

In addition, the following meteorological data were also captured at the CCDAS:

- Wind Speed, recorded by location of meteorological tower versus time. Figure 3 is a snapshot that shows the wind speed and direction
- Temperature, recorded by location of meteorological tower versus time. Figure 3 is a snapshot that shows the ambient temperature
- Atmospheric stability, recorded by location of meteorological tower versus time. Figure 4 is a snapshot that shows the atmospheric stability across the NTS
- Solar radiation, recorded by location of meteorological tower versus time. Figure 5 is a snapshot that shows the solar radiation
- Relative humidity, recorded by location of meteorological tower versus time. Figure 5 is a snapshot that shows the relative humidity

Figure 2: Spill pan, resting on load cells, that was used during the spill tests

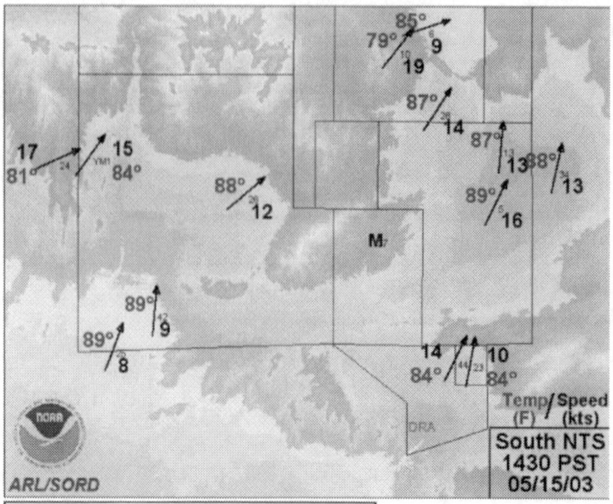

Stn	Wind (kts) Dir	Av	Mx	Mn	Temp (F)	Rel Humid	Wind Chill	Heat Index	Press
05	208	16	24	07	89	9%	-	85	904.3
06	252	09	18	03	85	2%	-	81	877.0
10	219	19	29	08	79	25%	-	-	-
13	185	13	23	06	87	9%	-	83	-
23	191	10	23	07	84	26%	-	82	883.7
24	246	17	27	07	81	21%	-	79	847.4
25	201	08	17	03	89	26%	-	85	915.2
26	231	12	19	06	88	-	-	85	-
27	Invalid								
28	213	14	23	09	87	6%	-	84	-
34	193	13	20	07	88	13%	-	83	-

Figure 3: Snapshot of atmospheric temperature, wind direction, and speed

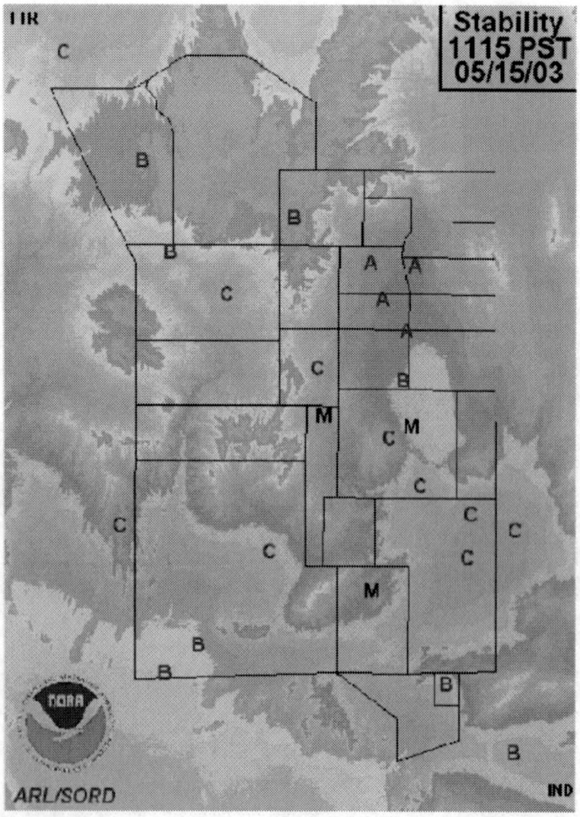

Figure 4: Snapshot of atmospheric stability across the NTS

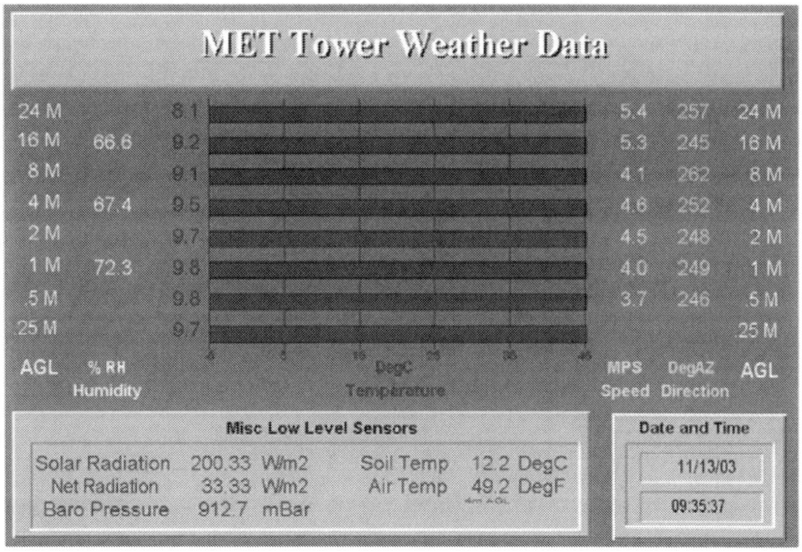

Figure 5: Snapshot of solar radiation, temperature, humidity, and wind speed

4. Spill Mitigation

Spill mitigation training was conducted during the spill tests using water fog, mid-expansion foam or "Spill-X-A®" agent. During a spill test, the trainees wear protective gear and learn to use the equipment for releasing and applying the mitigation agents. Figure 6 illustrates the mitigation using a water fog. A water stream or foam is never allowed to directly enter into the pan because of the violent reaction and potential for increased fuming. When CSA is in the pan, there is a potential for rapid-thermal-expansion explosions to occur. After the fuming is significantly reduced, a neutralizing agent is applied to the pan. After checking the pH, the material within the pan is shoveled out into 55 gallon drums and set aside for disposal.

Figure 6: Spill mitigation using a water fog

5. Atmospheric Measurement Methods

In addition to the spill mitigation training, data were also acquired during spill tests on the atmospheric dispersion behavior of the plume. The atmospheric data were obtained before the actual mitigation discussed in Section 5.

No continuous analyzers are available for measuring sulfuric acid mist concentrations. Therefore, to make the atmospheric measurements, wooden poles were used to dangle filter holders. The poles were then placed at five different distances along the centerline of pan in the 225 degree direction. The filter holders on the first two to three poles [located at 10, 12, 25, or 50 ft] were at a height of 35 to 40" above the ground. The filter holders at 100 and 250 ft were at a height of 67" above the ground. The filter holder on the last pole (at 500 ft) was located at a height of 98" above the ground.

During each test, the pre-weighed clean filters were placed in the filter holders and the gases sucked in at each point through Teflon® tubing connected to a sampling pump. The pump was stopped before the mitigation training was started for each spill. After the mitigation was completed the filters were retrieved and stored in clean glass bottles. The filters were later analyzed for sulfate and chloride ions, to determine the amount of sulfuric and hydrochloric acid.

In addition to filter samples, additional information was collected from pictures and videos of the plumes

6. Results

Figure 7 illustrates the dispersion behavior of the acid mist plume for one of the spill scenarios. Even under the low humidity conditions in Nevada, there is more than enough moisture in the atmosphere to convert the evolving SO_3 into H_2SO_4 mist. Atmospheric sampling data were collected along the centerline of the pan up to a distance of 500 feet.

Even though several spills were conducted, atmospheric data were collected only for a limited number of scenarios. Data relevant for modeling the spill behavior from a total of 18 spills conducted in 2003 and 2006 is provided in Tables 4 and 5, respectively.

The mean values of relative humidity, measured at a height of 1 m, varies anywhere between 6 and 21 % with a narrow standard deviation of 0 to 0.5 %. The solar radiation varied from 422 to 1092 Watts/m^2 with a standard deviation of 0 to 112 Watts/m^2. Most of the time, the Pasquill atmospheric stability was D ("neutral"), and C ("slightly unstable") in two of the cases. The atmospheric temperature, measured at a height of 8 m, varies anywhere between 25 and 31°C with a narrow standard deviation of 0 and 0.3^{o}C. Along the centerline of pan and the atmospheric samplers the wind direction would be 225 degrees. The actual direction of winds, measured at a height of 8 m, varied anywhere between 180 and 236 degrees, with a standard deviation of 0 and up to 28 degrees in some cases. The wind speeds, measured at a height of 8 m, varied anywhere between 4.5 and 8.5 m/s, with a standard deviation of 0 and up to 1.5 m/s in some cases.

Figure 7: Acid mist dispersion behavior

At various times, anywhere from 118 to 214 lb of either 65% Oleum or CSA was spilled as noted. In three cases, some amount of water was added to the pan before the chemical was spilled into the pan. The maximum mass in the pan is always higher than chemical plus the water initially added. It corresponds to the weight of the mixture based on the water absorbed from the atmosphere into the liquid phase. The surface area covered is the area of the liquid surface exposed to the atmosphere estimated based on the total amount spilled.

When 65% Oleum and CSA react with water in the pan or with moisture absorbed from the atmosphere, heat is generated in the liquid and the bulk temperature increases. The smallest increase in liquid temperature is 66 deg C and in one case the liquid temperature gets as high as 320 deg C. As the liquid pool vaporizes, the chemical vapors react with moisture in the atmosphere generating heat. The temperature of the acid mist plume is as low as 70 deg C and can be as high as 150 deg C.

Provided in Tables 4 and 5 are concentrations of H_2SO_4 and HCl at five different distances from the pan along the centerline (225 degrees direction). The acid mist particles collected on filters consist of H_2SO_4 and/or HCl depending on whether 65% Oleum or CSA is spilled into the pan.

Table 4: Field data collected during spill tests conducted in May 2003

Test	Date	Start Time	Duration	Relative Humidity	Solar Radiation	Atmospheric Stability	Ambient Temperature	Wind Direction	Wind Speed	Chemical Spilled	Chemical Mass Added Into Pan	Water Added Before Chemical	Maximum Mass in Pan	Surface Area Covered	Liquid Temperature	Plume Temperature	Sampler Location	H_2SO_4	HCl
			s	%	Watts/m^2		deg C	deg.	m/s		lb	lb	lb	ft^2	deg C	deg C	ft	mg/m^3	mg/m^3
1	5/13/03	15:24	328	7.5 ± 0.2	633 ± 67	D	28.0 ± 0.2	236 ± 11	7.5 ± 0.2	65% Oleum	132	0	154	36	117 ± 26	89 ± 15	10	491	
																	25	700	
																	50	21	
																	100	79	
																	500	5	
2	5/13/03	15:51	418	7.1 ± 0.2	539 ± 21	D	28.1 ± 0.2	236 ± 10	8.0 ± 0.8	65% Oleum	130	10	140	38	175 ± 44	116 ± 34	10	782	
																	25	360	
																	50	118	
																	100	6	
																	500	5	
3	5/13/03	16:21	262	7.0 ± 0.2	448 ± 96	D	28.1 ± 0.2	239 ± 10	8.0 ± 1.1	CSA	180	0	195	44	122 ± 47	85 ± 17	10	6536	7113
																	25	378	294
																	50	32	15
																	100	6	7
																	500	2	4
4	5/13/03	16:49	260	7.0 ± 0.2	460 ± 112	D	28.2 ± 0.1	228 ± 11	8.5 ± 1.3	65% Oleum	150	37	190	45	125 ± 33	130 ± 41	10	5418	
																	25	972	
																	50	209	
																	100	69	
																	500	2	
5	5/13/03	17:14	6	6.3 ± 0.1	422 ± 105	D	28.3 ± 0.3	230 ± 5	8.5 ± 1.0	65% Oleum	155	0	176	40	162 ± 59	92 ± 29	10	687	
																	25	426	
																	50	46	
																	100	8	
																	500	2	
6	5/15/03	12:24	740	21.0 ± 0.5	1063	C	25.3 ± 0.2	218 ± 13	6.8 ± 1.1	65% Oleum	208	0	224	46	91 ± 10	105 ± 10	10	137	
																	25	147	
																	50	21	
																	100	7	
																	500	1	
7	5/15/03	13:12	460	19.6 ± 0.4	1063	C	26.3 ± 0.3	213 ± 13	6.8 ± 0.9	CSA	214	0	231	47	106 ± 10	82 ± 1	10	216	328
																	25	147	373
																	50	13	13
																	100	3	6
																	500	1	1
8	5/15/03	14:41	288	17.1 ± 0.2	930	D	27.3 ± 0.3	190 ± 15	4.7 ± 1.1	CSA+Oleum	159 + 8	0	189	41	141 ± 62	88 ± 20	10	475	542
																	25	128	127
																	50	0	0
																	100	7	3
																	500	0	0
9	5/15/03	15:45	300	15.4 ± 0.2	788 ± 22	D	28.0 ± 0.1	225 ± 11	7.1 ± 0.8	CSA	234	0	245	47	139 ± 73	79 ± 11	10	1023	1053
																	25	209	392
																	50	43	22
																	100	14	10
																	500	4	2

Table 5: Field data collected during spill tests conducted in May 2006

Test	Date	Start Time	Duration	Relative Humidity	Solar Radiation	Atmospheric Stability	Ambient Temperature	Wind Direction	Wind Speed	Chemical Spilled	Chemical Mass Added Into Pan	Water Added Before Chemical	Maximum Mass in Pan	Surface Area Covered	Liquid Temperature	Plume Temperature	Sampler Location	H_2SO_4	HCl
			s	%	Watts/m²		deg C	deg.	m/s		lb	lb	lb	ft²	deg C	deg C	ft	mg/m³	mg/m³
10	5/2/06	12:45	120	13.9 ± 0.2	1090	D	28.3 ± 0.1	210 ± 12	4.7 ± 1.2	65% Oleum	125	0	136	35	130 ± 20	97 ± 7	12	376.13	
																	50	202.30	
																	100	11.67	
																	250	1.80	
																	500		
11	5/2/06	13:11	572	14.2 ± 0.2	1025 ± 11	D	28.3 ± 0.2	225 ± 28	4.5 ± 1.5	CSA	133	0	156	40	100 ± 26	83 ± 2	12	348	270
																	50	44	34
																	100	2	2
																	250	0	0
																	500	0	0
12	5/2/06	13:46	404	13.3	1092	D	30.5	225	8.2	CSA	132	0	155	40	92 ± 2	80 ± 1	12	363	122
																	50	67	171
																	100	5	5
																	250	0	0
																	500	0	0
13	5/2/06	14:14	610	13.3	1092	D	30.5	225	8.2	65% Oleum	139	0	171	43	146 ± 25	92 ± 7	12	2450	
																	50	67	
																	100	4	
																	250	0	
																	500	1	
14	5/2/06	14:40	356	13.3	1092	D	30.5	225	8.2	65% Oleum	136	0	166	41	144 ± 17	89 ± 4	12	913	
																	50	104	
																	100	11	
																	250	1	
																	500	1	
15	5/2/06	15:44	8	13.3	1092	D	30.5	225	8.2	65% Oleum	128	0	161	38	134 ± 24	90 ± 14	12	843	
																	50	88	
																	100	8	
																	250	1	
																	500	1	
16	5/3/06	14:48	12	13.5	1025	D	26.5	220	8.3	65% Oleum	140	0	155	43	103 ± 9	77 ± 6	12	1346	
																	50	103	
																	100	12	
																	250	1	
																	500	1	
17	5/3/06	15:16	310	13.7 ± 0.3	850 ± 73	D	26.9 ± 0.3	218 ± 8	8.3 ± 0.8	65% Oleum	118	0	136	33	142 ± 16	83 ± 8	12	494	
																	50	197	
																	100	18	
																	250	1	
																	500	4	
18	5/3/06	16:29	7	14	820	D	26.8	208	8.34	65% Oleum	150	118	300		263 ± 57	114 ± 35	12	1817	
																	50	261	
																	100	16	
																	250	2	
																	500	5	

7. Discussion and Summary

At the NPTEC facility, on a clear and bright afternoon when there are no thunderstorms the wind direction and speeds are always in a predictable range. The expected patterns develop after the lake bed heats up during the day. So the spill tests are allowed only in the afternoons provided there are no weather disturbances in the larger NTS site. For the days when the spill tests were conducted in May 2003 and 2006 the wind speeds ranged from a low of 3 m/s to a high of almost 10 m/s in a predictable direction 180 to 235 degrees direction. Even though the air is fairly hot and dry, the relative humidity of more than 6% provides enough moisture in the atmosphere to react with SO_3 and CSA in the pan and in the air.

The surface of the pan is generally uneven and typically covered by salt residues left after neutralization. For the different amounts of chemicals spilled, the surface area of the spill has been estimated and listed in Tables 4 and 5. This surface area affects the vaporization rate of the chemical spilled. For most other chemicals, the difference in weight of pan over a period of time would provide the average vaporization rate. However, since oleum and CSA are violently reactive with water, the weight in the pan seemed to fluctuate over the sampling time period. The maximum mass in the pan was found to be higher than the amount of chemical spilled into it. The only explanation for it appears to be that the increase in mass results from absorption of moisture from the atmosphere.

The reaction of the oleum and/or CSA with available water, as shown in Equations 1 and 2 previously, results in an increase in the bulk liquid temperature. The higher bulk liquid temperature results in a vaporization rate that would be higher than if the liquid were at ambient temperatures. The highest temperatures were observed when a known amount of water was added to the pan before the spillage of oleum or CSA. This represents the vaporization behavior that can be expected when these chemicals are spilled on to ground that is wet.

The acid mist that is formed when the vapors evolving from a pool react with the moisture in the atmosphere is also at a high temperature as shown in Tables 4 and 5. The higher than ambient temperatures would cause the acid mist plume to be buoyant under low wind speed conditions as shown in Figure 1. At high wind speeds observed during the tests the plume is bent more towards the ground and the ground-level concentrations are expected to be higher.

Tables 4 and 5 also show the concentrations obtained from analysis of filters located at five distances up to 500 ft from the edge of the pan along the centerline (225 degrees direction) for each case. Some differences are observed for the two sets of data obtained in May 2003 and May 2006. Detailed analysis and quality checks have yet to be conducted. These data are preliminary but do show that within a distance of 500 ft the concentrations are below ERPG-2 and AEGL-2 levels. Concentrations near the release location are much greater than ERPG-3 and AEGL-3. There were problems with instruments (i.e., less than expected variability in weather data for some cases) and analyses of filters from observations in May 2006.

In spite of problems typical of a field study, this work has shown that the reaction of oleums and CSA with water is very important and causes high bulk liquid and plume temperatures. It is very important for the pool vaporization and acid plume dispersion models to account for the associated effects (e.g., higher vaporization rates and plume buoyancy) resulting from the higher than expected temperatures. In future, more detailed analysis will be conducted of the data to develop a better understanding of the vaporization and dispersion behavior of plumes resulting from spills of 65% oleum and CSA.

8. References

[1] Grint, G., and G. Purdy, "Sulfur Trioxide and Oleum Hazard Assessment," Journal of Loss Prevention in the Process Industries, Volume 3, pp 177-184, 1990.

[2] Singh, M.P., M. Kumari, and S. Ghosh, "A Mathematical Model for the Recent Oleum Leakage in Delhi," Atmospheric Environment, Volume 24A, No. 4, pp 735-741, 1990.

[3] Kapias, T., and R.F. Griffiths, "A Model for Spills of SO3 and Oleum, Part I, Model Description," Journal of Hazardous Materials, Volume 62, No. 2, pp 101-129, 1998.

[4] Kapias, T., and R.F. Griffiths, "A Model for Spills of SO3 and Oleum, Part II, Results, Conclusion, and Discussion," Journal of Hazardous Materials, Volume 62, No. 2, pp 130-142, 1998.

[5] Kapias, T., Griffiths, R.F., and C. Stefanidis "REACTPOOL: A Code Implementing a New Multi-compound Pool Model That Accounts for Chemical Reactions and Changing Composition for Spills of Water Reactive Chemicals," Journal of Hazardous Materials, Volume A81, pp 1-18, 2001.

[6] HSE, *Modelling the behaviour of spillages of sulfur trioxide and oleum*, U.K. Health and Safety Executive (HSE) Report CRR217/1999, ISBN 0717624404, 1999.

[7] HSE, *Modelling the behaviour of spillages of sulfur trioxide and oleum: Further Work*, U.K. Health and Safety Executive (HSE) Report CRR257/2000, ISBN 0717617343, 2000.

[8] Aspen Custom Modeler, Software from Aspen Technology Inc., Cambridge, Massachusetts, U.S.A.

[9] *DuPont Sulfur Trioxide and Oleum: Properties, Uses, Storage, and Handling*, DuPont Chemical Solutions Enterprise, Customer Service Center, Barley Mill Plaza, Bldg. 23, Wilmington, DE 19898, 2002.

[10] Kirk-Othmer Encyclopedia of Chemical Technology, "Sulfuric Acid and Sulfur Trioxide, 4th Edition, Volume 22, John Wiley and Sons, 2003.

[11] Schrage, J., "Vapor-liquid Equilibrium of the System H2SO4-SO3: Part I. Vapor Pressure Measurements with a New Static Vapor Pressure Apparatus," Fluid Phase Equilibria, Volume 68, No. 30, pp 229-245, 1991.

[12] Nilges, J., and J. Schrage, "Vapor-liquid Equilibrium of the System H2SO4-SO3: Part II. Thermodynamic Description with Regard to the Formation of H2S2O7," Fluid Phase Equilibria, Volume 68, No. 30, pp 247-261, 1991.

[13] *Emergency Response Planning Guideline (ERPG) Handbook*, American Industrial Hygienists Association, U.S.A.

[14] *Acute Exposure Guideline Levels (AEGL)*, U.S. Environmental Protection Agency, (http://www.epa.gov/oppt/aegl/index.htm), 2007

LOSS
PREVENTION
SYMPOSIUM

Estimating the Consequences of Deflagration to Detonation Transition (DDT) in Hydrogen Explosions

Prankul Middha and Olav R. Hansen

GexCon AS
P. O. Box 6015 Postterminalen
NO-5892 Bergen Norway
Phone: +47-55-57-4318
prankul@gexcon.com

Prepared for Presentation at
American Institute of Chemical Engineers
2007 Spring National Meeting
41st Annual Loss Prevention Symposium
Houston, Texas
April 22–26, 2007

UNPUBLISHED

Estimating the Consequences of Deflagration to Detonation Transition (DDT) in Hydrogen Explosions

Prankul Middha and Olav R. Hansen
GexCon AS, P.O. Box 6015 Postterminalen
Bergen, N-5892, Norway

Abstract

Due to the development in computational resources, Computational Fluid Dynamics (CFD) has assumed increasing importance in recent years as a tool for predicting the consequences of accidents in petrochemical and process industries. CFD has also been used more and more for explosion predictions for input to risk assessments and design load specifications. The CFD software FLACS has been developed and experimentally validated continuously for more than 25 years. As a result, it is established as a tool for simulating hydrocarbon gas deflagrations with reasonable precision and is widely used in petrochemical industry and elsewhere. In recent years the focus on predicting hydrogen explosions has increased, and with the latest release the validation status for hydrogen deflagrations is considered good.

However, in many of these scenarios, especially involving reactive gases such as hydrogen, deflagration to detonation transition (DDT) may be a significant threat. In previous work, FLACS was extended to identify whether DDT is likely in a given scenario and indicate the regions where it might occur. The likelihood of DDT has been expressed in terms of spatial pressure gradients across the flame front. This parameter is able to visualize when the flame front captures the pressure front, which is the case in situations when fast deflagrations transition to detonation. Reasonable agreement was obtained with experimental observations in terms of explosion pressures, transition times, and flame speeds. The DDT model has now been extended to develop a more meaningful criterion for estimating the likelihood of DDT by comparison of the geometric dimensions with the detonation cell size. This article will discuss the new models to predict DDT, and compare predictions with relevant experiments.

1. Introduction

The CFD-simulator FLACS has been developed since 1980. Significant experimental validation and related computational development activity driven by a series of gas safety research programs and partly motivated by efforts to avoid serious accidents such as Piper Alpha [1] has contributed to the wide acceptance of FLACS as a reliable tool for risk assessment and explosive safety studies in real process areas offshore and onshore. For natural gas deflagration calculations several hundred experiments have been simulated using FLACS as part of validation exercises in the 1990s [2]. Similar studies have been carried out to ensure good performance of FLACS in other areas, including far-field blast propagation from deflagrations [3,4], dispersion of flammable gas and ignition of non-homogeneous gas cloud relevant for QRA-studies towards offshore oil and gas industry [5], dispersion of tracer gas [6] and dust explosion simulations with DESC, a tool based on FLACS [7].

Recently, there has been an increased focus on predicting the safety aspects of various hydrogen applications. One driving force is the increasing attention from the nuclear industry, particularly due to accidents like Chernobyl and Three Mile Island, and the potential increase in use of nuclear power. However, the main reason for a strong interest in hydrogen safety is the ongoing development of hydrogen transport vehicles and related infrastructure, such as refuelling stations. This development has been further accelerated by the growth in global awareness about the impact of greenhouse gases and the finite nature of fossil fuel reserves. Therefore, a strong effort has been made in recent years to learn more about hydrogen explosions and improve FLACS in that area. In 1998, a project funded by oil companies was carried out to study the effect of inert gas dilution on deflagrations, which also included some small-scale hydrogen experiments in a channel with obstacles [8]. In 1999, more hydrogen explosion experiments with varying gas concentration, obstacle height and number, ignition position, and confinement were carried out with partial funding from the Norwegian Research Council (NFR). In 2001, a 3-year research project with industry funding from Statoil, Norsk Hydro and IHI was established to improve the validity of FLACS when simulating hydrogen. A significant improvement in performance was obtained through a large number of varied small-scale experiments combined with simulations and model improvements. GexCon also got access to various large-scale hydrogen explosion tests such as FLAME facility experiments carried out by Sandia [9] and 20m hemispherical deflagration tests from Fraunhofer-ICT [10]. More details on the validation efforts are presented in Reference 11. There is still a major focus on hydrogen safety at GexCon, with involvement in the EU-sponsored HySafe network of excellence and IEA task 19 on hydrogen safety. In both these international projects, our main focus is carrying out applied CFD calculations and developing a comprehensive procedure for risk assessment of hydrogen applications.

Due to the high reactivity of hydrogen, a transition from deflagration to detonation (DDT) is likely in a variety of scenarios involving H_2-air mixtures and can be a significant threat and result in large-scale damage. Many of these situations are commonly employed in industrial settings. Although this phenomenon has been studied for around 70 years by various researchers (e.g. [12,13]) and many different explanations have been propounded, complete theoretical description is still lacking. Most commonly, DDT is understood to happen due to the merging of shock fronts caused by expanding burned gases and those resulting from autoignition of preheated unreacted mixture [9]. However, it has been shown by Oppenheim that in many cases, the transition can begin with a "local explosion" in a region of high turbulence, even though the compression heating of the unburned gas due to shock waves is not sufficient [14]. The interactions of flame with the wall have also been found to be important. Lee and co-workers have proposed the SWACER concept where chemical induction time gradients (associated with temperature and pressure gradients) produce a compression wave that can gradually get amplified into a strong shock wave. This shock wave can auto-ignite the mixture and produce DDT [15]. In addition to the mechanisms presented above, there are some other theories such as instabilities caused by sudden venting [16] which can lead to DDT.

Although the study of the phenomena is still an active research area, there is no doubt that it is very important to study this phenomenon from a process safety perspective as it can lead to some very adverse consequences. A lot of theoretical effort has been focused on development of criteria for DDT [17]. Researchers have tried to establish the initial and boundary conditions, and geometric and process parameters under which a transition to detonation can be expected. However, these criteria and scaling arguments are difficult to apply in a process setting where exact numerical results are much more valuable.

Although the validation of the current version of FLACS in simulating flame acceleration and high-speed deflagrations is good, a detonation model has been lacking. We are currently involved in an activity sponsored by NFR that aims at predicting the extent to which DDT may be expected using FLACS. As a part of this work, many practical scenarios have been simulated previously and the simulation results have been found to compare well with experimental data [18]. More details are presented in Section 3. The model has now been developed further and used to simulate various scenarios reported in literature where these high-speed deflagrations transition to detonation. Comparisons have been performed in terms of pressure traces, location and time of DDT, and the possibility of propagation of a detonation front. The main goal is to carry out hydrogen explosion simulations in various practical scenarios due to the increased need for understanding hydrogen safety.

2. Brief description of FLACS

A brief description of the various features of the CFD program FLACS is presented in this section. In principle most CFD-models are similar, with more or less the same conservation equations solved on a numerical grid, but the disparities in numerics can cause differences in accuracy and stability. In FLACS, second order schemes (Kappa schemes with weighting between 2nd order upwind and 2nd order difference, with delimiters for some equations to improve accuracy and stability) are used to solve the compressible conservation equations on a 3D Cartesian grid using a finite volume method [19,20]. The explosion experiments that were used to develop and validate FLACS initially are also published [21,22]. FLACS uses a standard k-ε model for turbulence with some important modifications, including a model for generation of turbulence behind sub-grid objects and a model for flame folding around them [23]. With the close coupling between sub-grid modelling and turbulence model, it is not believed that using a more advanced turbulence model with more equations and constants will improve accuracy significantly.

A model for development of the flame that describes how the local reactivity changes with parameters like concentration, temperature, pressure, turbulence, etc. is implemented for modelling combustion and explosion. A good description of geometry and the coupling of geometry to the flow, turbulence, and flame is one of the key elements in the modelling. The real flame area is described properly and corrected for curvature at scales equal to and smaller than the reaction zone. All flame wrinkling at scales less than the grid size is represented by sub-grid models, which is important for flame interaction with objects smaller than the grid size.

The representation of geometry using a distributed porosity concept is one of the key advantages of FLACS compared to several other CFD tools. FLACS can therefore be used to simulate a variety of complicated geometries (including cylindrical and other curvilinear elements) using a Cartesian grid. Large objects and walls are represented on-grid; smaller objects are represented sub-grid. Sub-grid objects contribute to flow resistance, turbulence generation, and flame folding in the simulation. For other CFD models without sub-grid representation of geometry, it is unrealistic to properly describe complex process areas.

3. Previous work on DDT at GexCon

The currently available commercial version of FLACS can predict fast deflagrations well, but lacks models to predict a transition to detonation. In the fast deflagration mode, the flame propagates because hot flame products heat and ignite the unburned gas ahead. Turbulence caused by obstructions increases the flame velocity and resulting overpressures. However, in situations with very fast flames and/or very strong pressure waves, DDT may change the propagation mode so that shock waves will ignite unburned gas ahead and lead to much faster flame propagation (1500-2000 m/s). This propagation mode is not handled by FLACS directly. For hydrogen explosions, DDT may be a significant threat. This is primarily a result of very high reactivity. As described above, the study of this phenomenon is very critical as the overpressures in a detonation front can cause losses which are much more severe as compared to those seen with blast waves produced by a deflagration.

In previous work that was presented in the last Loss Prevention Symposium in Orlando in April 2006, four practical systems were analysed for the likelihood of fast flames transitioning to detonation [18]. DDT was seen in all these systems experimentally and we extended the FLACS code to be able to predict the transition. Comparisons with experimental data in terms of the time and location of DDT were also performed. The likelihood of DDT was illustrated in terms of a parameter proportional to the spatial pressure gradient across the flame front (see Figure 1).

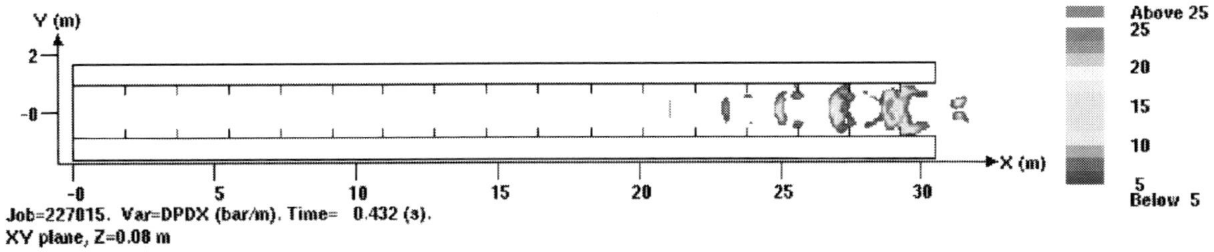

Figure 1. Spatial Pressure gradients in the XY plane for the FLAME facility [9] for test 22 indicating a possibility of DDT.

This parameter is able to visualize when the flame front captures the pressure front, which seems to correlate well with time for experimentally observed DDT. [18,24]. It was proposed that the presence of these spatial pressure gradients represents the indication of a possibility of the deflagration front transiting to detonation. Numerical simulations in diverse geometries in this work revealed that FLACS is able to pick up significant pressure gradients in areas where detonations have been observed in experiments. We have attempted to simulate other large-scale experiments available in literature such as the experiments in the FLAME facility performed at Sandia National Laboratories [9]. In general, the simulations are able to predict detonation pressures, transition times, and run-up distances quite well. The front speeds were found to be somewhat lower but still the right order of magnitude. We are therefore confident that this parameter can be used as an indication of the possibility of DDT in novel systems. More details can be found in Reference 18. It must be mentioned that there may also be other DDT mechanisms, for instance high-energy releases or local flame disturbances in connection to venting, that will not be picked up in the modelling.

4. Detonation Cell Size

Before presenting the current work on DDT modeling, the significance of detonation cell size in postulating the mechanisms and explanations for detonation initiation is discussed. It has been observed experimentally that the structure of a detonation front consists of a cell pattern, with a characteristic length scale that is the detonation cell size, λ. The detonation cell size is used as a measure of mixture sensitivity and has traditionally played a very important role in the development of theories to explain the onset of DDT in many different scenarios. This allows scaling of DDT conditions found for different mixtures and compositions. The cell size data themselves are available for many different conditions of temperature, pressure, and concentrations. However, it must be noted that there is a large uncertainty (up to a factor of 2 is not uncommon) in the experimental cell size data, and therefore, they should be used carefully. It should also be kept in mind that the detonation cell size cannot be considered as a fundamental mixture property (such as laminar flame speed). Therefore, its application as a scaling parameter needs to be validated experimentally. However, for systems involving hydrogen combustion (e.g. H_2-air, H_2-air-steam, H_2-air-CO_2 mixtures at both normal and elevated initial temperatures), the detonation cell size has been shown to be a relevant scaling parameter [18].

A detailed study of obstructed tubes (with diameters ranging from 5-30 cm) was conducted at McGill University in the 1980s with varying blockage ratio (BR), fuel type, and mixture composition [25,26]. In this study, the detonation cell size was used as a measure of the sensitivity of a given fuel mixture to detonation initiation. It was found that a size of the unobstructed passage, *d*, greater than 1λ was required for DDT to occur. However, this criterion was found to be necessary but not sufficient as large variations in the critical d/λ ratio were seen depending on the obstacle spacing and BR. Despite this limitation, it is still an important indication of the effect of scale on DDT. The cell size increases significantly as the mixture concentration deviates from stoichiometry, and hence a larger scale will support the occurrence of DDT for a larger composition range.

A large number of studies have been performed to formulate a criterion for the onset of detonation in typical geometrical enclosure found in process facilities [e.g., 27-29]. It has been proposed as $L > 7\lambda$, where L can be defined as a characteristic size of a room that is filled with a flammable mixture or the size of a gas cloud. Good agreement with the L/λ criterion has been seen by comparison with experimental data. However, there has been some ambiguity in the definitions of L, but in any case, this criterion has been found to be valid over a wide range of conditions.

Both the criteria described above are based on the comparison of a characteristic geometrical size with the characteristic mixture sensitivity scale (λ). This has been implemented in the FLACS CFD code and further details are presented below. The difference between the two criteria arises from different definitions of characteristic geometrical sizes. The first criterion ($d > \lambda$,) sets a limit on the transverse size of unobstructed passage in a channel for DDT to be initiated while the second criterion ($L > 7\lambda$) prescribes the minimum distance for detonation formation. Thus, the requirements can be considered as complementary for evaluating a particular geometrical configuration with regards to the possibility of a transition to detonation during the combustion of a reactive mixture such as near stoichiometric H_2-air. Thus, detonation cell size, λ, can act as a valuable parameter in making safety decisions. Further, recent experiments [30] have also shown that the initiation of spherical detonations by a turbulent jet can also be scaled using the detonation cell size. There have been many other studies verifying the above criteria and more details can be found in Reference 17.

5. Current Work

5.1 Introduction

In order to put the current DDT work into perspective, it must be noted that the description of DDT by FLACS can only be characterized as an "average" description. However, this is still valuable as an exact representation of DDT in a typical hazardous industrial scenario involves scales that differ by up to 10 orders of magnitude. An exact simulation for such a 3D system will take many years based on current computational power. Reasonably precise representations of the transition to detonation do exist (see [31] for a review) but these are mostly confined to small, 2D systems. Even this computation requires special, multidimensional numerical simulations that cannot be carried out for risk analyses that are required on an almost daily basis in process industry. We do not know of any other tool that can currently predict DDT in industrial systems with reasonable precision. As FLACS has been proven to be able to predict high-speed deflagrations with good accuracy, the description of DDT in terms of the region and time of occurrence and the resulting overpressures can be expected to be reasonable. However, the speed of the flame front may be too low, but the development of a shock ignition model is ongoing which is expected to resolve this discrepancy.

5.2 New models

As stated above, the spatial pressure gradient across the flame front has been used previously to indicate the likelihood of DDT. However, the value of this spatial pressure gradient was found to differ significantly (by up to an order of magnitude) depending on the scenario and geometry. It was also found to be dependent on grid resolution. An attempt has therefore been made to normalize and this key parameter and make it dimensionless. The normalization has been tested through FLACS simulations and the dimensionless spatial pressure gradient can be expressed as:

$$\left.\frac{dP}{dx}\right|_{normalized} = \left.\frac{dP}{dx}\right|_{actual} \frac{X_{CV}}{P_0} \qquad (1)$$

where P_0 is the initial pressure and X_{CV} represents the grid resolution. It is proposed with this scaling that a magnitude of the pressure gradients of order 1 indicates that DDT is likely, with values around 10 or larger indicate very strong possibility of DDT.

The previous model had some other shortcomings. It was missing any comparison of the detonation cell size with geometrical dimensions. Therefore, it was possible for the simulations to predict the occurrence of DDT in geometries that may have been too small for any detonation front to propagate. Also, it did not consider whether sufficient fuel was present for any possible DDT to develop into a detonation. This means that DDT and hence possible detonation would be indicated even if most fuel were exhausted. These shortcomings have been improved in the current work by introducing a parameter known as the "DDT Length Scale" (hereby referred to as DDTLS) that compares the relevant geometric length scale in FLACS (LSLIM) for a given geometry with the detonation cell size (which is based on the concentration of the fuel region).

Mathematically, $DDTLS = \dfrac{LSLIM}{\lambda}$ and based on the discussion in the previous section, it is proposed that a value of 7 or larger would ensure the propagation of any initiated detonation front (for obstructed tubes, a value of the order of 1 is probably sufficient). The advantage of this parameter is that it dynamically updates the concentration and hence the detonation cell size in the simulation domain as the fuel is consumed by the flame. Also, it is able to indicate a zero or very small value for DDTLS (and hence no possibility of DDT) where none or very dilute fuel is present. Therefore, this parameter along with the normalized spatial pressure gradient gives a much better indication of the likelihood of DDT.

In order to implement the calculation of DDTLS, we need an analytical model for detonation cell size, λ. There have been fairly extensive measurements of detonation cell size for hydrogen combustion systems carried out by various researchers. The results have been summarized in a detonation database at California Institute of Technology [32]. However, the experimental data is available only for some particular compositions and initial conditions. These measurements do not allow direct estimation of the cell size for any arbitrary composition and initial conditions. Therefore, analytical and numerical tools are necessary to give reliable values of detonation cell size for risk and accident analyses. One of the approaches has been based on fitting of experimental data by an analytic function and that is what has been used in FLACS. The analytic function can be used to interpolate experimentally available detonation cell size data. However, it must be realized that such a fit is valid only within the range of experimental data. Many different functions have been applied, but the one that gives the least deviation from the experimental data is presented below [33]:

$$\log(\lambda,\mathrm{cm}) = \left(a - m + \left(\frac{b}{\left(A - \frac{k}{B}\right)^{f}} + h\cdot(A - g\cdot B)^{2} + i\cdot(A - g\cdot B) \right)\cdot\left(1 + d\cdot C + e\cdot B\cdot C^{2}\right)\cdot\frac{j}{B} \right)\cdot \qquad (2)$$

$$\cdot(D-c)\cdot\left(\frac{1}{0.1-c} + n\cdot(D-0.1)\right) + m$$

Parameters:

$a = -1.13331E{+}00$	$b = 4.59807E{+}01$	$c = -1,57650E{-}01$
$d = 4.65429E{-}02$	$e = 3.59620E{-}07$	$f = 9.97468E{-}01$
$g = -2.66646E{-}02$	$h = 8.74995E{-}04$	$i = -4.07641E{-}02$
$j = 3.31162E{+}02$	$k = -4.18215E{+}02$	$m = 2.38970E{+}00$
$n = -8.42378E{+}00$		

Variables:

A: Dry hydrogen concentration, vol. %

B: Initial temperature, *T*, Kelvin

C: Steam concentration, vol. %

D: Initial pressure, MPa

This function is used in FLACS for the calculation of the detonation cell size, and hence the DDT length scale in order to estimate the likelihood of DDT.

5.3 Calculation Results

Next, we present some results of using the current approach. All the simulations have been carried out with the latest available commercial version of FLACS, FLACS v8.1, which has been modified to include the above, mentioned models. Grid embedding and other recommended guidelines have been followed and extensive QA has been performed to ensure the accuracy of the results.

5.3.1 McGill Detonation Tubes

It would be instructive to apply the new models to the calculations presented in our previous work to demonstrate the methodology. Therefore, we have resimulated the detonation tube experiments carried out at McGill University, which are briefly described in the previous section [26]. As presented in Reference 18, these experiments were performed in three long (18m each) tubes of circular cross-section (diameter, *D* = 5cm, 15cm, and 30cm, respectively) closed at both ends. Ignition was at one of the closed ends. The tubes contained an array of circular orifice rings placed at the inner wall of the tube and running the entire length (spacing one diameter; blockage ratio (BR) = 0.43, 0.39, and 0.43 for *D* = 5, 15, and 30cm respectively). The previous results are summarized in Figure 2 where the spatial pressure gradients in the XY plane are presented for all three tube diameters. The results were found to be representative of experimental observations in the sense that large pressure gradients were seen very close to the ignition end for the cases in which a transition to detonation was observed. However, quite different magnitudes were seen for the pressure gradients for the three different diameters (up to 500 or larger).

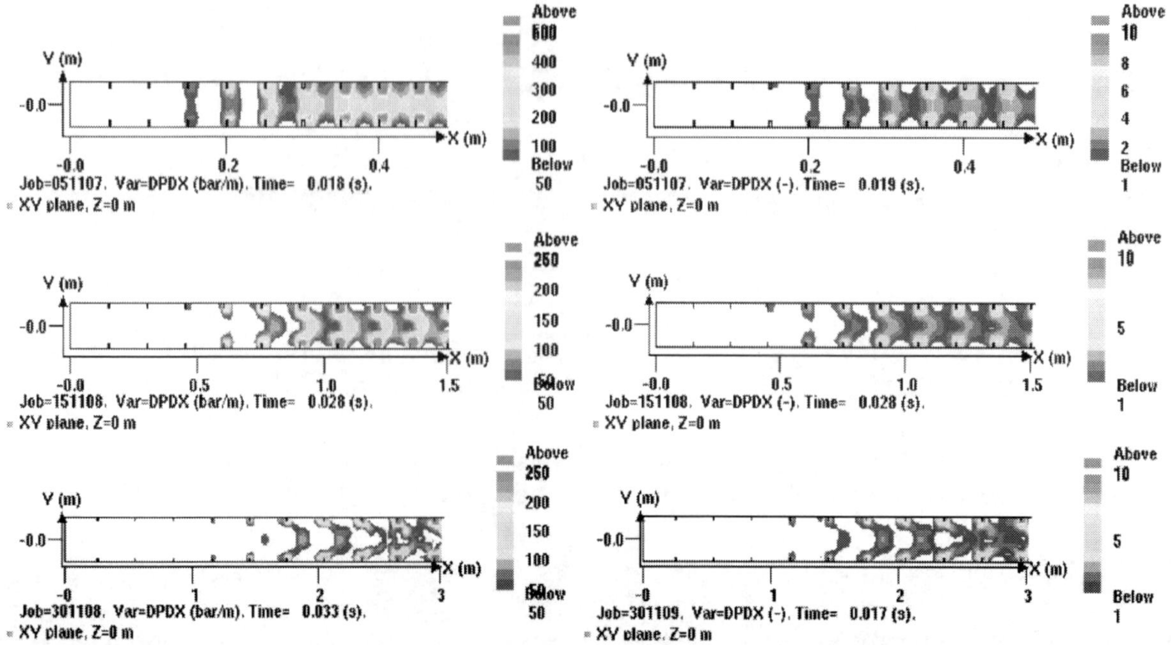

Figure 2. Pressure gradients in the XY plane in the detonation tube [26] for tube diameter of 5 cm (top), 15 cm (middle), and 30 cm (bottom). The left figures show the results from our previous work and the right figures show the normalized pressure gradients based on the current model.

Figure 2 also presents the same simulations carried out with the current model for calculating the spatial pressure gradient (right figures) and it can be seen that the values for all three cases lie in the range from 1 to 10 as proposed above. Strongest likelihood of DDT is seen for the 30 cm tube near the baffles. Hence, it can be concluded that the scaling of the pressure gradient is appropriate.

Next, we demonstrate the calculation of the DDT Length Scale (DDTLS) illustrated above for the 30 cm detonation tube (see Figure 3) for two different times. It can be seen from that the values of DDTLS are larger than 10 where fuel is present but they are set to zero as the fuel is consumed. Thus, it can be concluded that DDT is likely in this case and the resulting detonation front can easily propagate. It can also be seen that the maximum values of DDTLS change from around 20 for the 30 cm tube to less than 4 for the 5 cm tube. Although the transverse size of the unobstructed passage, *d* is sufficient for the 5 cm tube (as shown by DDTLS > 1), it can be shown that the characteristic distance *L* which prescribes the minimum distance for formation of a detonation front is somewhat less than 7λ in this case (It has been shown that for obstructed tube, the characteristic distance that scales the DDT experimental observations best is given by $L = D/(1 - d/D)$ which can be calculated based on BR and DDTLS values). Thus, the formation of three dimensional cell structure seen in detonation fronts is somewhat difficult for the smallest tube, and a detonation if formed may decay to deflagration.

We have also simulated other experiments carried out in obstructed tubes. Comparisons between simulations and experimental data have been carried out for the two different tube experiments of Kuznetsov et al. [34,35] and Chan & Dewit [36]. In general, reasonable results are seen and some details have been published previously [18] for the experiments described in Reference 34.

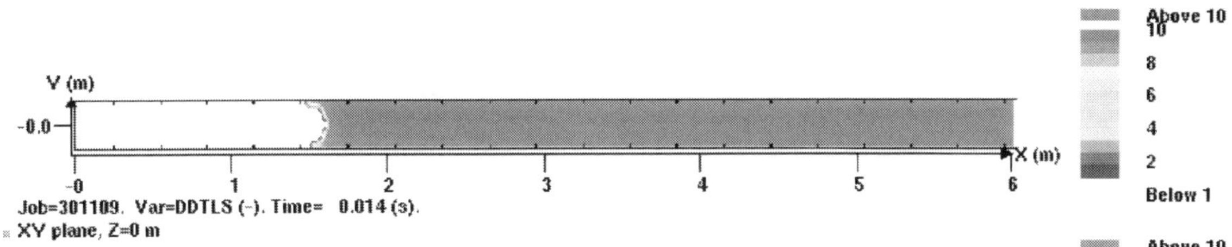

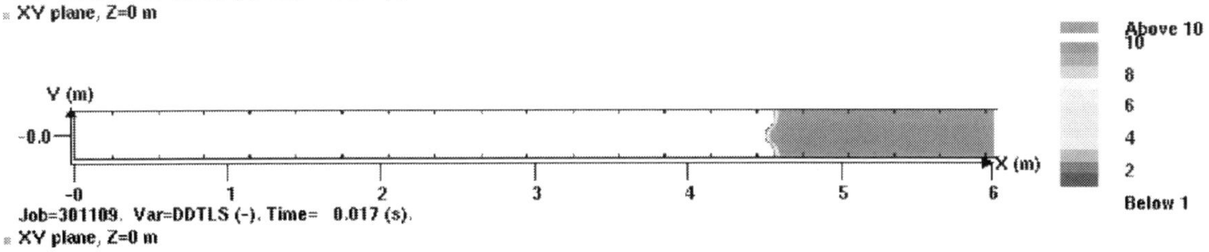

Figure 3. DDT Length Scale in the XY plane in the detonation tube [26] for tube diameter of 30 cm for two different times: 14 ms (top) and 17 ms (bottom).

5.3.2 GexCon 1.44m Explosion Channel

In principle, it is possible to apply this approach to any geometrical configuration. Next, we present some results from small-scale channel experiments conducted at GexCon in early 2003. The channel is a 1.44 × 0.3 × 0.3 m^3 rectangular channel with one open end. Baffles can be placed along the channel to generate turbulence [37]. The number of baffles could be varied, and the results presented here simulated the set-up with 4 baffles. Ignition was at the closed end of the channel. A schematic view of the experimental set-up with all the pressure sensors is presented in Figure 4. The simulations were carried out with a "fine" grid resolution of 1.67 cm, which corresponds to 18 grid cells across the width and height of the channel. Two different hydrogen concentrations were considered for the simulations (Equivalence ratio = 0.7 and 1.0). The results from this channel serve to validate the methodology further for the cases where DDT is not observed experimentally, but high-speed deflagrations are seen.

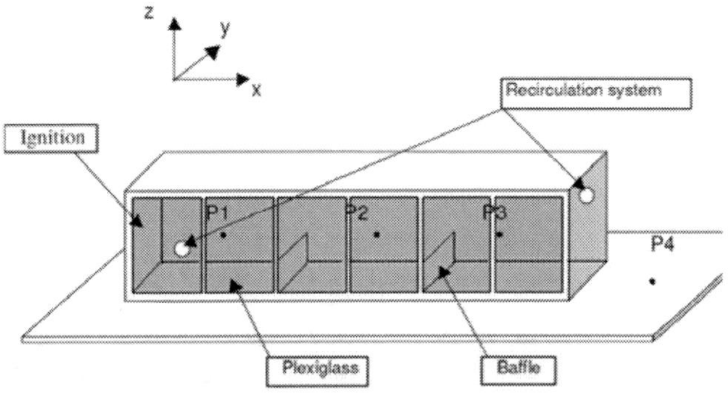

Figure 4. Schematic view of the 1.44m explosion channel [37]. The 4 pressure sensors are also shown.

The simulation results for the lean hydrogen mixture are shown in Figure 5. The normalized spatial pressure gradient in the XZ plane at the centre of the channel is shown in the top picture at 16 ms. It can be seen that the values of the pressure gradients seen in this case are significantly smaller than 1 and the maximum values are around 0.25. The plot of DDTLS at the same location and time shows that it is possible to propagate a detonation front if it is initiated but it is unlikely in this case due to the fact that the pressure gradients across the flame front are small. This agrees with the experimental observations where no indications of a detonation front were seen, although quite high pressures (around 3.5 barg) were observed.

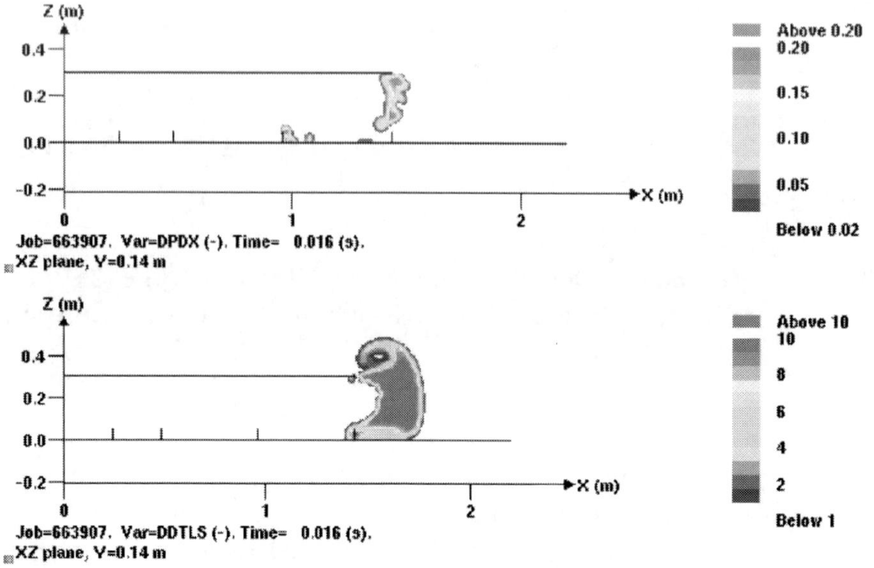

Figure 5. Normalized spatial pressure gradients and the DDT length scale in the XY plane at t = 16ms for the explosion channel [37] with 4 baffles for H2-air mixture (ER = 0.7).

For the stoichiometric mixture, the spatial pressure gradients are somewhat larger, but still much smaller than 1. Therefore, DDT is expected to be not very likely in this case as well, although it is possible. Similar calculations as above can be drawn from the DDTLS plot. In general, the simulation results are able to capture the experimental measurements very well in terms of the calculated overpressures. Most importantly, the normalized spatial pressure gradient is able to correctly predict the experimental observations.

5.3.3 Fh-ICT Lane Experiments

This section presents the simulations of the experiments carried out at Fraunhofer Institut Chemische Technology (Fh-ICT) in 1984. These concerned the transition of a deflagration into a detonation in jet ignited hydrogen air mixtures within a partial confinement [38,39]. Similar experiments carried out by Dorofeev and colleagues [16] have also been modelled and reported previously [18]. These experiments are extremely significant as they are large-scale and can be expected to consider the physical phenomena that are important in real-life situations and accidents. The test set up consisted of a driver section that was a rectangular container (3 m × 1.5 m × 1.5 m). There was a square spaced opening in the front wall of the driver section with blocking ratio 0.1 (tests IA1, IA2, IA3) and 0.3 (tests IA4 and IA5). The container was followed by a "lane" which consisted of 2 parallel walls at a distance of 3 m with a length of 12 m and a height of 3 m (see Figure 6 for details). The whole volume was filled with H_2-air mixture. The mixture was ignited at the rear side of the container with 5 distributed pyrotechnic igniters. Two vertical tubes (diameter 14 cm) were installed with a distance of 5 cm from the wall each in the middle and at the end of the lane for all tests except test IA1. The relevant scenario parameters are presented in Table 1.

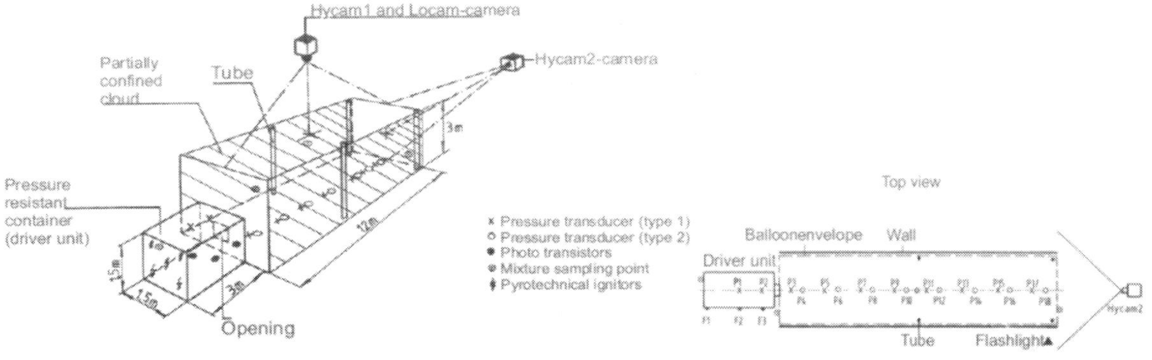

Figure 6. Fh-ICT Lane Experiments test facility, with details of sensors [38]. The tubes were not installed in test IA1.

Quite high pressures (greater than 10 bar) and a transition to detonation (in 2 of the tests) was observed in the experiments due to the high turbulence generated by a jet flame shooting into a large, reactive gas cloud followed by reflections of the high speed combustion front from the ground and the walls. The experiments observed DDT for 21% hydrogen concentration, but not for less sensitive mixtures. However, no DDT was seen for test IA1 where no tubes were installed even if the concentration was larger than 21% H_2. Also, DDT was observed to occur near the tube for test IA2 but near the ground for test IA4. No detonation was seen for test IA5. Four of the tests were simulated using the CFD program FLACS (except test IA3).

Table 1. Relevant scenario parameters for the four tests considered.

Parameters	IA 1	IA 2	IA 4	IA 5
Ambient Temp. (K)	279.2	281.5	293	293.4
Ambient pressure (bar)	0.988	0.991	0.993	0.996
H_2 concentration in driver unit (%)	21.9	20.8	22.3	19.9
H_2 concentration in lane (%)	21.0	21.1	22.5	20.0
DDT observed	No	Yes	Yes	No

The simulation domain was resolved by around 3.25 million grid cells with a grid resolution of 5 cm. The grid was according to the FLACS guidelines for explosion simulations. The simulations were carried out on a single processor of a LINUX PC with 3-4 GB RAM, and took 2 days to complete. It is seen that the simulations agree reasonably well with measurements in terms of the pressure measured at the sensors. The maximum simulated pressure in test IA2 was 10.2 barg at sensor 12, compared to 9.2 barg in the experiments also at sensor 12. The flame arrival times were calculated to be 50 and 90 ms at photo transistors F2 and F3, compared to the observed values of 50 and 80 ms. For test IA4, the maximum simulated pressure was 9 barg at sensor 8, compared to around 12 barg in the experiments also at sensor 8. In this case, the flame arrival times at F2 and F3 were calculated to be 35 and 52 ms, compared to observed values of 38 and 49 ms. The maximum pressure in the driver section was also found to compare very well with experimental results for all tests. More details on the comparison of pressure traces are presented in Reference 40.

2D snapshots of the pressure field at the ground, along with the flame and the spatial pressure gradient for test IA2 at different times with a detailed description of various different stages during the simulation are shown in Figure 7. The figure serves to show that the simulations are able to capture the physical phenomena in a reasonable manner. As the hot flame jet shoots into the lane, it creates a lot of turbulence and strong mixing of hot products with the unreacted mixture. The hot products act as ignition sources, and a large amount of the unburnt mixture is simultaneously ignited. This results in a very fast pressure rise (2 to 20 barg in 3 ms). In the shear layer near the wall, still higher turbulence levels and reflections are seen, with very high pressures and maximum likelihood of DDT. The shock wave is seen to sustain a pressure of 20 barg before it decays as seen in the last picture in Figure 7. This decay is expected, as FLACS does not have a detonation model. However, an "average" shock ignition model, which is expected to replicate a detonation front, is currently under development. 2D plots of the variable DDTLS in the XY plane are shown in Figure 8 at the two instances where DDT likelihood is indicated in Figure 7. It can be seen from this figure that any initiated detonation should be able to propagate through the geometry.

Similar snapshots for test IA4 are presented in Figure 9. In this case also, a quick pressure rise is seen as the hot flame jet shoots into the lane. The first indication of DDT in the form of a spatial pressure gradient larger than one is seen much quicker in this case (59 ms instead of 93 ms) which is in agreement with experimental data. In this case, DDT was observed near the ground. The simulations also predicted that DDT is possible near the ground close to a sidewall of the lane (see Figure 9). However, the magnitude of the pressure gradients and the dimension of the region where they are seen are smaller in this case compared to test IA2, and hence DDT is less likely in this case. Also, the pressure was found to decay much more quickly in this case (not shown). The values of DDTLS were found to be more than sufficient in this case as well for the propagation of a detonation front.

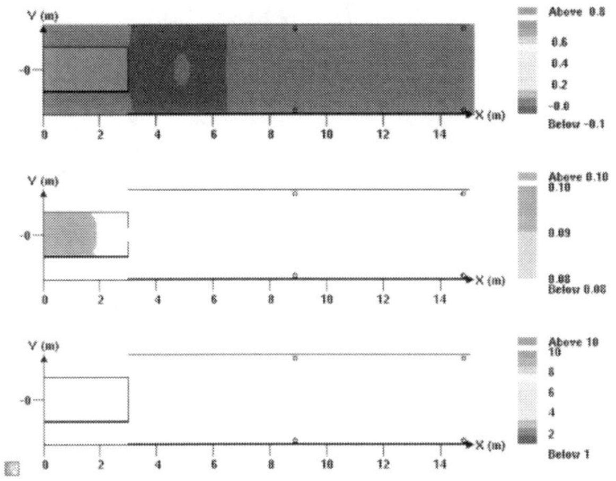

Initial phase (flame inside driver unit, P=0.8barg)

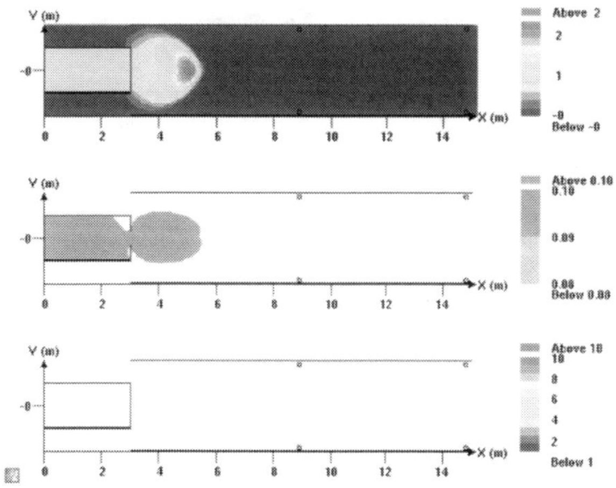

90ms (Flame jet shooting into the lane, P=2 barg)

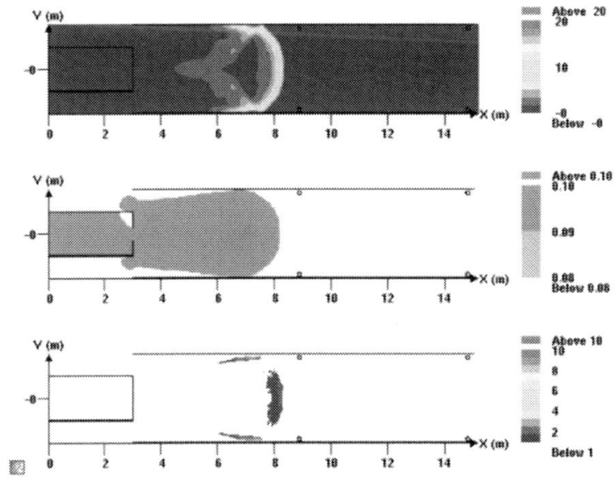

93ms (First indication of possible DDT, 20 barg)

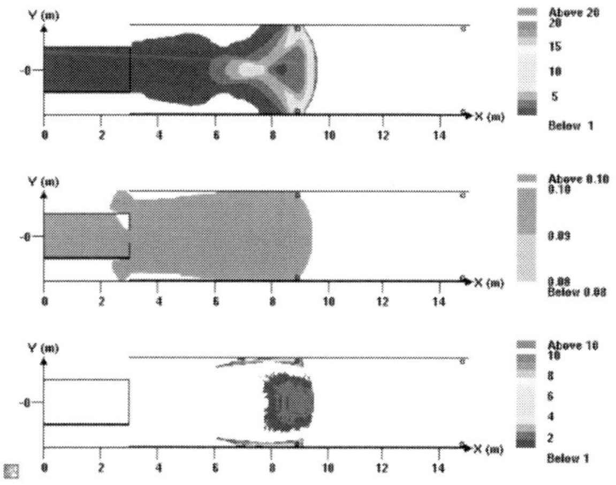

95ms (DDT likelihood maximum near pipes, 39 barg)

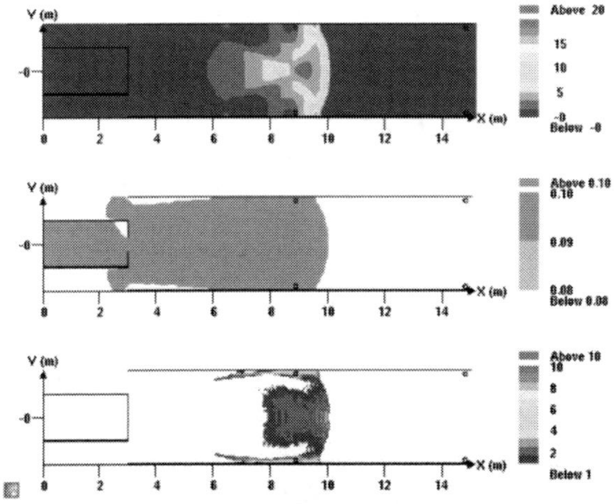

96ms (Propagation of shock/detonation front)

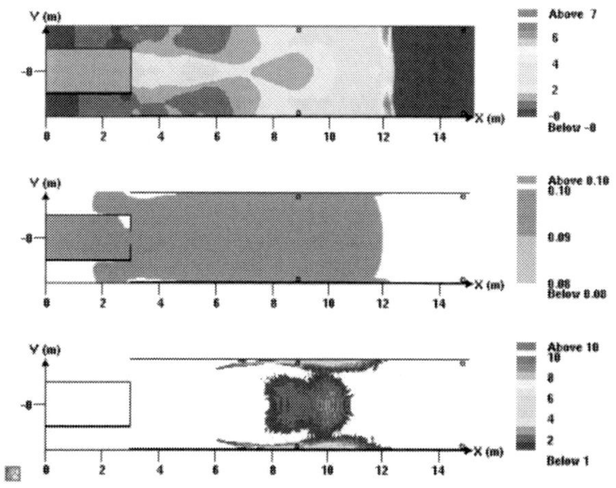

98ms (Shock front has decayed, P=7 barg)

Figure 7. 2D snapshots of simulation results (test IA2): Pressure (top), flame (middle), DDT indication parameter DPDX (bottom).

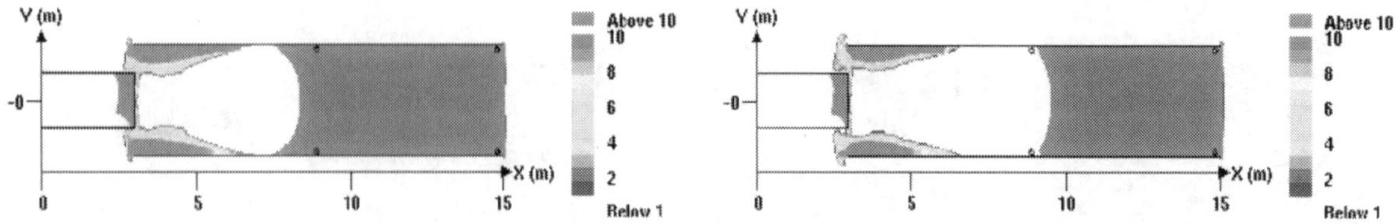

Figure 8. 2D snapshots of DDT length scale for test IA2 at two instances when DDT is indicated: 93 ms (left) and 95 ms (right).

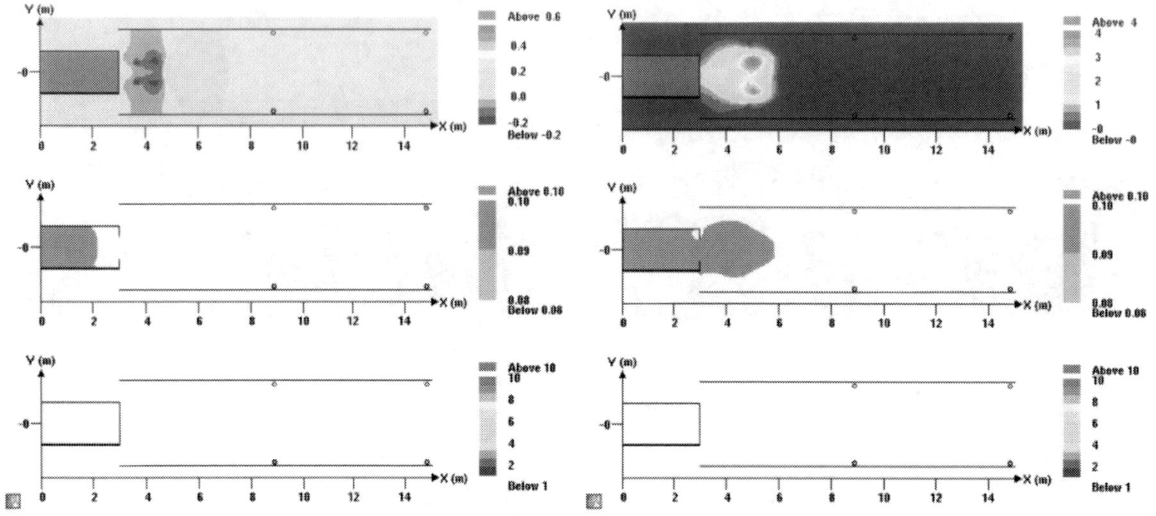

Initial phase (flame inside driver unit, P=0.6barg) **57ms (Flame jet shooting into the lane, P=4 barg)**

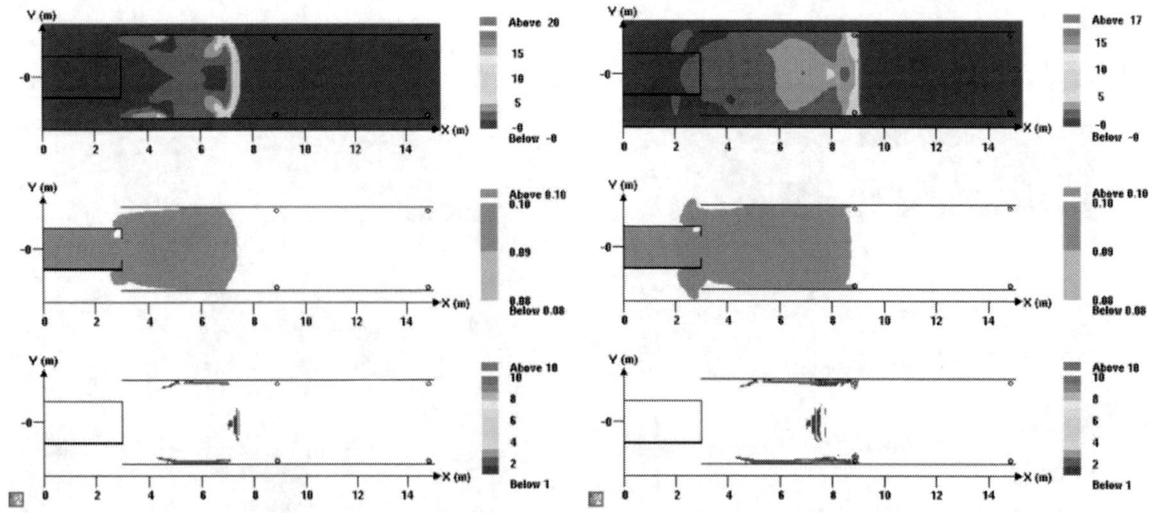

59ms (First indication of possible DDT, 20 barg) **61ms (DDT likelihood max. near sidewalls, 17 barg)**

Figure 9. 2D snapshots of simulation results (test IA4): Pressure (top), flame (middle), DDT indication parameter DPDX (bottom).

The maximum simulated flame speed was seen to be 1208 m/s in test IA2 and 1374 m/s in test IA4 (compared to 1651 m/s and 1740 m/s in the experiments). These are somewhat lower than those observed in the experiments but the development of a shock ignition model is ongoing which is expected to resolve this difference. Simulations of test IA1 revealed significant pressure gradients near the ground at a location that is similar to that of test IA4. This can be explained by the fact that the concentrations in both the driver section and the lane are higher for this test than for test IA2. Hence, the simulations indicated the likelihood of DDT but it was not observed experimentally. However, for test IA5, very small values of DPDX were seen, and hence DDT was not predicted to be likely which is in agreement with experimental observations. In general, the modelling results are able to capture the experimental observations, including pressure traces and locations of DDT, reasonably well.

6. Summary and Conclusions

FLACS has become increasingly accepted for risk assessment and explosion safety studies in offshore oil platforms worldwide. This is due to accurate and fast numerics and very extensive validation against available experimental data. Due to the increasing focus on hydrogen safety, significant use of the tool is seen in evaluating the viability of new and existing establishments in terms of probability and consequences of accidental events. It is expected that FLACS will be used in carrying out risk assessments for hydrogen risk with the same frequency as that for oil and gas applications. As a result of a dedicated effort towards hydrogen safety the CFD-tool FLACS can now demonstrate an improved performance and validation simulating hydrogen explosions.

The possibility of DDT represents a severe danger for hydrogen explosions due to its high reactivity. However, the currently available commercial version of FLACS can only model fast deflagrations, and a detonation model is lacking. This article presents the continuing work to enable FLACS to indicate the possibility, location, and time of occurrence of a transition to detonation. As a part of this work, the spatial pressure gradient has been used previously to predict the likelihood of DDT. The current work has developed and refined the model further, and comparisons of simulations with experimental data show that the CFD program is able to correctly replicate many aspects of the observations. The dimensions of a given geometry have been compared with the detonation cell size in order to estimate whether the propagation of any initiated detonation front is possible.

However, this model is still under development. An especially important limitation is the absence of a shock ignition model, which will be able to represent a propagating detonation front and predict the correct speed. But in any case, this work represents an important step forward for the development of a numerical tool to predict this highly complex phenomenon in real-scale geometries, even if it is on an "average" basis.

7. References

[1] Cullen, Lord W. D., 1990, The Public Inquiry into the Piper Alpha Disaster. Department of Energy, London, UK, October 1990.

[2] Hansen, O. R., Storvik, I. E., and van Wingerden, K., 1999. Validation of CFD-models for gas explosions, where FLACS is used as example - model description, experiences and recommendations for model evaluation. Proceedings European Meeting on Chemical Industry and Environment III, Krakow, Poland, available at the Department of Environmental Systems Engineering, Faculty of Process and Environmental Engineering, Technical University of Lodz, Wolczanska 175, 90-924 Lodz, Poland, pp. 365–382.

[3] van Wingerden, K., Hansen, O. R., and Teigland, R., 1995, Prediction of the strength of blast waves is the surroundings of vented offshore modules, ERA-report 95-1173, London, UK.

[4] van Wingerden, K., Hansen, O. R., and Foisselon, P., 1999. Predicting blast over pressures caused by vapor cloud explosions in the vicinity of control rooms. *Proc. Safety Prog.* **18**, 14–17.

[5] Hansen, O. R., Renoult, J., Bakke, J. R., 2001. Explosion risk assessment: how the results vary with the approach chosen. Fall Symposium Proceedings, Mary Kay O' Connor Process Safety Center, Dept. of Chemical Engineering, Texas A&M University, 3122 TAMU, College Station, TX 77843-3122, pp. 395–410.

[6] Dharmavaram, S., Hanna, S. R. and Hansen, O. R., 2005, Consequence Analysis-Using a CFD Model for Industrial Sites. *Proc. Safety Prog.*, **24**, 316-327.

[7] Skjold, T., Arntzen, B. J., Hansen, O. R., Taraldset, O. J, Storvik, I. E. and Eckhoff, R. K., 2005, Simulating Dust Explosions with the First Version of DESC. *Proc. Safety Env. Protec.*, **83**, 151-160.

[8] Hansen, O. R., van Wingerden, K. 1998, Modeling the Effects of Water Deluge and Inert Gas Dilution on Gas Explosions in Congested Geometries, 9th International Symposium of Loss Prevention and Safety Promotion in the Process Industries, Barcelona, Spain, Addenda to Proceedings, pp. 23-32.

[9] Sherman, M. P., Tieszen, S. R. and Benedick, W. B, FLAME Facility, The Effect of Obstacles and Transverse Venting on Flame Acceleration and Transition to Detonation for Hydrogen-Air Mixtures at Large Scale, Sandia National Laboratories, Albuquerque, NM 87185, USA, NUREG/CR-5275, SAND85-1264, R3, April 1989.

[10] Becker, T. and Ebert, F., 1985, Vergleich zwischen experiment und theorie der explosion grober, freier gaswolken. *Chem.-Ing. Tech.*, **57**, 42-45.

[11] Hansen, O. R., Renoult, J., Sherman, M. P., and Tieszen, S. R. 2005, Validation of FLACS-Hydrogen CFD Consequence Prediction Model Against Large Scale H_2 Explosion Experiments in the FLAME Facility, Proceedings of International Conference on Hydrogen Safety, Pisa, Italy, September 2005.

[12] Lee, J. H. S. and Moen, I. O. (1979). The mechanism of transition from deflagration to detonation in vapour cloud explosions. *Proc. Energy Comb. Sci.*, **6**, 359-389.

[13] Lee, J. H.S., Soloukhin, R. I., and Oppenheim, A. K. (1969). Current views on gaseous detonation. *Astron. Acta*, **14**, 565-584.

[14] Oppenheim, A. K., 1970, On the inadequacy of the gas dynamic processes for triggering the transition to detonation. *Comb. Flame*, **14**, 13-20.

[15] Lee, J. H. S., Knystautas, R., and Yoshikawa, N., 1978, Photochemical initiation and gaseous detonations. *Acta Astron.*, **5**, 971-972.

[16] Dorofeev, S. B., Bezmelnitsin, A. V., and Sidorov, V. P., 1995, Transition to detonation in vented hydrogen-air explosions. *Comb. Flame*, **103**, 243-246.

[17] Breitung, W., et al., 2000, Flame Acceleration and Deflagration to Detonation Transition in Nuclear Safety. State-of-the-Art Report, OECD Nuclear Energy Agency, Ref. NEA/CSNI/R (2000) 7.

[18] Middha, P, Hansen, O. R., and Storvik, I. E., 2006. Prediction of deflagration to detonation transition in hydrogen explosions. Proceedings of the AIChE Spring National Meeting and 40th Annual Loss Prevention Symposium, Orlando, FL, April 23-27, 2006.

[19] Patankar, S.V., 1980, Numerical Heat Transfer and Fluid Flow, Hemisphere Publications, New York, NY.

[20] Hjertager, B. H., 1985, Computer simulation of turbulent reactive gas dynamics. *J. Model. Identification Control*, **5**, 211–236.

[21] Hjertager, B. H., Bjørkhaug, M., and Fuhre, K., 1988. Explosion propagation of non-homogeneous methane-air clouds inside an obstructed 50 m^3 vented vessel. *J. Haz. Mater.*, **19**, 139–153.

[22] Hjertager, B., Fuhre, K., and Bjorkhaug, M., 1988, Gas explosion experiments in 1:33 and 1:5 scale offshore separator and compressor modules using stoichiometric homogeneous fuel/air clouds, *J. Loss Prevention Proc. Ind.*, **1**, 197-205.

[23] Arntzen, B. J.., 1998. Modeling of turbulence and combustion for simulation of gas explosions in complex geometries, PhD Thesis, NTNU, Trondheim, Norway.

[24] Tegnér, J. and Sjögreen, B., 2002. Numerical investigation of deflagration to detonation transitions. *Comb. Sci. Tech.* **174**(8): 153-186.

[25] Guirao, C. M., Knystautas, R., and Lee, J. H. S., 1989, A summary of hydrogen-air detonations for reactor safety, Report NUREG/CR-4961, Sandia National Laboratories/McGill University.

[26] Peraldi, O., Knystautas, R., and Lee, J. H.S., 1986, Criteria for transition to detonation in tubes. *Proc. Comb. Inst.*, **21**, 1629-1637.

[27] Dorofeev, S. B., Efimenko, A. A., Kotchourko, A. S., and Chaivanov, B. B., 1994, Evaluation of the hydrogen explosions hazard, *Nuclear Engineering and Design*, **148**, 305-316.

[28] Dorofeev, S. B., Sidorov, V, P., Breitung, W., and Kotchourko, A. S., 1997, Large-scale combustion tests in the RUT facility: Experimental study, numerical simulations, and analysis on turbulent deflagrations and DDT, *Transactions of the 14th International Conference on Structural Mechanics on Reactor Technology,* Lyon, France, Vol. 10, 275-283.

[29] Dorofeev, S. B., 1996, Turbulent combustion and DDT events as an upper bound for hydrogen mitigation techniques, OECD/NEA/CSNI workshop on the implementation of hydrogen mitigation techniques, Winnipeg, Manitoba, Canada, AECL-11762, NEA/CSNI/R(96)8, 415-433.

[30] Bezmelnitsyn, A. V., Dorofeev, S. B., and Yankin, Yu G., 1997, Direct comparison of detonation initiation by turbulent jet under confined and unconfined conditions, *Proceedings of 16th International Colloquium on Dynamics of Explosions and Reactive Systems,* Krakow, Poland, 222-225.

[31] Oran, E. S. And Gamezo, V. N., 2007, Origins of the deflagration-to-detonation transition in gas-phase combustion. *Comb. Flame*, **148**, 4-47.

[32] Kaneshige, M. and Shepherd, J. E., 1997, Detonation Database, Explosion Dynamics Laboratory Report FM97-8 (revised September 1999), California Institute of Technology, Pasadena, CA, USA.

[33] Dorofeev, S. B., Efimenko, A. A., and Besmelnitsyn, A. V., 1997, DDT scaling for severe accidents. Detonation cell size data as a function of composition and initial conditions. Report RRC "Kurchatov Institute" RRC KI 80-05/15.

[34] Kuznetsov, M., Alekseev, V., Matsukov, I., and Dorofeev, S., 2005, DDT in a smooth tube filled with a hydrogen-air mixture. *Shock Waves*, **14**, 205-215.

[35] Kuznetsov, M., Alekseev, V., Matsukov, I., and Dorofeev, S., 2000, Comparison of critical conditions for DDT in regular and irregular cellular detonation systems. *Shock Waves*, **10**, 217-233.

[36] Chan, C. K. and Dewit, W. A., 1996, Deflagration-to-detonation transition in end gases. *Proc. Comb. Inst.*, **26**, 2679-2684.

[37] Renoult, J. and Wilkins, B. A., 2003, Hydrogen Explosion Safety Phase 2 – Small scale explosion experiments, GexCon-03-F46201-2, GexCon, Bergen, Norway (confidential).

[38] Pförtner, H., Schneider, H., Tests with Jet Ignition of Partially Confined Hydrogen Air Mixtures in View of the Scaling of the Transition from Deflagration to Detonation, Final Report for Interatom GmbH, Bergisch Gladbach, Germany, Oct. 1984, Fraunhofer ICT Internal Report.

[39] Schneider, H., 2005, Deflagration and deflagration to detonation transition within a partial confinement similar to a lane Proceedings of International Conference on Hydrogen Safety, Pisa, Italy, September 2005.

[40] Middha, P, Hansen, O. R., and Schneider, H., Deflagration to Detonation Transition (DDT) in Jet Ignited Hydrogen-Air Mixtures: Large Scale Experiments and FLACS CFD Predictions. Paper accepted for presentation at 12th International Loss Prevention Symposium, Edinburgh, UK, May 22-24, 2007.

LOSS
PREVENTION
SYMPOSIUM

Use of Adiabatic Calorimetry and Aging Test for Safe Storage Study of Hydroxylamine Nitrate

Lijun Liu[1], Maria Papadaki[2], Yingchun Zhang[1], Yanjun Wang[1], William J. Rogers[1], and M. Sam Mannan[1]

(1) Artie McFerrin Department of Chemical Engineering, Texas A&M University, TAMU 3122, College Station, TX 77840, (2) Department of Environmental and Natural Resources Management, University of Ioannina, Agrinio, GR30100, Greece
Tel: (979) 862-3985
mannan@tamu.edu

Prepared for Presentation at
American Institute of Chemical Engineers
2007 Spring National Meeting
41st Annual Loss Prevention Symposium
Houston, Texas
April 22–26, 2007

UNPUBLISHED

Use of Adiabatic Calorimetry and Aging Test for Safe Storage Study of Hydroxylamine Nitrate

Lijun Liu[1], Maria Papadaki[2], Yingchun Zhang[1], Yanjun Wang[1], William J. Rogers[1], and M. Sam Mannan[1]

(1) Artie McFerrin Department of Chemical Engineering, Texas A&M University, TAMU 3122, College Station, TX 77840, (2) Department of Environmental and Natural Resources Management, University of Ioannina, Agrinio, GR30100, Greece
mannan@tamu.edu

Abstract

Since 1970s hydroxylamine nitrate (HAN) has been involved in several incidents at several military facilities including a major explosion at the Plutonium Reclamation Facility at Hanford in 1997 [1]. Following these incidents extensive research on HAN has been conducted at MKOPSC [2-3] and other institutes [4-9]. However, most of these research efforts were mainly focused on understanding the unstable behaviors or combustion behavior of HAN.

On-line monitoring, safe storage and handling criteria of HAN are still far from well studied and developed. In the current work, aging test and adiabatic calorimetry, inverse-based fitting methods are combined to elucidate the aging and storage behaviors of HAN. Aqueous solutions of HAN, 18 mass%, have been subjected to aging tests and adiabatic decomposition test. The experimental data are used to elucidate and deduce the mass loss kinetics parameters. These data could be used to monitor the storage behavior of HAN.

1 Introduction

1.1 Reactive chemicals

Reactive chemicals are substances that can explode or enter into violent reactions releasing large amounts of light, heat and gases. They have to be handled very carefully to avoid the violent explosion, fires or release of large amounts of toxic materials during transport, storage and usage. Reactive chemicals can be self-reactive or reactive with other materials. Self-reactive chemicals normally have a finite shelf time (maximum storage time). Several incidents related to self-reactive chemicals indicate inadequate understanding of their properties and behavior [1].

Reactive chemicals should be stored under proper conditions to minimize the potential hazards associated with them. Currently there are some general storage guidelines for chemicals, which can be referenced from OSHA and US EPA. For example chemicals should be stored separately according to their categories, such as solvent, bases, oxidizers and poisons etc. However, these general guidelines are not adequate for self-reactive chemicals, which are undergoing spontaneous decomposition process [10]. A quantitative model is eagerly needed to help the decision-making for storage operation and has the following benefits.

- Minimize the risk of incidents during storage, production and processing procedure.
- Fully utilize the service ability (life) of reactive chemicals to enhance the economical efficiency.
- Provide qualified material for the usage of its application.
- Minimize the cost during quality control and safety control with fewer measurements.
- Detect faulty measurements or abnormal phenomena when the measures against the model predictions.

1.2 Quantitative reactive chemicals storage study

Under the requirement of public safety and the lure of above benefits, some research of reactive chemicals storage is being carried out worldwide. However, current research on reactive chemical storage is mainly focused on the qualitative rating of reactivity, storage tank design, new storage material and critical storage conditions or high temperature stability study [11-16].

Currently storage study of reactive chemicals can be classified as reaction kinetics based model and statistics based model. Reaction kinetics based model is a fully predictive model, which is built based on the complete understanding of the reaction kinetics beneath the apparent chemical decomposition process. For reactive chemicals, kinetics based model is more suitable than the statistical based model, which is based on the best-fit functions. Because statistical based model doesn't have any chemical implication inside, the risk, errors and uncertainty with the application of statistical based model are much higher than the kinetics based model. However, complete understanding of reaction mechanism for chemicals is still rare case since too many possible reaction pathways may exist in the reality. Furthermore, the reaction mechanism may change under different conditions, because some other reaction steps may become dominant in specific conditions. Simplification and best-fit functions are still necessary to assist the development of kinetics-based model.

1.3 Storage study on HAN

The storage behavior of reactive chemicals is significantly different from each other considering their variety. This work is focused on the self-catalytic reactive chemicals, whose potential hazards were normally underestimated in normal industrial practice. Based on the incidents data and reactivity, HAN is chose to be studied. HAN is a salt of hydroxylamine and nitric acid, which is also named as hydroxylammonium nitrate. It is widely used in nuclear industry as reductant and liquid propellant for gun/rocket.

During the past 40 years, this compound has been involved in several incidents in different locations, such as Hanford and Savannah River Site (SRS) [1]. The poor understanding of its properties and self-catalytic behavior caused most of the past incidents related to HAN, where either nitric acid concentration exceeded the safe limit, caused by wrong operating procedure or water evaporation, or excess heat was introduced into the storage facilities or piping systems. To safely store reactive chemicals, currently some criteria such as self-accelerating decomposition temperature (SDTA) were already proposed [16]. However, those empirical criteria in their natural approaches are not adequate to reach the requirement of current practice, especially for self-catalytic chemicals. The decomposition of self-catalytic chemicals is a dynamic process since its composition and properties change with time. For example, the T_{SDTA} of HAN decreases during the storage process. Such a criterion without consideration of decomposition mechanism may even mislead the operator and cause potential accidents. Overall, any new storage model of self-catalytic chemicals is imperative to include decomposition mechanisms and monitoring criteria associated.

1.4 Decomposition mechanism of HAN

The decomposition mechanism of HAN is so complex that many factors, such as transition metal ions, acid concentration, pressure and solution concentration, may affect the decomposition mechanism. Research has been carried out in different institutes and gives several different decomposition mechanisms. All of those mechanisms could explain some observed phenomena and agree with each other in some detail reaction steps, especially the initial step [4-7]. However it is not appropriate to use them directly for the reason of different experimental conditions and HAN sources, where impurities are significantly different. Furthermore the decomposition mechanism changes along the storage process that it requires huge efforts on the mechanism development to describe the decomposition process correctly at different stages.

Another way to describe the decomposition mechanism is to utilize global reaction rates. Although it also changes along the storage process, the effort of deriving summarized reaction mechanism normally is much less than that of obtaining the detail decomposition mechanism. A series of reaction steps and parameters could be developed for the practical conditions with small costs. They could be very useful within the initial developing scope. Currently this method has been applied to several rockets/gun propellants, such as ε-Hexanitrohexaazaisowurtzitane (ε-HNIW, ε-CL20) and hydrazinium nitroformate (HNF) etc [17-19]. Considering the cost and application, the summarized reaction mechanism and their parameters of HAN decomposition have been derived according to the mass loss data and literature survey.

2 Experimental

2.1 Sample

HAN (18 mass% water solution, FLUKA 93726) was used in the tests without further purification.

2.2 Aging test (isothermal test)

Aging test (isothermal test) aims to get the isothermal kinetics parameters to describe the normal storage behavior of HAN. The reaction vessel with HAN/water sample was loaded in the reaction cell and the cell was placed in the calorimeter. In this research Automatic Pressure Tracking Adiabatic Calorimeter (APTAC) is used as an isothermal calorimeter [20]. All experiments are performed in a closed environment to minimize the influence of evaporation of water and nitric acid generated. To further depress the evaporation of water, nitric acid, and HAN, reaction vessel is pressurized to around 50psi at the beginning of the run. Given the decomposition process happens in the condense phase, the change in pressure does not affect the kinetics parameters significantly. The sample is heated and kept isothermally at specific temperature for a few to several hours to decompose isothermally. After desired isothermal time, the mixture is cooled down and the sample mass is measured to obtain the mass loss data. During the iso-aging reaction pressure, temperature and a number of other quantities are recorded.

3 Results and discussion

3.1 Mass loss evaluation

The mass loss of HAN decomposition is elucidated to derive the mass loss kinetics parameters by utilizing the species conservation equations, which are based on the summarized global reaction steps of HAN [4-7, 9]. The following reduced global reaction steps were proposed by Lee to describe the decomposition process of HAN based on the literature survey [5].

$$HAN \rightarrow NH_2OH + HNO_3 \quad \text{(R1)}$$

$$NH_2OH + HNO_3 \rightarrow HONO + HNO + H_2O \quad \text{(R2)}$$

$$NH_2OH + HONO \rightarrow N_2O + 2H_2O + HNO_3 \quad \text{(R3)}$$

$$NH_2OH + HNO \rightarrow N_2 + 2H_2O + HNO_3 \quad \text{(R4)}$$

$$3HONO \rightarrow 2NO + HNO_3 + H_2O \quad \text{(R5)}$$

$$HONO + HNO_3 \rightarrow 2NO_2 + H_2O \quad \text{(R6)}$$

$$2HNO \rightarrow N_2O + H_2O \quad \text{(R7)}$$

$$HNO + HNO_3 \rightarrow 2HONO \quad \text{(R8)}$$

It is generally accepted that the initial steps of HAN decomposition are composed of the proton transfer to produce NH_2OH and HNO_3 (R1) and the formation of HONO and HNO (R2). Furthermore HONO and HNO appear in many reaction steps and pathways [3], which demonstrate these two unstable species play crucial roles in the decomposition process of HAN. So reaction R3 and R4 are kept to describe this process.

Recently Amariei et al. found that primary gas decomposition products are N_2, NO and N_2O [8], which contribute to majority of the mass loss in the decomposition process. Furthermore HNO_3 is most likely to disassociate completely in the dilute solution. Based on these two facts, the reaction generating NO_2 (R6) is skipped. Reaction (R8) is also skipped for the similar reason of HNO_3 disassociation. In diluted solution, the probability of HNO reacting with NH_2OH (R4) is much larger than the dimerization process (R7). Although HONO is also an intermediate species that its concentration is very low, reaction (R5) is still kept to explain the formation of NO, one of the major decomposition gas components. So the first five reactions, R1, R2, R3, R4 and R5, are used to describe the decomposition mechanism of diluted HAN solution.

3.2 Computational procedure

Around 5g 18%wt HAN aqueous solution is used in each run. Considering the small quantity used and the small solubility of NO, N_2 and N_2O in water, the mass of gases dissolved in the condense phase are too small to be measured. So it is assumed they enter the gas phase once they are generated. The evaporation of water, nitric acid and HAN are already eliminated by the initial pressurization and the closed system setup. The intermediate species, such as HONO and HNO, are not considered for they are highly reactive and their concentrations are very small in the solution. Overall the physical evaporation phenomenon is not considered to reduce the complexity of optimization.

Therefore the mass conservation of the jth species in condense phase is expressed as follows.

$$\frac{d(m_j)}{dt} = \sum_{i=1}^{n} (v_{j,i} k_{MLi} \prod_{j=1}^{m} C_j^{v_{j,i}}) \times MW_j \times V \quad (1)$$

where m_j is the mass of species j in condense phase, $v_{j,i}$ is the stoichiometric coefficient of reactants in the ith reaction, k_{MLi} is the mass loss reaction rate constant of the ith reaction, C_j is the concentration of species j in condense phase, MW_j is the molecular weight of specie j, n is the total number reaction involved in this model, m is the total number of species in condense phase, and V is the condense phase volume. The mass loss, ML, of the decompostion process is given as:

$$ML = \sum_{g=1}^{l} (\int_0^t v_{g,i} k_{MLi} \prod_{j=1}^{m} C_j^{v_{j,i}} \times MW_g \times V) \quad (2)$$

where $v_{g,i}$ is the stoichiometric coefficient of gas component g, l is the total number of reaction generating non-soluble gases, MW_g is the molecular weight of gas g. The initially guessed reaction rate coefficients are used to calculate the mass loss theoretically according to the species conservation equations. Then the following objective function is minimized with non-linear least-square method.

$$\chi = \sqrt{\sum_{i=1}^{l} (\frac{ML_i^e - ML_i^c(k)}{\sigma(ML^e)})^2}$$

Where superscripts e and c represent the experimental measurements and regression results respectively, k represents the kinetics parameters under determination, $\sigma(ML^e)$ represent the standard deviation of the experimental measurements, ML represents the mass loss data of the decomposition process, χ represents the objective function of parameter optimization. The optimization is terminated when the difference of objective function is less than 10^{-5}.

3.3 Mass loss of HAN and the corresponding mass loss kinetics parameters

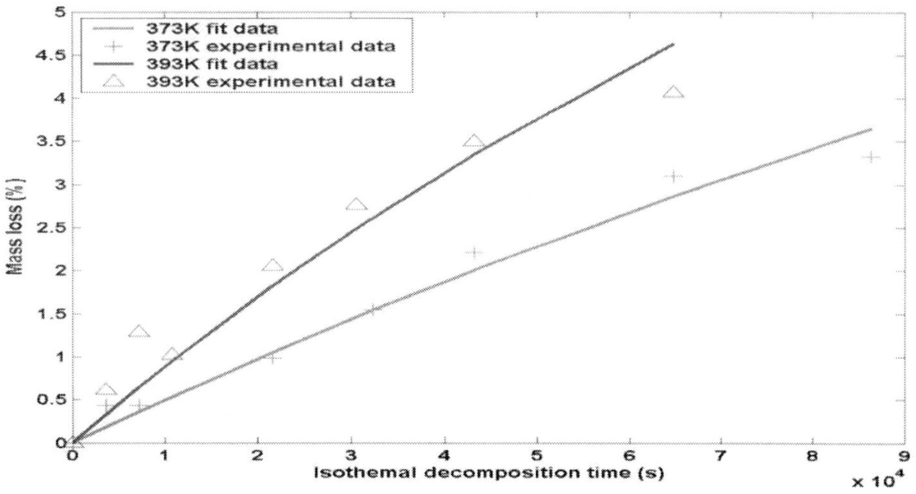

Figure 1. Mass loss description of isothermal decomposition test of HAN

The mass loss data of HAN are shown in Figure 1 as the function of isothermal time. The result indicates the mass loss rate of HAN increases with the increasing storage temperature. The mass loss kinetics parameters are obtained through above computational procedure, which is given as Table 1.

Table 1. Mass loss kinetics parameters of isothermal decomposition of HAN

Temp (K)	373	393
$K_{ML,1}$ *	4.54e-6	8.10e-6
$K_{ML,2}$ *	43.7	41.3
$K_{ML,3}$ *	3.52e-3	3.09e-4
$K_{ML,4}$ *	46.3	45.9
$K_{ML,5}$ **	1e-16	1e-16
χ	0.4167	0.6843

Unit of K_{ML}: * $(dm^3/mol)/sec$, ** $(dm^3/mol)^2/sec$.

Original mass loss method proposed by Bohn [17-19] is only applicable to those simple reactions, such as first order or second order reaction. However, this method may be applied to a more complex reaction network with advanced numerical optimization method. In this research, a simplified global reaction network is used to describe the isothermal decomposition process of HAN. Obtained mass loss data are used to deduce the mass loss kinetics parameters of HAN decomposition. Furthermore mass loss data could be converted to measurable variable, material density, to monitor the storage process of energetic materials. However the mass loss kinetics parameters obtained from mass loss information introduce high uncertainty into the optimization process. To build accurate reaction rate database, more data, such as concentration of NH_2OH and HNO_3, are needed to assist mass loss data to finish this job.

4 Conclusions

Based on the mass loss information of HAN, its mass loss kinetics parameters are deduced, which can be used to monitor the storage process of HAN. This work demonstrates that mass loss information could be used as one input for complex reaction. It also could provide a bridge connecting the complex reaction network and storage operation with an easily monitoring variable, material density. In the future, a model of HAN decomposition based on easily measurable variables will be proposed to meet the practical storage application. The lifetime of energetic material will be expressed as a function of several variables, such as density and storage temperature history profile. This function is eligible to predict the important criteria, such as the induction time, based on above inputs.

5 Acknowledgement

This research was supported by the Mary Kay O'Connor Process Safety Center at Texas A&M University.

6 Nomenclature

m_j	Mass of species *j* in condense phase
$v_{j,i}$	Stoichiometric coefficient of reactants in the *i*th reaction
k_{MLi}	Mass loss reaction rate constant of the *i*th reaction
C_j	Concentration of species *j* in condense phase
MW_j	Molecular weight of specie *j*
n	The total number reaction involved in the decomposition model of HAN
m	The total number of species in condense phase
V	The condense phase volume
ML	Mass loss of isothermal decomposition process
$v_{g,i}$	The stoichiometric coefficient of gas component *g*
l	The total number of reaction generating non-soluble gases
MW_g	The molecular weight of gas *g*
e	Experimental measurements
c	Regression results
k	The kinetics parameters under determination
$\sigma(ML^e)$	The standard deviation of the experimental measurements
χ	Objective function of parameter optimization

7 Reference

[1] D. G. Harlow, R. E. Felt, S. Agnew, G. S. Barney, J. M. McKibben, R. Garber, M. Lewis, Technical Report of Hydroxylamine Nitrate, U. S. Department of Energy, (1998).

[2] C. Wei, W. J. Rogers, M. S. Mannan, Thermal decomposition hazrd evaluation of hydroxylamine nitrate, *Journal of Hazardous Materials,* 130 (2006), 163-168.

[3] C. Wei, W. J. Rogers, M. S. Mannan, Detection of autocatalytic decomposition behavior of energetic materials using APTAC, *Journal of Thermal Analysis and Calorimetry*, 83 (2006), 125-130.

[4] H. Lee, T. A. Litzinger, Chemcial kinetics study of HAN decomposition, *Combustion and Flames*, 135 (2003), 151-169.

[5] E. S. Kim, H. S. Lee, C. F. Mallery, S. T. Thynell, Thermal decomposition studies of energetic materials using confined rapid thermolysis / FTIR spectroscopy, *Combustion and Flames*, 110 (1997), 239-255.

[6] J. C. Oxley, K. R. Brower, Thermal decomposition of hydroxylamine nitrate, *SPIE*, 872 (1988), 63-69.

[7] N. Klein, Twenty-Seventh JANNAF Combustion Subcommittee Meeing, *CPIA Publ.* 557, Vol. 1, 1990, 443.

[8] D. Amariei, L. Courthéoux, S. Rossignol , C. Kappenstein, Catalytic and thermal decomposition of ionic liquid monopropellants using a dynamic reactor Comparison of powder and sphere-shaped catalysts, *Chemical Engineering and Processing* 46 (2007), 165–174.

[9] V. A. Rafeev, Yu I. Rubtsov, Kinetics and mechanism of thermal decomposition of hydroxylammonium nitrate, *Russian Chemical Bulletin*, 42 (1993), 1811-1815.

[10] Center for Chemical Process Safety, *Guidelines for Safe Storage and Handling of Reactive Materials*, AICHE, (1995), 14-42.

[11] S. Mukherjee, Understanding atmospheric storage tanks, *Chemical Engineering*, 113 (2006), 74-84

[12] U. Eberle, G. Arnold, R. von Helmolt, Hydrogen storage in metal–hydrogen systems and their derivatives, *Journal of Power Sources*, 154 (2006), 456-460.

[13] B. Roduit, Ch. Borgeat, B. Berger, P. Folly, B. Alonso and J. N. Aebischer, The prediction of thermal stability of self-reactive chemical From milligrams to tons, *Journal of Thermal Analysis and Calorimetry,* 80 (2005), 91-102.

[14] C. M. Bosch, E. Velo, F. Recasens, Safe storage temperature of peroxide initiators: prediction of self-accelerated decomposition temperature based on a runaway heuristics, *Chemical Engineering Science*, 56 (2001), 1451-1457.

[15] Y. Liu, W. J. Rogers, S. M. Mannan, Screening reactive chemicals hazards, *Chemical Engineering Progress*, 102 (2006), 41-47.

[16] T. Kotoyori, The self-accelerating decomposition temperature (SDTA) of solids of the quasi-autocatalytic decomposition type, *Journal of Hazardous Materials*, 64 (1999), 1-19.

[17] M. A. Bohn, Determination of the kinetic data of the thermal decomposition of energetic plasticizers and binders by adiabatic self-heating, *Thermochimica Acta*, 337 (1999), 121-139.

[18] M. A. Bohn, Kinetic description of mass loss data for the assessment of stability, compatibility and aging of energetic components and formulations exemplified with ε-CL20, *Propellants, Explosives, Pyrotechnics*, 27 (2002), 125-135.

[19] M. A. Bohn, Thermal stability of HNF investigated by mass moss and heat generation in the temperature range 50°C to 80°C and lifetime predictions, the 36th International Annual Conference of ICT and the 32nd International Pyrotechnics Seminar, (2005), 167-1 to 167-25.

[20] S. Chippett, P. Ralbovsky, R. Granville, The APTAC: a high pressure, low thermal inertia, adiabatic calorimeter, in International Symposium and Runaway Reactions, Pressure Relief Design, and Effluent Handling, 1998, New Orleans: America Institute of Chemical Engineers.

LOSS PREVENTION SYMPOSIUM

Industry-Government Alliances Share Information to Help Prevent Workplace Injuries, Illnesses and Fatalities

Scott Berger, Director (Author)
Center for Chemical Process Safety
AIChE
Three Park Avenue
New York, NY 10016
212-591-7237
scotb@aiche.org

Lee Anne Jillings, Director (Presenter)
Office of Outreach Services and Alliances
Occupational Safety and Health Administration
200 Constitution Ave, NW
Washington, DC 20210
(202) 693-2124
jillings.leeanne@dol.gov

Prepared for Presentation at
American Institute of Chemical Engineers
2007 Spring National Meeting
41st Annual Loss Prevention Symposium
Houston, Texas
April 22–26, 2007

UNPUBLISHED

Industry-Government Alliances Share Information to Help Prevent Workplace Injuries, Illnesses and Fatalities

Scott Berger, Director (Author)
Center for Chemical Process Safety
AIChE
Three Park Avenue
New York, NY 10016
212-591-7237
scotb@aiche.org

Lee Anne Jillings, Director (Presenter)
Office of Outreach Services and Alliances
Occupational Safety and Health Administration
200 Constitution Ave, NW
Washington, DC 20210
(202) 693-2124
jillings.leeanne@dol.gov

Abstract

Through the Alliance Program, OSHA works with groups committed to safety health, including businesses, trade or professional organizations, unions and educational institutions, to leverage resources and expertise to develop compliance assistance tools and resources and share information with employers and employees to help prevent injuries, illnesses and fatalities in the workplace. Existing Alliances address a wide range of occupations, specific safety competencies, and general industry concerns. Since March 2004, AIChE's Center for Chemical Process Safety has been a signatory to the OSHA and Chemical Reactivity Hazards Management Alliance to raise awareness of chemical reactivity hazards. The OSHA and Chemical Reactivity Hazards Management Alliance has provided input into the development of an OSHA Safety and Health Topics page on Chemical Reactivity Hazards that provides useful references and tools from the Center for Chemical Process Safety and other Alliance signatories. This paper describes the OSHA and Chemical Reactivity Hazards Management Alliance-developed resources, and highlights other compliance assistance materials of interest to practicing safety and health professionals in the chemical and petroleum industries.

1. The OSHA Alliance Program

1.1 Mission of the Alliance Program

The Occupational Safety and Health Administration (OSHA) works with the public to promote safety and health in the workplace by offering compliance assistance services and programs to businesses and organizations. These services and programs help organizations work collaboratively with the Agency and comply with the Occupational Safety and Health Act of 1971.

In March 2002, OSHA created the Alliance Program. Through the Alliance Program, OSHA works with groups committed to safety health, including businesses, trade or professional organizations, unions and educational institutions, to leverage resources and expertise to develop compliance assistance tools and resources and share information with employers and employees to help prevent injuries, illnesses and fatalities in the workplace.

1.2 Benefits of an Alliance

There are many benefits to participating in an Alliance with OSHA. The agreements, which are signed for two years, help organizations to:

- Build trusting, cooperative relationships with the Agency.
- Network with others committed to workplace safety and health.
- Leverage resources to maximize worker safety and health protection.
- Gain recognition as proactive leaders in safety and health.

Organizations may be cooperating with OSHA for the first time, or they may be continuing existing relationships with the Agency established through other cooperative programs.

1.3 Getting An Alliance Started

The Alliance Program supports OSHA's strategic areas of emphasis by addressing the plan's occupational trends, emerging issues and the Agency's priorities. There are few formal program requirements for Alliances and the agreements do not include an enforcement component. However, OSHA and the participating organizations must define, implement and meet a set of short and long-term goals that fall into three categories:

- Training and Education, such as developing training and education programs and seminars aimed at reducing workplace hazards, providing the OSHA Training Institute with educational and training materials on specific safety issues upon request, and providing peer review of OSHA training curricula.
- Outreach and Communication, such as sharing the most up-to-date ergonomic information for educational purposes, promoting participation in OSHA's cooperative programs, and providing information in Spanish and other languages.
- Promoting the National Dialogue on Workplace Safety and Health, by sharing data on safety and health hazards, participating in various forums and groups to discuss ways of improving

workplace safety and health programs, and demonstrating the effectiveness of safety and health programs.

1.4 Alliance Implementation

OSHA and the Alliance participants form implementation teams. The teams, consisting of representatives from OSHA and the organizations, are responsible for developing strategies and implementing programs or processes that meet the defined goals.

1.5 Products and Activities

Current Alliances have developed a number of products and undertaken many activities, including:

- Assisting OSHA in developing interactive, web-based training tools on occupational safety and health topics;
- Helping develop Safety and Health Topics pages;
- Appearing at trade shows and conferences.

F. Information About Alliances

For national Alliances, contact OSHA's Office of Outreach Services and Alliances at 202-693-2340. For regional or local Alliances, contact your area's Regional OSHA Office. Visit OSHA's website at www.osha.gov for more information about the Alliance Program and the Regional Offices.

2. The Chemical Reactivity Hazard Management Alliance

2.1 Formation and Current Membership

In 2002, the US Chemical Safety and Hazard Investigation Board (CSB) concluded an investigation into "Reactive Chemical" incidents[1] [1]. One of CSB's key findings was that the hazards which caused a large majority of the 167 incidents studied were published in readily available references. CSB postulated that the companies involved in the incidents did not have sufficient awareness of reactivity hazards to drive them to consult these references. CSB issued a range of recommendations aimed at raising awareness of chemical reactivity hazards.

CSB recommended to CCPS to publish a basic guidance book on managing chemical reactivity hazards, and "Essential Practices for Managing Chemical Reactivity Hazards" [2] was released in April 2003. In the ensuing months, CCPS collaborated with OSHA, US EPA, the American Chemistry Council, and the Synthetic Organic Chemical Manufacturer's Association to make an electronic version

[1] In other words, incidents that could be traced back to some kind of chemical reactivity hazard.

of "Essential Practices" available on the web at no cost to users. This program proved to be quite successful, attracting more than 2,000 readers in the first three months.

Buoyed by this success, the above group decided to organize under the Alliance Program in March 2004. Additional parties were invited to participate, bringing the current membership of the Chemical Reactivity Hazard Management Alliance (CRHMA) to include:

- U.S. Occupational Safety and Health Administration
- U.S. Environmental Protection Agency (EPA)
- AIChE's Center for Chemical Process Safety (CCPS)
- American Chemical Society (ACS)
- American Chemistry Council (ACC)
- Chlorine Institute, Inc. (CI)
- National Association of Chemical Distributors (NACD)
- Synthetic Organic Chemical Manufacturers Association (SOCMA)
- Texas A&M University's Mary Kay O'Connor Process Safety Center (MKOPSC)

2.2 CHRMA's Accomplishments

Effective immediately upon signing the Alliance agreement, CHRMA partners began or continued to promote the free on-line use of "Essential Practices", as well as other free publications available from CCPS, the U.S. EPA., and the National Oceanographic and Atmospheric Association (NOAA). Links to these and links or references to other key resources were consolidated over the ensuing months, and made available on the Chemical Reactivity Hazards Topics page in the OSHA website (http://www.osha.gov/SLTC/reactivechemicals/) in December 2004.

CHRMA developed a one-half day workshop targeted at familiarizing non-experts with chemical reactivity hazards. To minimize the cost of the workshop, it is designed to be piggy-backed onto other conference events. Its first formal presentation was in September 2005, immediately following a Chlorine Institute conference, and was given again in conjunction with a CCPS meeting in January 2006. A slightly modified version, planned for an ACS meeting in March, 2007, should be completed by the date this paper is presented. Anyone interested in bringing this workshop to an organization's event may contact the author for more details.

CHRMA has also maintained an exhibit booth to distribute its materials at several conferences, including meetings of CCPS/AIChE, CI, NACD, and MKOPSC.

3. Other Alliances of Interest to Process Safety Professionals

As of November 30, 2006, OSHA has more than 450 national, regional and area office Alliances. Readers are encouraged to review the web page http://www.osha.gov/dcsp/alliances/index.html to see the extensive range of Alliances and their activities and products. In addition to the CRHMA, some other Alliances of particular interest to process safety professionals include:

- OSHA and The Dow Chemical Company Alliance on process safety and ergonomics

- OSHA and Safe Tank Alliance
- OSHA and Abbott Alliance on emergency preparedness and communicating the business value of safety

4. Conclusion

Through the Alliance Program, OSHA works with groups committed to safety health, including businesses, trade or professional organizations, unions and educational institutions, to leverage resources and expertise to develop compliance assistance tools and resources and share information with employers and employees to help prevent injuries, illnesses and fatalities in the workplace. The Alliance Program is an excellent way for organizations that support safety and health to develop and disseminate important safety and health information and tools to employers and employees that need it.

5. References

[1] U.S. Chemical Safety and Hazard Investigation Board, *Improving Reactive Hazard Management*, Washington, DC, 2002

[2] Johnson, R.W., S. Rudy, and S. Unwin, *Essential Practices for Managing Chemical Reactivity Hazards*, American Institute of Chemical Engineers CCPS Concept Series, New York, 2003

Communicating RAGAGEP Responsibility Beyond PSM and RMP Coverage

Angela S. Blair, P.E.

Investigator
US Chemical Safety and Hazard Investigation Board
2175 K ST. NW Ste 400
Washington, DC 20037
202-261-3607
202-974-7603 (fax)
Angela.Blair@csb.gov

Prepared for Presentation at
American Institute of Chemical Engineers
2007 Spring National Meeting
41st Annual Loss Prevention Symposium
Houston, Texas
April 22–26, 2007

UNPUBLISHED

Communicating RAGAGEP Responsibility Beyond PSM and RMP Coverage

Angela S. Blair, P.E.

Investigator
US Chemical Safety and Hazard Investigation Board
2175 K ST. NW Ste 400
Washington, DC 20037
202-261-3607
202-974-7603 (fax)
Angela.Blair@csb.gov

Abstract

Many of the facilities at which the CSB investigates incidents are not covered by the OSHA PSM rule or the EPA RMP rule. In some cases, the CSB finds that existing consensus codes and standards, if properly applied, might have prevented or mitigated the incident. The difficulty lies in communicating the responsibility of engineers in these facilities to consult and apply such standards when there is no clear-cut regulatory requirement to do so.

The PSM and RMP rules both require facility owners to document that their equipment complies with *Recognized and Generally Accepted Good Engineering Practices,* known by the acronym RAGAGEP. However, no other OSHA or EPA rule specifically assigns responsibility for RAGAGEP compliance. OSHA's General Duty Clause can be used to enforce the RAGAGEP standards, only when OSHA can establish that the employer could reasonably be expected to be aware of the hazard.

This paper examines the available literature, state licensing laws, and published codes of ethics for engineers, to establish a citable link between the engineer at a "non-covered" facility and the responsibility to adhere to RAGAGEPs.

1.0 Origins of RAGAGEP

OSHA's process safety management standard (29 CFR 1910.119) introduced a new term to the technical vocabulary: RAGAGEP. RAGAGEP stands for "Recognized And Generally Accepted Good Engineering Practices." In the PSM standard, OSHA requires that employers with covered processes document that equipment complies with RAGAGEPs [1]. The EPA's Risk Management Rule (40 CFR Part 68.112(r)) contains the same requirement for large, high-risk facilities—the Program 3 facilities [2].

OSHA didn't invent RAGAGEP; they simply codified it. Engineers have always known of the existence of engineering standards and other generally accepted practices for designing and maintaining process equipment. Engineers recognize that adherence to these standards and practices saves them work and helps ensure process operability and safety.

2.0 What are RAGAGEPs?

Recognized and generally accepted good engineering practices are analogous to the legal term "standard of care". In the case of professionals, the standard of care refers to the level of care that a reasonable professional in the same or similar circumstances would take to prevent harm or injury to another person [3].

There is no established or official list of what constitutes a RAGAGEP. RAGAGEPs encompass the whole body of guidelines, standards, generally accepted principals (both taught in school and learned from others) that establish the ways in which responsible engineers accomplish their tasks. Here are but a few examples:

- Trade association[1] recommended practices
- ASTM Standards
- Piping, mechanical and electrical codes
- Professional society standards (ASME Boiler & Pressure Vessel Code)
- Fire codes and standards
- Methods and rules in engineering texts
- Lessons learned communicated in scientific and technical literature and at conferences.

It is not this author's intent to engage in a discussion of what is, or is not, considered a RAGAGEP by OSHA. The reader can consult Schuler (1997) and Salot (1997) for two researched opinions on the subject [4,5].

3.0 RAGAGEP Responsibility Beyond PSM and RMP

Many of the incidents the CSB investigates occur in facilities not covered by the PSM or RMP rules. West Pharmaceutical Services, Hayes Lemmerz, CTA Acoustics and Marcus Oil are a few of these non-PSM/RMP facilities. Without the PSM and RMP regulations to codify compliance with RAGAGEPs, what other means exist in these facilities to encourage their use?

3.1 Engineering Code of Ethics

Licensed professional engineers are bound by regulations established by the states in which the engineers are registered. A sampling of states reveals that most states include codes of ethics and/or principles of engineering practice into the state laws for professional engineers. For example, the Texas regulation states:

[1] For example, the Chlorine Institute, the American Petroleum Institute, the International Institute of Ammonia Refrigeration, the American Chemistry Council.

> "Engineers shall not perform any engineering function which, when measured by generally accepted engineering standards or procedures, is reasonably likely to result in the endangerment of lives, health, safety, property, or welfare of the public." [6]

Not every practicing engineer is legally required to obtain a license. Many states, including California, Texas, Vermont and Mississippi[2] do not require a license for engineers employed by a manufacturing company. For these license-exempt engineers, their connection to codes of ethics is through applicable professional societies such as the American Society of Mechanical Engineers.

The Accreditation Board for Engineering and Technology (ABET) is the body charged by the engineering societies (ASME, AIChE, ASCE, etc.) with establishing requirements for and monitoring compliance with engineering curricula. ABET-accredited engineering curricula include instruction on engineering ethics. [7]

The Association of Engineering Societies is the holder of record of the Engineering Code of Ethics, although the most thorough statement of the code is promoted by the National Society of Professional Engineers. The entire code is available from the NSPE website [8], but the section most applicable to this discussion is in the Rules of Practice, Section II (b): "Engineers shall approve only those engineering documents that are in conformity with applicable standards."

3.2 The Engineer's Responsibility

The connection between the non-licensed engineer working in a non-PSM-covered facility and RAGAGEP is *responsibility,* as established in the Engineering Code of Ethics. Each engineer has a responsibility to practice only within his/her area of expertise and to approve only those designs that comply with applicable standards.

According to Michael Davis of Center for the Study of Ethics in the Professions at the Illinois Institute of Technology, knowledge of the codes and standards applicable to ones work is a hallmark of engineering competency [9, 10]. Therefore, the code of ethics binds engineers to seek out, become familiar with and adhere to the standards that apply to their work.

4.0 Case Studies

The major sources on engineering ethics consulted for this paper present the information on codes of conduct through case studies [11, 12]. The following are examples from CSB investigations, where failure to consult and adhere to established codes or standards contributed to an accident. In most cases, the engineers were unaware of the applicable standards. These are only a few examples; many other CSB investigations revealed failure to research or comply with applicable codes and standards as causal to the incident.

---

[2] Mississippi's exemption applies only insofar as the engineer is not engaged in original process or equipment design.

4.1 West Pharmaceutical Services, Inc., Kinston, NC

On January 29, 2003, a massive explosion, fueled by polyethylene powder that had accumulated above a suspended ceiling, virtually destroyed the Kinston, NC facility [13]. Six workers died and 37 others were injured; some were permanently disabled due to debilitating burn injuries.

The CSB found that the engineers who designed the building and equipment used for rubber compounding did not consult or apply the requirements of NFPA 654[3], *Standard for the Prevention of Fire and Dust Explosions from the Manufacturing, Processing, and Handling of Combustible Particulate Solids*, 2000. The building was not designed to minimize combustible dust accumulation; a suspended ceiling in the manufacturing area allowed dust to accumulate out of sight and uncleaned for nearly 20 years.

4.2 Hayes Lemmerz International, Huntington, IN

One employee died and another was seriously burned on October 29, 2003 when aluminum exploded in a dust collector near the foundry furnaces for Hayes Lemmerz at their Huntington, IN wheel casting plant [14]. The explosion propagated through ducting to engulf employees working near one of the furnaces.

The CSB found that, when engineers modified a system for remelting scrap aluminum chips, they did not design the dust collector according to NFPA 484, *Standard for Combustible Metals, Metal Powders, and Metal Dusts* (2002). The dust collector was placed too close to occupied areas of the building, and the system was not adequately maintained to prevent dangerous accumulations of aluminum dust in the dust collector, ducting and on surfaces in the work areas.

4.3 Marcus Oil and Chemical, Houston, TX

A tank filled with melted polyethylene failed on December 3, 2004, injuring nearby residents and firefighters and causing offsite structural damage [15]. The failed tank was used to reprocess off-color or residual wax. The CSB found that the tank failed because of sub-standard welding during a modification of the tank, which was actually a converted pressure vessel and operated at pressures of several atmospheres.

American Society of Mechanical Engineers (ASME) codes and standards for pressure vessels and National Board of Boiler and Pressure Vessels Inspectors (NBBI) codes for modifying pressure vessels were not followed at Marcus Oil. Further, although this lack was not causal to the incident, the wax vessels were not equipped with pressure relief, as required by ASME code.

The State of Texas is one of eleven in the United States that has not adopted the ASME pressure vessel codes into state law. When, as in this case, the facility was not covered by the PSM and RMP regulations and Texas had no state pressure vessel code, the responsibility falls on engineers involved in the design or modification of equipment to adhere to RAGAGEPs under their own code of ethics.

---

[3] Although the 2000 edition of the standard is cited, NFPA 654 was first issued in 1943.

5.0 Conclusion

Many of the accidents investigated by the CSB have a common causal factor of failure to comply with codes, standards and other Recognized and Generally Accepted Good Engineering Practices (RAGAGEPs). In some cases, the facility is not covered under OSHA's Process Safety Management rule or the EPA Risk Management Plan Rule, the two federal regulations that mandate RAGAGEP compliance. For these and situations where state or local law do not force RAGAGEP compliance, the working public must rely on the codes of ethics to establish the engineer's responsibility to know and adhere to RAGAGEPs.

Within the Engineering Code of Ethics is a requirement for engineers to practice only within their area of expertise, and to ensure their designs meet standards and established practices. This code of ethics is a legal requirement for licensed engineers. All accredited engineering curricula include ethics instruction, which convey the responsibility for any engineer to adhere to the code of ethics.

6.0 References

[1] Occupational Safety and Health Administration regulation *Process Safety Management of Highly Hazardous Chemicals,* 29 CFR 1910.119, May 1992.

[2] U.S. Environmental Protection Agency regulation Risk Management Plan Rule, Chemical Accident Prevention Program, 40 CFR 68, 112 (r), June 1996.

[3] Lexus-Nexus online glossary of legal terms http://support.lexis-nexis.com/academic/record.asp?ArticleID=academic_legal_glossary, consulted February 2007.

[4] Schuyler, Roy L, 1997. "Factors Impacting the Identification and Application of RAGAGEP", presented to the AIChE 1997 Spring National Meeting, Houston, TX, March 1997.

[5] Salot, W.J., 1997. "GEP and RAGAGEP in Resource-Limited Facilities", presented to the AIChE 1997 Spring National Meeting, Houston, TX, March 1997.

[6] Texas Engineering Practice Act and Rules Concerning the Practice of Engineering and Professional Engineering Licensure, Title 6 §137.55, December 2006. http://www.tbpe.state.tx.us/downloads/law_rules1206.pdf

[7] Accreditation Board for Engineering and Technology (ABET), "Criteria for Accrediting Engineering Programs", October 20. 2005. http://www.abet.org/Linked%20Documents-UPDATE/Criteria%20and%20PP/E001%2006-07%20EAC%20Criteria%205-25-06-06.pdf

[8] National Society of Professional Engineers. NSPE Code of Ethics for Engineers, http://nspe.org/ethics/eh1-code.asp, consulted February 2007. http://nspe.org/ethics/eh1-code.asp

[9] Davis, Michael, 1991. "Thinking Like an Engineer", Center for the Study of Ethics in the Professions at Illinois Institute of Technology, Princeton University Press, 1991.

[10] Davis, Michael, 1992. "Technical Decisions: Time to Rethink the Engineer's Responsibilities", Business & Professional Ethics Journal, Vol. 11, Nos. 3 & 4., Princeton University Press, Spring 1992.

[11] Martin, Mike W., and Schinzinger, Roland, 1996. Ethics in Engineering, Third Edition, McGraw-Hill, New York, NY. 1996.

[12] Whitbeck, Caroline. 1998. Ethics in Engineering Practice and Research, Cambridge University Press, New York, NY, 1998.

[13] The US Chemical Safety Board (CSB), 2004, Investigation Report, Dust Explosion, West Pharmaceutical Services, Inc., U.S. Chemical Safety and Hazard Investigation Board (CSB), September 2004.

[14] CSB, 2005. Investigation Report. Aluminum Dust Explosion, Hayes Lemmerz International-Huntington, Inc., September 2005.

[15] CSB 2006. Case Study, Polyethylene Wax Processing Facility Explosion and Fire, Marcus Oil and Chemical, June 2006.

LOSS PREVENTION SYMPOSIUM

Education Materials for Universities and Industry

Joseph Louvar
Wayne State University
and CCPS Consultant
jlouvar@ameritech.net

and

Dennis Hendershot
Chilworth Technology Inc.
and CCPS Consultant
d.c.hendershot@att.net

Prepared for Presentation at
American Institute of Chemical Engineers
2007 Spring National Meeting
41st Annual Loss Prevention Symposium
Houston, Texas
April 22–26, 2007

UNPUBLISHED

Education Materials for Universities and Industry

Joseph Louvar
Wayne State University
and CCPS Consultant
jlouvar@ameritech.net

and

Dennis Hendershot
Chilworth Technology Inc.
and CCPS Consultant
d.c.hendershot@att.net

Abstract

The Safety and Chemical Engineering Education (SACHE) program is a cooperative effort between the Center for Chemical Process Safety (CCPS) members and engineering schools to provide education materials that bring elements of process safety into the education of students studying chemical engineering, mechanical engineering, and chemistry. Although these educational materials were originally developed for universities, they can also be used for engineers in industry.

Effective education requires effective educational materials. The ultimate benefit of using effective educational materials is the elimination of accidents when handling hazardous materials, including chemicals and dusts. The targeted industries include chemical, pharmaceutical, biological, microchip, petroleum, etc. Additionally:

- These Internet accessible materials include the most important safety concepts and safety technology; and are in a format that is effective for teaching, because they are developed by industrial experts who understand and the principals of process safety, and professors who understand the principals of developing educational materials. Some examples of available products include: a) Dust Explosion Control, b) Design for Overpressure and Underpressure Protection, c) Movies on the Safety in Chemical Process Industries, d) Emergency Relief System Design for Single and Two-Phase Flow, e) Chemical Reactivity Hazards, f) Mini Case Histories, and g) Inherently Safer Design. These materials are ready to use by educators in various computer formats including PowerPoint presentations, Excel spreadsheets, and DVD movies. They can be downloaded and modified to fit the educator's style of teaching and to provide additional material to customize the content to better suit the audience.
- The SACHE products are available, internationally, to any SACHE member. A university membership allows all departments in the university to access and use the

products. A company membership allows any company facility (internationally) to access and use the products.

A very reasonable membership fee entitles members access to the specific year's products. Other years' products can be obtained for the membership fee for that year. The membership fee is a fraction of the value of the products. This article gives a complete list of available web-based SACHE products, and some descriptions.

1. Education versus Training

In a luncheon speech at the 2nd Global Congress on Process Safety in Orlando, Florida in April 2005, Mr. Glenn Erwin, Project Director of the Triangle of Prevention (TOP) Program at the United Steel, Paper and Forestry, Rubber, Manufacturing, Energy, Allied Industrial and Service Workers International Union (USW) strongly differentiated between "training" and "education." Glen commented "I train my dog, I educate my children." It important to recognize that everyone in our chemical plants and laboratories needs training and education. The following explanation should help you to know when and how to train and educate.

In a 2003 presentation [1], Dr. Andrew L. Carney, a surgeon, provided an excellent description of the difference between training and education. Training seeks "to impart a set of established facts and skills and to obtain a uniform predictable behavior from the trainees without the necessity of their understanding why they should act is the prescribed manner". This contrasts with education, which "seeks to have the student learn skills and to understand why actions are taken or not," and to "observe, analyze and question, to formulate hypotheses and make conclusions and then to act, live and modify their actions according to these conclusions."

Thus, training is the process that gives workers a skill or skill set. It requires the modification of certain employee behaviors. The effectiveness of a training program is measured by assessing how well the workers perform their daily tasks and obtain the desired results. Training means that you teach people to respond in a particular way to a situation - if the temperature gets too high, close this valve.

In Joseph Conrad's *Heart of Darkness* (1899) [2], Marlowe, the narrator and protagonist, gives an excellent example of an operator who has been trained, not educated. He describes the native fireman on river steamboat which took him up the Congo River:

> "A few months of training had done for that really fine chap....He was useful because he had been instructed; and what he knows was this - that should the water in that transparent thing disappear, the evil spirit inside the boiler would get angry through the greatness of his thirst, and take a terrible vengeance."

The fireman knows what to do in response to a specific piece of data – the water level in the boiler gage glass. However, he has no understanding of why this is important, and therefore is unable to use this knowledge more broadly.

Education modifies behaviors and thought processes. The effectiveness of an educational process is measured by noting the amount of information the learners have retained and the different uses to which they put the acquired information. The application of knowledge to different uses is based on understanding the fundamental principles. This is especially important when applying knowledge to identifying and eliminating hazards. H. H. Fawcett said. "To know is to survive and to ignore fundamentals is to court disaster" [3]. Education means understanding not just what to do, but why. Both training and education are necessary for the safe and efficient operation of chemical process plants. Many times, actions need to be taken quickly and automatically in a given situation, and good training along with frequent practice and drills will help to ensure proper response. But, in many cases, we need more education, particularly for engineers and plant operators.

2. SACHE Products for Training and Educating

The Safety and Chemical Engineering Education (SAChE) program, initiated in 1992, is a cooperative effort between the Center for Chemical Process Safety (CCPS) and engineering schools to provide teaching materials and programs that bring elements of process safety into the education of undergraduate and graduate students. These materials are also excellent for training and educating in an industrial setting.

3. Description of the Major SACHE Products

The SACHE products that are available on line for SACHE members are listed in Table 1. Complete descriptions of all of the products are available on the Internet via the SACHE Newsletters [4] for the specific year of development. A description of a few of the especially interesting products follows:

A. Dust Explosion Prevention and Control; This SACHE product covers the fundamentals and guidelines for preventing and mitigating the effects of dust explosions. It includes three sections:

Section 1: A PowerPoint presentation describes the basic concepts for understanding dust explosions (deflagrations, detonations, necessary conditions for causing dust explosions, and the consequences of dust explosions).

Section 2: This section includes two movies called Deadly Dust II and III. Deadly Dust II was developed in the 1980s and it emphasizes the technology to prevent dust explosions. Deadly Dust III was developed in the early 2000s and emphasizes the injuries caused by dust explosions. Both movies include case histories. Although the movies were specifically developed for the grain industry, everything in this product is relevant to the prevention of any dust explosion.

Section 3: This section covers the fundamentals for eliminating dust ignitions from the build-up of static electricity. The topics which are described in this PowerPoint presentation include: a) conditions for the accumulation of static charges, b) specific types of static electricity discharges, c) calculation methods for estimating the potential for ignitions, d) case histories, and e) design methods for eliminating ignitions.

This product supports the U.S. Chemical Safety Board's emphasis concerning the hazards of handling dusts, as identified in their November 2006 investigation report on dust explosion hazards in the United States [5]. This emphasis is due to the continued accidents and injuries that are the result of dust explosions – 281 incidents occurred in industry between 1980 and 2005, resulting in at least 119 fatalities and 718 injuries.

B. Design for Overpressure and Underpressure Protection: This product helps faculty, students, and professionals incorporate proper overpressure and underpressure protection in their process designs. This package helps students:

- Understand the technology, special engineering devices, and methods that are used for protection against overpressure and underpressure (vacuum) incidents,
- Understand the root causes of overpressure and underpressure incidents, and
- Design plants with the appropriate features to protect against overpressure and underpressure incidents.

This SACHE product contains two PowerPoint presentations that can be copied and modified to fit specific teaching/learning objectives. It includes detailed instructions covering a) emergency relief systems, b) runaway reactions, and c) safeguards to prevent accidents. This product can be used in design, fluid flow, or control courses. Some background information is included to assist the student in understanding these general concepts. This package can be used for three half-hour lectures or self-study exercises.

C. Safety in the Chemical Process Industries This movie series entitled "Safety in the Chemical Process Industries" presents an excellent introduction to the application of chemical process safety technology in an actual chemical facility. All of the movies were taped in the Chemical Engineering Department of the BASF Corporation. It includes demonstrations with actual process equipment.

This series is designed as instructional material for undergraduate students in chemical or mechanical engineering as well as industrial engineers or chemists who are being introduced to industrial safety. This series provides significant supplementary material for an existing undergraduate chemical engineering course on chemical process safety. A Study Guide and Instructor's Guide were written to accompany the movies.

The teachings of this product can also be effectively used in industry, e.g. it contains demonstrations for the vessel entry procedure.

D. Emergency Relief System Design for Single and Two-Phase Flow: This design module covers the principles and procedures for sizing emergency relief systems for both single-phase (gas or liquid) and two-phase flow, including relief devices as well as inlet and discharge piping. The Homogeneous Direct Integration method for two-phase flow is used; it is simpler, more general, and more rigorous than methods previously presented. The material is in a form which could be incorporated into courses on applied fluid mechanics, process safety, or process design, or it could be presented separately as a "stand alone" topic. It is assumed that the student or engineer has a familiarity and working knowledge of the macroscopic conservation laws for mass, energy, and momentum, and Newtonian fluid flow through pipes and fittings, including incompressible as well as compressible flow up to and including choked flow. A PowerPoint presentation and spreadsheet are given for describing the concepts and for making the calculations respectively. Example problems and solutions are also included.

E. Chemical Reactivity Hazards: This web-based instructional module contains about 100 web pages with extensive links, graphics, videos, and supplemental slides. It can be used either for a classroom presentation or as a self-paced tutorial. The module is designed as a junior or senior chemical engineering course or for engineers, by showing how uncontrolled chemical reactions in industry can lead to serious harm. Key concepts are given for avoiding unintended reactions. The five main sections in the module cover (a) three major incidents that show the potential consequences of uncontrolled reactions, (b) how chemical reactions get out of control, (c) data and lab testing resources used to identify reactivity hazards, (d) four approaches to make a facility inherently safer with respect to chemical reactivity hazards, and (e) strategies for designing facilities both to prevent and to mitigate uncontrolled chemical reactions. The module concludes with a ten-question informative quiz. An extensive glossary and bibliography are directly accessible from any page. This product is also available for purchase on a CD as a part of AIChE's publication venture with Wiley [6].

F. Mini-Case Histories: This product emphasizes learning from history or being doomed to repeat it. It includes three sections: a) Mini-Case Histories, b) Process Safety Beacon, and c) Safety Concepts as described below:

a) Mini-Case Histories: There are eight PowerPoint presentations, with notes, that cover the accidents at Bhopal, Monsanto, Phillips, Flixborough, Morton, Tosco, Hickson, and Sonat. The root causes of each accident are given, such as, poor designs, lack of training, and poor management. The information summarized in these case histories was taken from SACHE products and reports by the United States Chemical Safety and Hazard Investigation Board.

b) Process Safety Beacon: The Center for Chemical Process Safety is publishing monthly, one-page case histories or lessons learned. Currently the Beacon is sent by email to subscribers free of charge, and CCPS is planning on making past Beacons available on the Internet and as a SACHE product in the near future. A few of the Beacons are included in this product to remind readers that these lessons are available.

c) Safety Concepts: This product includes six PowerPoint presentations: i) Relief Valves, ii) Prevent Runaway Reactions, iii) Manage Design, Construction, and Operations, iv) BLEVEs, v) Explosions, and vi) Prevent Explosions. The concepts covered are those that are especially relevant to chemical plants.

The Mini-Case Histories and Safety Concepts are deliberately developed in a PowerPoint format to give the users the opportunity to add slides or lessons concerning their own interest.

G. Inherently Safer Design: This module introduces the concept of inherently safer design in the chemical processing industry. Inherently safer design (ISD) is receiving more attention in recent years from industry, community groups, and government. For example, the Contra Costa County, CA Industrial Safety Ordinance requires an evaluation of inherently safer process options for covered facilities in the county. When feasible, ISD can be a more robust and reliable approach to managing risk in the chemical industry. Following the terrorist attacks of September 11, 2001, ISD has also been promoted as a means to reduce chemical hazards, making chemical handling facilities less attractive terrorist targets. In 2005, the New Jersey Department of the Environment required that facilities covered by the state Toxic Catastrophe Prevention Act (TCPA) review the practicality of adopting inherently safer technology as an approach to reducing the potential impact of a terrorist attack. Similar requirements have been considered by the United States Congress, but have not yet been included in any legislation to date. It is important for chemical engineers to understand ISD, including its potential benefits and limitations, both to implement ISD where feasible and also to participate in public discussions about the application of ISD in the chemical industry.

Table 1: List of All SACHE Products on the SACHE Web-Site

Period	Number	Title	Author
2003	57	Mini Case Histories	Louvar
	58	The Hazards of Hydroxylamine; Case History	Christine Kelly/John Birtwistle
	59	SHE Text for Textbooks	Louvar, Horsch and Horsch
	60	Practical Design Practices for Relief Valve Sizing	Sterling/Parvin
2004	61	Consequence Modeling	Wagner
	62	Transportation Case Histories	Willey
	63	ioMosaic	
	64	Green Engineering	Schonnard
	65	Metal Structured Packing Fires	Mark Roberts
	66	Improving Communication Skills	Bethea
2005	67	Emergency Relief System Design for Single and Two-Phase Flow	Darby
	68	Electronic Version: Problem Set – Vol. 1	Wagner
	69	Electronic Version: Problem Set – Vol. 2	Willey, Crowl, Welker, and Darby
	70	Mach II (Digital Solutions)	
	71	Experiments for Runaways	Darby
	72	Chemical Reactivity Hazards	Johnson
	73	BioSafety	Schwartz
2006	74	Dust Explosion Control	Schoeff/Louvar
	75	Design for Overpressure and UnderPressure Protection	Grossel/Louvar
	76	Student AIChE Design Problem Solution (2002 Product)	Horsch, Louvar, and Wehman

Period	Number	Title	Author
	77	Chemical Process Safety – Videos on DVD	Reed and Crowl
	78	Inherently Safer Design	Hendershot
Planned for 2007	79	Explosion Prevention Research: Venting of Low Strength Enclosures	Howard and Louvar
	80	Beacon	Hendershot
	81	DOW's Static Electricity Control	Coe and Kennan
	82	Static Electricity	Rothschild
	83	Emergency Shutdown Systems	Art Dowell
	84	Fundamentals of Atm. Dispersion	Spicer
	85	Risk Assessment	Ralph Pike

4. Benefits of SACHE Products

The ultimate benefit of using the SACHE products is the elimination of accidents when handling hazardous materials, including chemicals and dusts. The targeted industries include chemical, pharmaceutical, biological, microchip, petroleum, etc. Additionally:

A. These materials are developed for chemical, mechanical, control, instrument and safety engineering students, and chemistry students; or young engineers and young chemists. These teaching materials and programs bring elements of process safety into the education of undergraduate and graduate students, and the education of young engineers who handle hazardous chemicals and/or dusts.

B. These materials include the most important safety concepts and safety technology; and are in a format that is effective for teaching, because they are developed by industrial experts (who understand the targeted industries and the principals of safety that must be included to prevent major accidents), and professors (who understand the principals of developing educational materials that can be used effectively in classrooms).

C. These materials are ready to use by educators (e.g. PowerPoint presentations, Excel programs, and DVD movies), however, in most cases they can be downloaded and modified to fit the educators' style of teaching.

D. The materials are on the Web; therefore they can be accessed by educators and engineers internationally.

E. The cost of the materials is a fraction of the value of the products. Committee members produce the materials with SACHE funds, and they are distributed via the Web, therefore distribution costs are minimized.

F. The SACHE products are available, internationally, to any SACHE member. A university membership allows all departments in the university to access and use the products. A company membership allows all plants (internationally) to access and use the products.

5. SACHE Committee

The SACHE committee is comprised of representatives from academe and industry in addition to American Institute of Chemical Engineers (AIChE) staff. University members have a specific interest in chemical process safety and the industrial members are process safety experts.

The committee chooses the products to be developed each year (ideas for products are solicited from the SACHE membership), and they are typically developed and written by committee members. The SACHE committee reviews the products extensively before being placed on the web-site with the ultimate aim of safer operations.

6. SACHE Funding

Universities (academic departments) and industry (companies) fund SACHE through annual membership fees. SACHE has over 100 members. These membership fees are periodically subsidized by CCPS to a maximum of $30,000 for a specific year. The extra funding is needed for Workshops or especially expensive products. The membership fee entitles members access to the specific year's products via the web-site (SACHE.org). Other years' products can be obtained for that year's cost of membership.

7. SACHE Deliverables

- The SACHE membership allows the members to download most of the products.
- The SACHE Newsletter (which can be downloaded from www.sache.org) can be searched for safety related topics, gives summaries of new products, and includes safety news in general.
- The SACHE members are invited to periodic Faculty Workshops (about every two years). SACHE members (one person from the membership organization) are placed in front of the attendance list compared to nonmembers. However, the organizers of the Workshop select the final list of attendees, based on diversity, etc. The SACHE Faculty Workshops give members a unique opportunity to learn more about safety in an industrial setting; and
- Links are provided that focus on especially effective safety training and education materials.

8. Summary

A. Engineers and students need to be appropriately educated in the concepts of chemical process safety to eliminate accidents and their consequences.

B. The education of students and engineers requires excellent educational materials.

C. The SACHE materials are especially effective, because they are developed by professors who know the requirements for effective learning, together with practicing engineers who know the practical side of chemical process safety.

D. The SACHE materials are available for a fraction of the cost (via a membership fee for the specific year) of a comparable commercial product.

E. The SACHE WEB based educational materials are available to university members (any department within a university), and industrial members (any plant internationally).

9. References

[1] Carney, A. L., "Factors in Instructional Design: Training versus Education," Instructional Technology Lab Seminar – Spring 2003, University of Illinois at Chicago, March 17-18, 2003. (http://www.uic.edu/depts/accc/itl/conf2003/usetech2enhance/carney2.pdf)

[2] Conrad, Joseph, *Heart of Darkness*. Ria Press Edition, November 2005, page 17 (originally published in 1899), http://riapress.com/riapress.

[3] Fawcett, H. H. and Wood, W. S., *Safety and Accident Prevention in Chemical Operations*, 2nd Edition. (New York: John Wiley and Sons, 1982), p. 1.

[4] SACHE Web Site, www.SACHE.org.

[5] Investigation *Report: Combustible Dust Hazard Study*. United States Chemical Safety and Hazard Investigation Board, November 2006, http://www.csb.gov.

[6] http://www.aiche.org/Publications/pubcat/listings/0470036648.aspx

CSB Safety Videos: A New and Effective Communication Tool for Accident Investigation Findings

Daniel Horowitz, US Chemical Safety Board, Sandy Gilmour, Sandy Gilmour Communications LLC

US Chemical Safety and Hazard Investigation Board
2175 K ST. NW Ste 400
Washington, DC 20037
202-261-7613
202-974-7603 (fax)
Daniel.Horowitz@csb.gov

Prepared for Presentation at
American Institute of Chemical Engineers
2007 Spring National Meeting
41st Annual Loss Prevention Symposium
Houston, Texas
April 22–26, 2007

UNPUBLISHED

CSB Safety Videos: A New and Effective Communication Tool for Accident Investigation Findings

Daniel Horowitz, US Chemical Safety Board, Sandy Gilmour, Sandy Gilmour Communications LLC

US Chemical Safety and Hazard Investigation Board
2175 K ST. NW Ste 400
Washington, DC 20037
202-261-7613
202-974-7603 (fax)
Daniel.Horowitz@csb.gov

Abstract

Since 1998 the U.S. Chemical Safety Board (CSB) has issued detailed written reports based on investigations of major U.S. chemical accidents. While the reports are indispensable written records of the investigations, in many cases the audiences for these lengthy technical products have been limited. In 2005, the CSB introduced a new product: short safety videos of five to ten minutes that summarize an investigation. The videos have been distributed primarily as streaming content over Internet and also in DVD and downloadable formats. A centerpiece of each video is a 3-D computer animation sequence that recreates a complex chemical accident. The videos also have included CSB board members and investigators discussing good industrial safety practices to prevent similar accidents. In the first eight months of the video program, the audience for CSB accident investigations increased by more than one order of magnitude, and the results of the investigations are now being used for training, hazard analysis, and plant engineering in a variety of industries worldwide. The presentation will include excerpts from recent safety videos, discuss their use by industry, and examine possible reasons why their distribution is so broad.

1. Results

Computer animations, which have been used for some time to present complex scenarios to juries in civil litigation, are now finding use in safety videos designed to inform chemical industry audiences about the prevention of major accidents. Short safety videos that incorporate computer animation are proving to be among the most effective vehicles for communicating accident scenarios and findings to managers, engineers, operators, and emergency responders.

Figure 1. CSB animation of BP Texas City refinery disaster that occurred during unit startup March 23, 2005.

In October 2005, an investigative team from the U.S. Chemical Safety Board first used a 3-D computer animation to explain the BP Texas City refinery disaster to an audience at a public hearing. The two-minute video sequence showed how flammable gasoline components overfilled a distillation tower during unit startup and then overflowed from a nearby vent stack. The animation sequence proved to be an invaluable tool for explaining the complex startup process and the failures that led to the accident.

The CSB further enhanced the animation by adding a narration by the lead investigator, vapor cloud and explosions models, and aftermath footage. The resulting six-minute video became the most widely known CSB product ever, viewed more than 410,000 times from the CSB website during its first 14 months online. International companies have translated the video into French, German, Spanish, Mandarin, and Cantonese. Surprisingly, the video also found significant audiences outside the traditional oil and chemical sectors, such as in health care, construction, and even space exploration. Boards of directors, top government officials, and entire petrochemical facility workforces have viewed the video.

Part of the success of the video program stems from the ease of distributing video materials through the Internet, through downloading, file-sharing, and emailing. The CSB did not actively market the BP video during the first few weeks of its release. Nonetheless, as word of the video evidently spread, the number of hits increased exponentially, peaking at over 24,000 hits in a single day on January 5, 2006. The steep climb reflected the phenomenon known as viral marketing, where a document or video spreads rapidly through existing social networks, in the manner of a communicable illness. In this case, the video spread among networks of operators, engineers, and safety professionals.

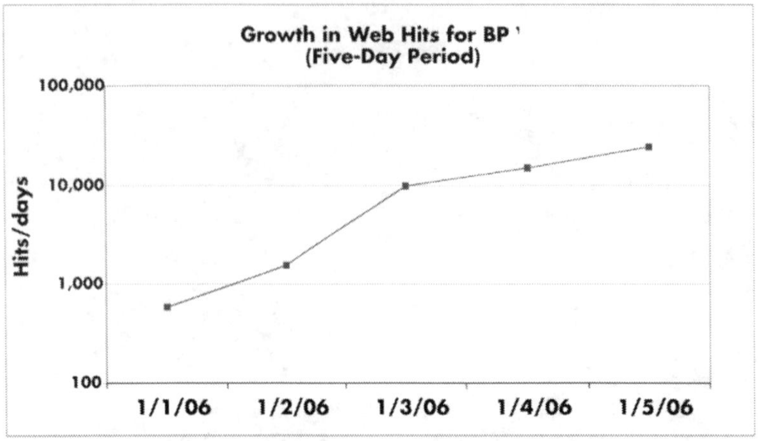

Encouraged by the widespread use of the BP video, the CSB developed and distributed five additional safety videos between January and November 2006, and two more videos were planned for release in March 2007. Unlike the BP video, which presented preliminary results from an ongoing investigation, the new videos described CSB investigations that had been completed and where root causes had been determined.

The new videos feature on-camera appearances by CSB board members and investigators discussing those root causes. All the videos use sophisticated 3-D computer animations to explain accident scenarios in just two to four minutes. For most audience members, this represents a substantial time saving compared to reading several technical chapters in an accident report. Unlike the CSB's written reports, the safety videos emphasize general lessons from accidents that can be applied more broadly throughout industry.

For example, a January 2006 safety video used a fatal accident at a New Jersey acetylene facility to illustrate the importance of appropriate operating procedures and safe venting designs when handling highly flammable gases. A March 2006 video used an ethylene oxide explosion in California to underscore the need for comprehensive training before authorizing operators to bypass automatic safety systems.

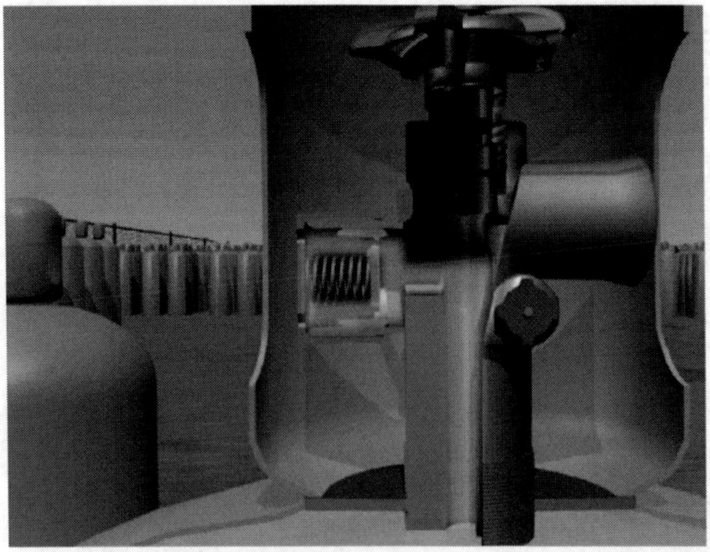

In June 2006, it was the flawed design of common propylene cylinder relief valves that were the subject of yet another CSB safety video. The video not only the depicted the failure mode of the valves, but also used dramatic video from the emergency response to show the severity of gas cylinder fires. It was followed by a July 2006 safety video that looked at fireproofing and isolation practices that can prevent catastrophic fires in petrochemical facilities.

But the most successful and widely viewed of the series, after the BP video, was a November 2006 safety video that examined a common hazard not only in the oil and chemical industry but throughout many sectors: confined space entry. The video focused on an accident at a Delaware refinery that tragically took the lives of two workers from oxygen deprivation. It showed how easy it is to underestimate the hazards of entry into an oxygen deficient atmosphere, and it explained the deadly hazards of attempting rescue in a confined space without proper procedures and equipment. The video also included a noted occupational physician describing the physiological effects and the human experience of oxygen deprivation.

Unlike the depictions of major fire and explosions, this video needed to present a series of mistaken human actions and decisions, and thus it was the first CSB video to use humanoid figures in a computer simulation. Surprisingly, computer animation techniques were at least as effective in portraying human beings as they were in illustrating equipment issues. To date, the video has generated more than 60,000 web hits and downloads in just over three months online.

2. Conclusion

Compared to conventional training videos – that often rely heavily on actor reenactments and extensive text and narration – CSB safety videos represent a new approach for conveying accident findings and providing useful elements for worker training programs. They use the power of actual events and interviews with the people who investigated them in order to present a few key messages in ways that viewers find compelling.

The videos are formatted and presented so as to capitalize on the nature of Internet, making extensive and costly marketing unnecessary, and in the near future the agency plans to use podcasting and posting to file sharing sites to further broaden distribution. Compared to conventional accident reports, which often have audiences of only a few thousand experts, the safety videos have broad appeal among personnel at all levels of specialization.

3. Acknowledgments

The authors thank Don Holmstrom, James Lay, Stephen Selk, Randy McClure, Robert Hall, John Vorderbrueggen, Chris Kirkpatrick, Bill Hoyle, Carolyn Merritt, John Bresland, Gary Visscher, Robert Newton, Ines Perez-Thompson, and Dr. Tee Guidotti for participating in the development of safety videos.

LOSS PREVENTION SYMPOSIUM

New Tools and Approaches to Expand Global Process Safety Capabilities

Scott Berger, Director
Center for Chemical Process Safety
American Institute of Chemical Engineers
New York, NY 10016
212-591-7237
scotb@aiche.org

and

Karen Person, Sponsor Support Engineer
Center for Chemical Process Safety
American Institute of Chemical Engineers
New York, NY 10016
212-591-7319
karep@aiche.org

Prepared for Presentation at
American Institute of Chemical Engineers
2007 Spring National Meeting
41st Annual Loss Prevention Symposium
Houston, Texas
April 22–26, 2007

UNPUBLISHED

New Tools and Approaches to Expand Global Process Safety Capabilities

Scott Berger, Director
Center for Chemical Process Safety
American Institute of Chemical Engineers
New York, NY 10016
212-591-7237
scotb@aiche.org

and

Karen Person, Sponsor Support Engineer
Center for Chemical Process Safety
American Institute of Chemical Engineers
New York, NY 10016
212-591-7319
karep@aiche.org

Abstract

From 1985 and 2000, large western chemical and petroleum companies, collaborating through the Center for Chemical Process Safety (CCPS), developed a sophisticated level of process safety knowledge and shared this knowledge in the western world through conferences and English-language books. This approach proved satisfactory to address the process safety needs of a large proportion of global industry existing at the time. However, since 2000, the industry has been expanding significantly in developing regions of the world. Communications in this region are complicated by language, time zone differences, and culture.

Meanwhile, as reported by the US Chemical Safety Board, many smaller companies, even in the west, are often unaware of the considerable resources that have been developed. Finally, new engineers entering the workplace around the world are generally unable to travel, and due to the Internet resources available to them during their formative years are less enthused about plowing through hardcover books to retrieve information. It is clear that traditional communication means do not successfully reach any of these groups.

CCPS has recognized that the communication paradigm must be changed and expanded. This paper is addresses means that CCPS has used to close physical and language gaps, through on-site participation and translation of key materials, and to utilize exciting new Internet tools to facilitate interaction across distances and meet the needs of new engineers.

1. Introduction

1.1. The formation of CCPS

In the months following the tragic chemical release accident in Bhopal India, industry executives were quick to realize that similar accidents could happen to them. While it might have been possible to dismiss Bhopal as a one-of-a-kind tragedy, the fact that Bhopal came just after a major LPG conflagration at a Pemex terminal in Mexico the month before, and a chemical tanker explosion in San Francisco Bay the month before that presented a long paragraph of catastrophe ending with a double exclamation point.

The industry executives started by discussing their companies' process safety programs, which each felt was quite good, though perhaps not as good as the program of Union Carbide. (Note: at the time, the terms loss prevention, plant safety, and process safety were used by different companies to mean what we now call Process Safety. We use the term Process Safety here throughout for consistency). As they dug deeper, they came to realize that their colleagues' Process Safety programs contained useful features that they had not previously known about. Perhaps more importantly, each executive realized that some of their colleagues had experienced accidents similar to the accidents they had recently experienced, and that had they known about the earlier accident, their own accident could have been avoided.

The conclusion was obvious. Process Safety could not be treated safety as a trade secret. The executives realized that their managers and technical employees had to share knowledge, practice, and lessons learned broadly with their colleagues around the world in order to protect their capital investments, their workers, and neighbors in their surrounding communities.

Of course the governments of the world, and particularly the United States government, frown upon companies sharing trade secrets. However, the anti-trust laws do provide a mechanism for sharing certain information for the public good, involving certain not-for-profit educational organizations to act as the mechanism for sharing. The not-for-profit becomes the owner of the shared information, assures that all sharing that takes place is for the public good, and makes the knowledge shared and developed in final form available at a reasonable price to the public at large.

And so the executives came to the American Institute of Chemical Engineers, long known for sharing chemical engineering knowledge, with a request to create a "Center for Plant Safety". Intensive discussions occurred between mid-February and mid-March, 1985, during which time the final name "Center for Chemical Process Safety (CCPS)" was chosen. On March 25, 1985, CCPS was born.

Since then, the goals of CCPS have remained relatively constant:

- Advance the state of the art in process safety technology and management practices
- Serve as the premier resource for information on process safety
- Foster process safety learning
- Promote process safety as a key industry value

1.2. Communicating Expert Knowledge, 1985-2001

Over CCPS' first 15 years, nearly all of these goals were realized through the development of Guideline and Concept books, or tied directly back in some way to books. In general, this was satisfactory for reaching the major chemical, petroleum, and pharmaceutical companies in the US, and indeed was the way that engineers of those days preferred to access their information. This approach unfortunately missed smaller US companies and almost all non-US companies. But it proved pareto-effective in addressing the biggest opportunities for process risk. However, by the turn of the century, globalization had begun to seriously take root. Large companies with overseas operations became large multinational companies, small companies were placed under more pressure than ever to survive, and new companies emerged in parts of the world which previously had limited chemical industry. It was no longer sufficient to publish a book, distribute a catalog, and hope that customers around the world would buy it. Indeed, the recent classes of chemical engineers graduating from universities were beginning to operate under the assumption that the Internet, rather than books, would be the source of all state-of-the-art knowledge. Clearly some changes were needed.

CCPS recognized this trend, and over the last six years, has been striving to use the latest technologies to make its materials more useful to its traditional constituents, to adapt to the expectations of emerging constituents, and to reach out to smaller companies in the US, as well as to companies of all sizes around the world.

2. Broadly Disseminating the Message

2.1 The Process Safety Beacon

November 2001 marked the first edition of the Process Safety Beacon, a monthly process safety communication to operating personnel. In just one page per month, the Beacon dispenses process safety lessons in easily absorbable doses in a simple format:

- Describe an actual incident or situation
- Explain how and why it happened
- Explain what the reader needs to know to prevent similar accidents

An example Beacon is included as Figure 1.

Beacons are used in many ways, including safety meetings, toolbox talks, and communications from the plant management. The Beacon has even been seen on bathroom and locker-room walls, unmarked by graffiti.

Early in the development of the Beacon, there was significant discussion among the members of the Beacon publishing subcommittee about whether the Beacon should be posted on a website, distributed by e-mail, or both. The subcommittee concluded that posting on a website is too passive; that readers could easily get out of the habit of looking for it every month. Instead, the Beacon is e-mailed to registered recipients each month, in a format that encourages recipients to forward it to their colleagues. To enable ease of receiving and forwarding, the Beacon is designed with low resolution graphics so that it is easy to transmit and receive by e-mail, even for recipients using dial up access.

A statistical study, conducted in 2004, showed that on the average, each Beacon recipient forwards their copy to 10 others, and each of these secondary recipients forward their copies to another 10, for a distribution multiple of 100. Consequently, the primary 7,000 Beacon e-mails transmitted each month reach an end-user audience of nearly three-quarters of a million readers.

To avoid leaving readers hanging without guidance about how to address the kinds of issues identified in each month's Beacon, the transmittal e-mail references a CCPS book that addresses the subject. The cover e-mail also makes the readers aware of other CCPS news of interest.

Possibly the biggest contribution the Beacon has made to the global expansion of process knowledge comes from its availability in many global languages. In 2003, CCPS began receiving offers from all around the world from volunteers willing to translate the Beacon. As of this writing, the Process Safety Beacon is now available in the 19 languages listed in table 1. Of course, many major world languages remain unavailable, and CCPS would welcome any volunteers willing to assist in making more languages available.

Figure 1: The Process Safety Beacon in Thai and English

Table 1: Available Languages for the Process Safety Beacon

Arabic	Farsi	Hindi	Portuguese (Brazil)
Chinese (Mandarin)	French	Italian	Spanish
Chinese (Traditional)	German	Japanese	Swedish
Dutch	Gujarati	Korean	Thai
English	Hebrew	Portuguese (Portugal)	

One drawback of the e-mail distribution method is that new recipients may find it difficult to obtain back-issues of the Beacon. CCPS has been providing CD's containing the Beacon archives at a modest cost, but in 2007 will launch a new web application which will enable users to search and browse past Beacons. However, the main distribution method will remain e-mail.

2.2 Website Content

Prior to 2001, the majority of content available on the CCPS website was, like most of the Internet, marketing related. In 2002, however, CCPS began adding short publications called "Safety Alerts" to its website. Safety Alerts were intended to provide basic awareness about important process safety topics, along with beginner level how-to information, and refer readers to more in depth resources.

This trend accelerated in late 2005, when the CCPS website underwent a major overhaul. One of the objectives of the overhaul was to put in place a structure in which many more short publications could be posted in a public "Knowledge Base" in a topic-driven format. When this effort is complete, we expect that engineers whose plants have fairly basic process safety needs will be able to meet them using this content. Engineers whose plants have more extensive process safety needs will be able to use this content to understand where in the CCPS universe of content to go for additional resources. Of course, tools that help CCPS meet its outreach goals, like the Process Safety Culture Evaluation Toolbox http://www.aiche.org/CCPS/PSCulture.aspx, which is useful for companies of all sizes, will continue to be available as appropriate.

As of this writing, CCPS is in discussions with the US Occupational Safety and Health Administration and others to establish an Alliance Program focused on process safety for small companies. We expect that this Alliance will tap into the public knowledge base to broadly disseminate general awareness of process safety hazards and methods.

The CCPS Knowledge base will also maintain more detailed information suitable for more complex operations and more skilled engineers. Because this more advanced content really needs to be used by experienced engineers who operate within a community of other experienced users, this level of content will be available mainly to the CCPS member community, who can interact with each other to better understand and continually update and perfect the content.

2.3 Electronic Books

It is logical for CCPS to follow the trend towards publishing technical e-books, following the emerging market of young engineers who prefer electronic references. However, CCPS pursued an electronic book strategy well ahead of this trend with a different goal in mind: to put the entire CCPS book collection at the fingertips of every engineer in a corporation. CCPS began converting its books into electronic form in mid-2003in partnership with Knovel Corporation. By early 2004, a substantial portion of the CCPS library was available in a fully searchable on-line format, and today every CCPS book currently in print is available at http://www.Knovel.com (Figure 3).

The CCPS Knovel library enables engineers to type in keywords, and be directed to the exact pages that these subjects are discussed no matter where in the CCPS library this information may be found.

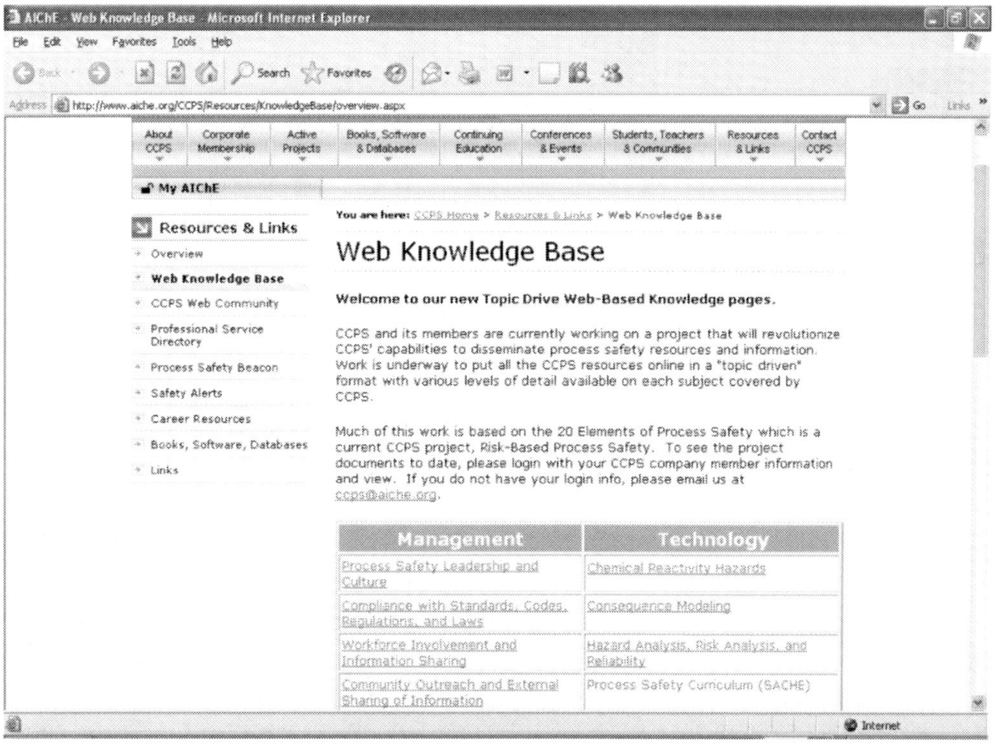

Figure 2. CCPS topic-drive Knowledge Base

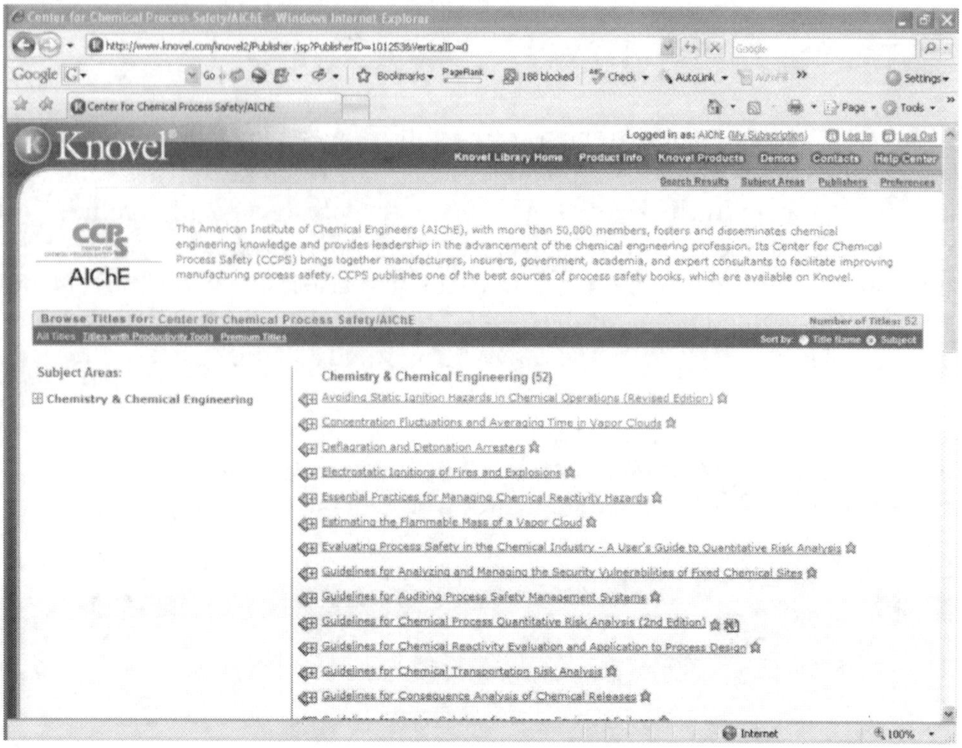

Figure 3: CCPS Library on Knovel.com

This form of access to CCPS books has tended to increase the usage of many of CCPS' lesser-selling books, including research reports (e.g. Estimating the Flammable Mass of a Vapor Cloud) as well as hands-on guidelines (e.g. Guidelines for Process Safety Fundamentals).

CCPS electronic books are also available in collection form and individually through Information Handling Services (titles before 2005 only) and CCPS' current hardcover book publisher John Wiley and Sons (titles after 2004 only).

3. Creating a Global Electronic CCPS Community

3.1 Electronic Meetings

During the last 6 years, would-be CCPS participants have been under intense pressure to reduce travel budgets, even as their jobs require them to travel more broadly around the world. These factors conspired to make it much more difficult for participants to attend 3 – 4 Technical Steering Committee meetings per year as well as numerous project subcommittee meetings, as was typical in the 1980's and 1990's.

By 2002, CCPS project teams were spending more of their project time on teleconferences than meeting face-to-face. While some of the camaraderie aspects of a project were lost, in some ways subcommittee momentum improved, because subcommittee members could meet more frequently for shorter periods of time. In the past few years, web conferencing has been added; helping participants collaborate as closely as it they were present in person.

In 2004, CCPS held its first Technical Steering Committee by web conference. TSC web conferences last only 3 hours, rather than the usual 1 - 2 days, but we have demonstrated over the past 2+ years that nearly all of the functions normally carried out at a face-to-face TSC meeting, such as project reports, technical presentations, voting, and workshops can be carried out in this format. In addition, participants from all over the world who might have otherwise been unable to travel to a TSC meeting have been able to participate. Unfortunately, we have not yet found a way to reproduce the informal technical and social interactions that are so valuable in face to face meetings.

3.2 Internet Community

While we have not found a way to create technical and social interactions in a web conference format, in 2006 we launched an application that allows CCPS members to have such interactions online. On the CCPS blogsite, members may post technical papers for discussion and comment, pose questions about issues important to them, and solicit and receive feedback on important issues. CCPS member companies are not limited in the number of their employees who may blog, chat, and browse. Confidential discussion groups may also be created related to relevant CCPS topics.

As CCPS starts its latest slate of projects, the CCPS Internet Community is also being used to share and project documents and debate important project issues with each other. Subcommittees have the option to make their project discussion open to other CCPS members to view, in order to follow progress, as well as to invite the general CCPS member community to provide input. This allows the CCPS membership to stay even more connected than ever to current developments.

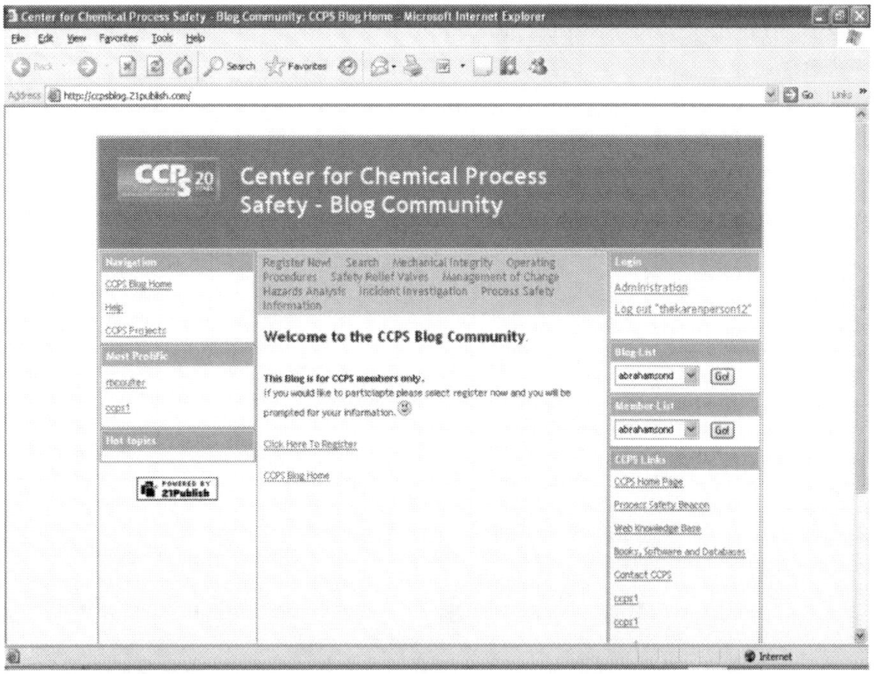

Figure 4: CCPS Process Safety Blog Home Page

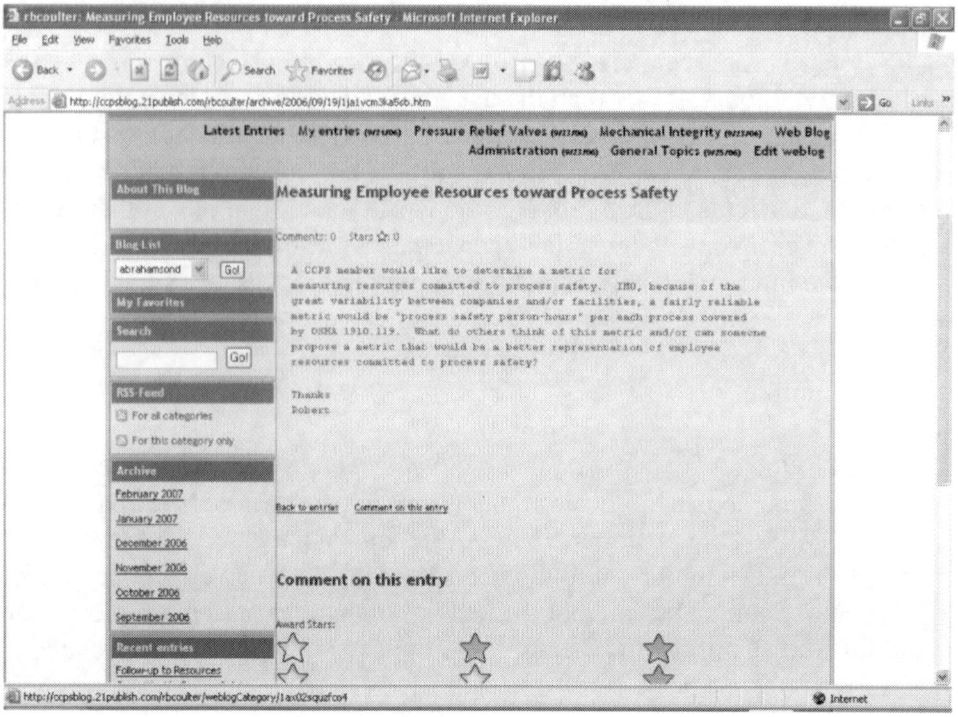

Figure 5: Sample Blog on Employee Resources

3.3 File Sharing and Collaboration

CCPS has also imitated project file sharing using Sharepoint technology which works like an FTP site but with more sophisticated capabilities. Users can upload project documents and edit them online in real time as others watch, run surveys and benchmarks, or use the General Discussion feature to hold newsgroup-style discussions on topics relevant to the project team.

CCPS members are currently using Sharepoint for posting meeting minutes and project announcements, performing document editing feedback and peer review as well as offering a secure working environment for detailed project updates.

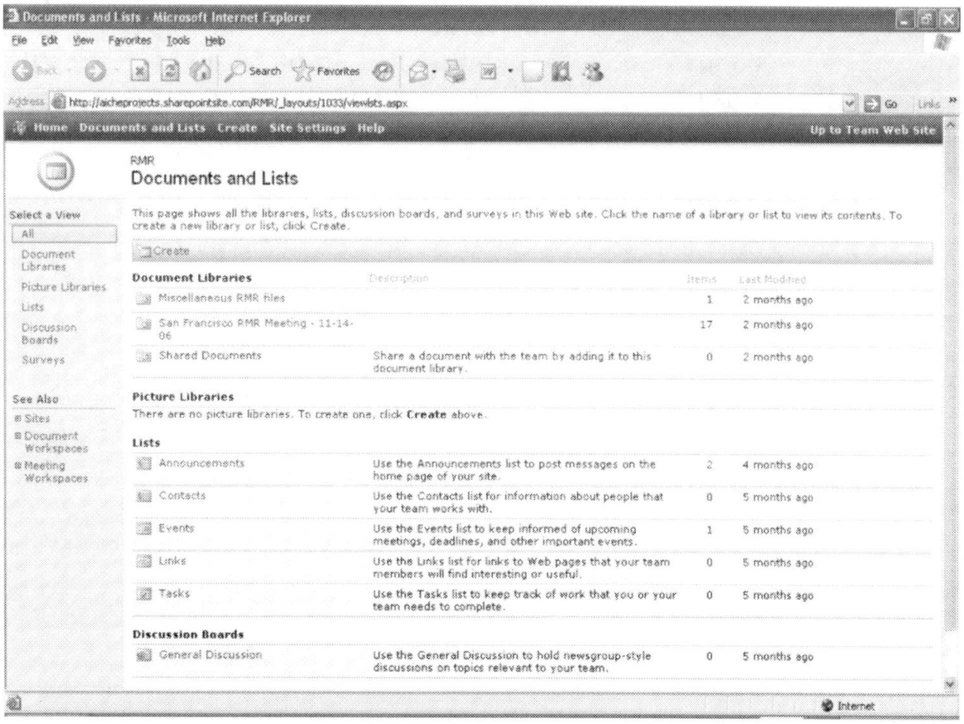

Figure 6: Sharepoint Technology

4. The Global CCPS Organization

The founders of CCPS always intended for CCPS' to have global impact. In its first full year of operation, CCPS called its first conference an international conference, held it in London, and invited global participation. Subsequent conferences, though all in North America, retain a significant international participation to this day. Likewise, CCPS books are known and used all around the world. Nonetheless, the membership of CCPS remained largely US-based, perhaps due to the difficulty in making multiple trips to the US every year in order to maintain active participation.

CCPS' new electronic books and tools remove many of these obstacles, and as a result, in 2006 new membership from outside the US exceeded new membership from inside the US for the first time. This new international perspective caused CCPS to take a good look at its mission, and come to the realization that while it could not back off on its first two goals, it needed to focus increased attention on its last two goals of fostering process safety learning and promoting process safety as a key industry value. This was found to be particularly in the two fastest growing economies, China and India.

In 2006, CCPS held two major workshops in Qingdao, China, and Mumbai, India to gauge the interest and level of expertise of engineers in these countries. Over 100 engineers from a broad range of companies large and small attended each workshop and provided an enthusiastic reception to the information presented by CCPS.

Buoyed by this experience, CCPS expects to have established, by the publishing date of this paper, franchise operations in these locations, named respectively the Center for Chemical Process Safety, China Section (CCPS-CS), and the Indian Institute of Process Safety (IIPS). These two arms will serve as CCPS' primary conduit for professional and pre-professional learning in these regions. We expect that additional country / regional operations will follow in the coming years.

While training will be the initial focus of CCPS-CS and IIPS, we fully expect that once the learning curve is surmounted, the experts we develop in these regions will be able to function as partners in advancing all of CCPS' goals.

5. Conclusion

There are many areas of business, technology development, and production, for which secrecy and confidentiality are essential to success, but Process Safety is not one of them. To the contrary, no accident that has happened at one plant should be allowed to occur at another plant, even a competitor's plant, because the lessons learned from the first accident weren't shared. Taken to the ultimate, the collective body of process safety knowledge must be captured in an organized set of concept textbooks explaining the technology and guidelines explaining how to put this technology into practice.

CCPS has filled this role since 1985, always with an eye to collecting this knowledge and disseminating it to a global audience. Recognizing that the world of business and the world of technology has been changing substantially since the turn of the century, CCPS is embracing new communication technologies and new business approaches to enable it to carry out its mission on a broadening scale.

Notes in the CCPS archives show opinions stated as early as 1995, and continuing to the present, that CCPS' work was nearly finished. The truth is quite the contrary. Every time we hear news of another major process safety incident, of workers injured, property damaged, business destroyed, and communities disrupted, it is tempting to ask ourselves whether we are making any progress at all. In truth, eliminating significant process safety incidents is a constant battle: as technologies evolve, as productivity pushes ever upward, and as business expands in another region of the world, new opportunities for incidents are created even as we eliminate the old ones.

CCPS has embraced new technologies and new business structures to help it communicate expert knowledge to technical communities around the world. If we have any hope to realize the ultimate goal of CCPS to eliminate process safety incidents, we must continue to exploit these new technologies and approaches, master them, and embrace the new ones as they emerge. The stakes are high, but with your help we will succeed.

Explosion Hazards During Fuel Transition in Combustion Equipment

Scott Davis, Ph.D.
Exponent
21 Strathmore Road
Natick, MA 01760 U.S.A.
Phone: 508-652-8557
Fax: 508-647-1899
daviss@exponent.com

Abstract

Combustion equipment capable of burning both gaseous and/or liquid fuels allows facilities to transition between various fuels based on the relative availability and price of the fuels. Explosion hazards, however, exist when transitioning from one fuel to another fuel in this equipment during normal operation. This paper describes a recent explosion in a single burner, positive pressure, thermal fluid heater.

The explosion occurred while transitioning the combustion fuel from natural gas to #6 fuel oil. During the transition the amount of combustion air was not sufficient to accommodate the addition of the #6 fuel oil resulting in an unstable, pulsating flame. The flame instability resulted in a partial or total extinction of the flame, followed shortly thereafter by a re-ignition of the unburned fuel/air mixture. The subsequent re-ignition created an overpressure that blew out the convection section of the heater. This paper reviews the several lessons that can be learned from this incident in order to prevent future incidents, including a review of guidelines and standards applicable to similar combustion equipment that outline control strategies for safely transitioning between fuels.

1. Background

This paper describes a recent investigation performed by Exponent, where an explosion occurred in a thermal fluid heater during a transition from natural gas to #6 fuel oil. The details of the incident are discussed in the first section, along with a review of guidelines and standards applicable to similar combustion equipment. In the second section, a detailed analysis of the incident is provided which includes operator procedures, process data recorded and the most likely scenario for the incident. The paper will conclude with lessons learned and recommendations to prevent similar incident from occurring.

1.1 Incident

Exponent Failure Analysis Associates (Exponent) was asked to investigate the explosion that occurred in a thermal fluid heater at 2:00 pm on the day of the incident. The explosion occurred while transitioning the combustion fuel from natural gas to #6 fuel oil. The incident heater was a vertical-cylindrical process heater with a radiant section that was approximately 43 feet in height followed by a 12-foot convective section (see Figure 1). The thermal fluid heater had a BTU input of 54.0 mmbtu/hr, and was fired by a single burner capable of firing either natural gas or # 6 fuel oil. The burner can simultaneously fire both fuels during a transfer from natural gas to #6 fuel oil. Under normal operation, the stack has a slight positive pressure (0.2" w.c.). The damage was mostly limited to the convective section, which ruptured due to an overpressure inside the heater as shown in Figure 2.

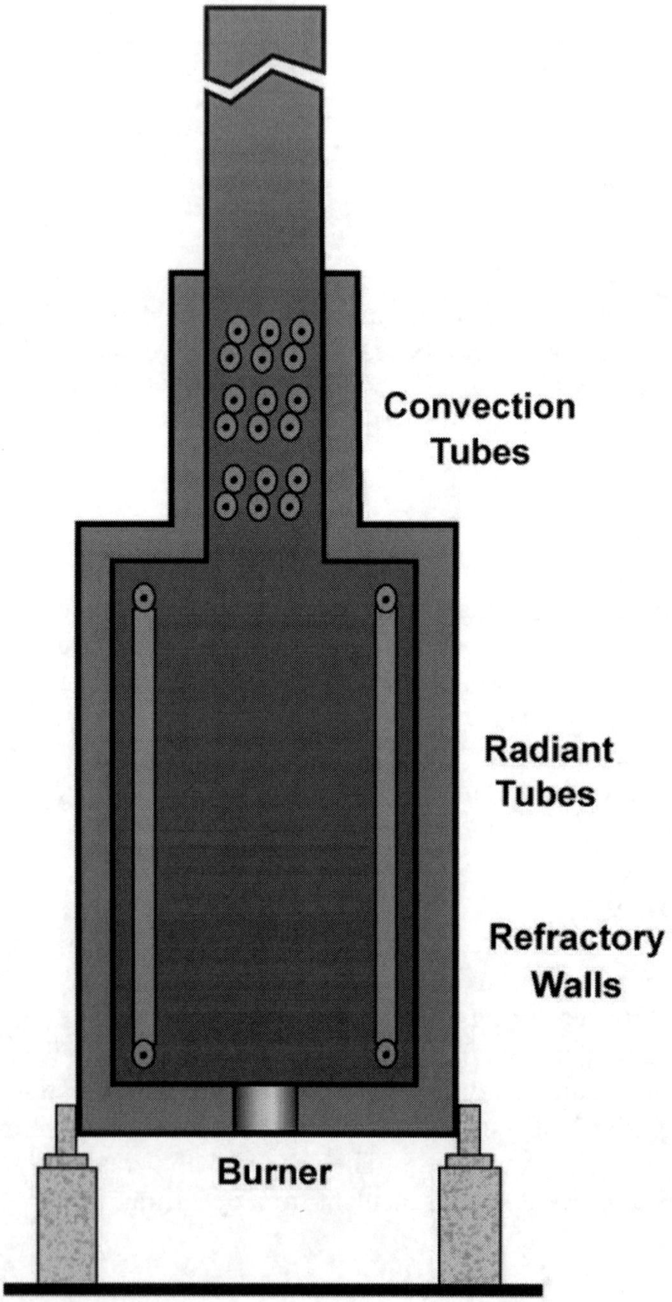

Figure 1: Vertical Cylindrical Process Heater

Figure 2: Ruptured Convection Section

At this facility, the fuel in the burner was transitioned manually from natural gas to #6 fuel oil. The process data for the thermal fluid heater around the time of the incident were recorded on strip charts and included: combustion-air flow rate, natural-gas flow rate, #6 fuel-oil flow rate, recirculation fuel-oil flow rate, percent oxygen in the flue gas, stack temperature and thermal-fluid flow rate, pressure and outlet temperature. A limited temporal sampling of data was also recorded on the Distributed Control System (DCS) for the thermal fluid heater, but this data does not include the combustion-air flow rate.

1.2 Codes and Standards

Thermal fluid heaters are not currently covered by an NFPA standard and are specifically excluded in the scope of NFPA 85 "Boiler and Combustion Systems Hazards Code".[1] However, guidance provided in Chapter 5 of NFPA 85 and in Factory Mutual Data Sheet (FM DS) 6-4 on single burner boilers that are capable of simultaneously firing natural gas and oil is helpful. Section 5.7.4.2 "Subsequent Fuel Light-Off — Oil" specifically addresses guidelines associated with a manual transition from natural gas to oil. A brief review of section 5.7.4.2 from NFPA 85 will be given next. The reader is referred to NFPA 85 (2004) section 5.7.4.2 for the actual text.

The general procedure for transitioning manually from natural gas to fuel oil as outlined by NFPA 85 includes the following steps: (1) ensure all controls are in the manual position, (2) reduce the control for the gas-firing rate to the light-off position, (3) place the oil control to the light-off position and (4) place the fuel transfer into combination oil/gas mode. Manual shutoff off valves, capable of adjusting the firing rate, shall be provided at the burner downstream of the safety shutoff valves for both natural gas and fuel oil. Once all of the oil safety interlocks are satisfied, the fuel transfer is performed using the manual gas and oil valves mentioned above and

[1] Currently an NFPA working group is developing a proposed standard for thermal fluid heaters.

not by the using control valves that are normally used to adjust firing rate. The procedure entails slowly opening the manual oil valve to light the oil flame from the gas flame. Once the oil flame is lit, the oil-firing rate is slowly increased while simultaneously lowering the gas-firing rate, trying to maintain a constant heat input of the combined fuel flow to the burner.

This procedure continues until the manual oil valve is fully open and the manual gas valve is completely closed. During transition, excess air should be maintained at all times by continuously monitoring the air-to-fuel ratio, oxygen indicator or opacity indicator. Also the airflow must be maintained at a constant rate.

2. Analysis

2.1 Typical Transfer Procedure at the Facility

The procedure employed at the facility for transitioning from natural gas to #6 fuel oil at the time of the incident is summarized as follows. All controls were set to "manual" and the fuel transfer was set to combination gas/oil. Guidelines at the facility instructed operators to lower the gas-firing rate to the low-fire position during fuel transfer. These guidelines were based on the burner manufacturer's specifications. However, operators indicated that it was not possible to lower the gas-firing rate to the low-fire position due to loss of flame during the transition. Therefore, the gas-firing rate needed to be maintained at a typical gas-firing rate just prior to transfer. The oil-firing rate was set to a minimum level to maintain a stable oil flame. The manual oil valve was opened and the oil flame was established. The transfer was performed at the control valves by lowering the gas-firing rate and gradually increasing the oil-firing rate to maintain a stable flame.

2.2 Strip Chart and DCS data

The data recorded on the DCS can be seen in Figure 3, while Figure 4 shows the time shifted data obtained from the strip charts. The time coordinate indicated on the original strip charts is generally not the same as the actual time of day and can differ from one strip chart to another. Furthermore, the time shift between pens is approximately 5 minutes. The data in Figure 4 was compiled by matching the time coordinate for the natural gas flow rate so that the changes in the natural gas flow rate profile on the strip charts matched those recorded on the DCS. Next, the time coordinate for combustion air flow rate and the #6 fuel flow rate were shifted relative to the natural gas flow rate by adding 5 minutes and subtracting 5 minutes, respectively. A time shift similar to the natural gas flow rate was also performed for the percent excess oxygen.

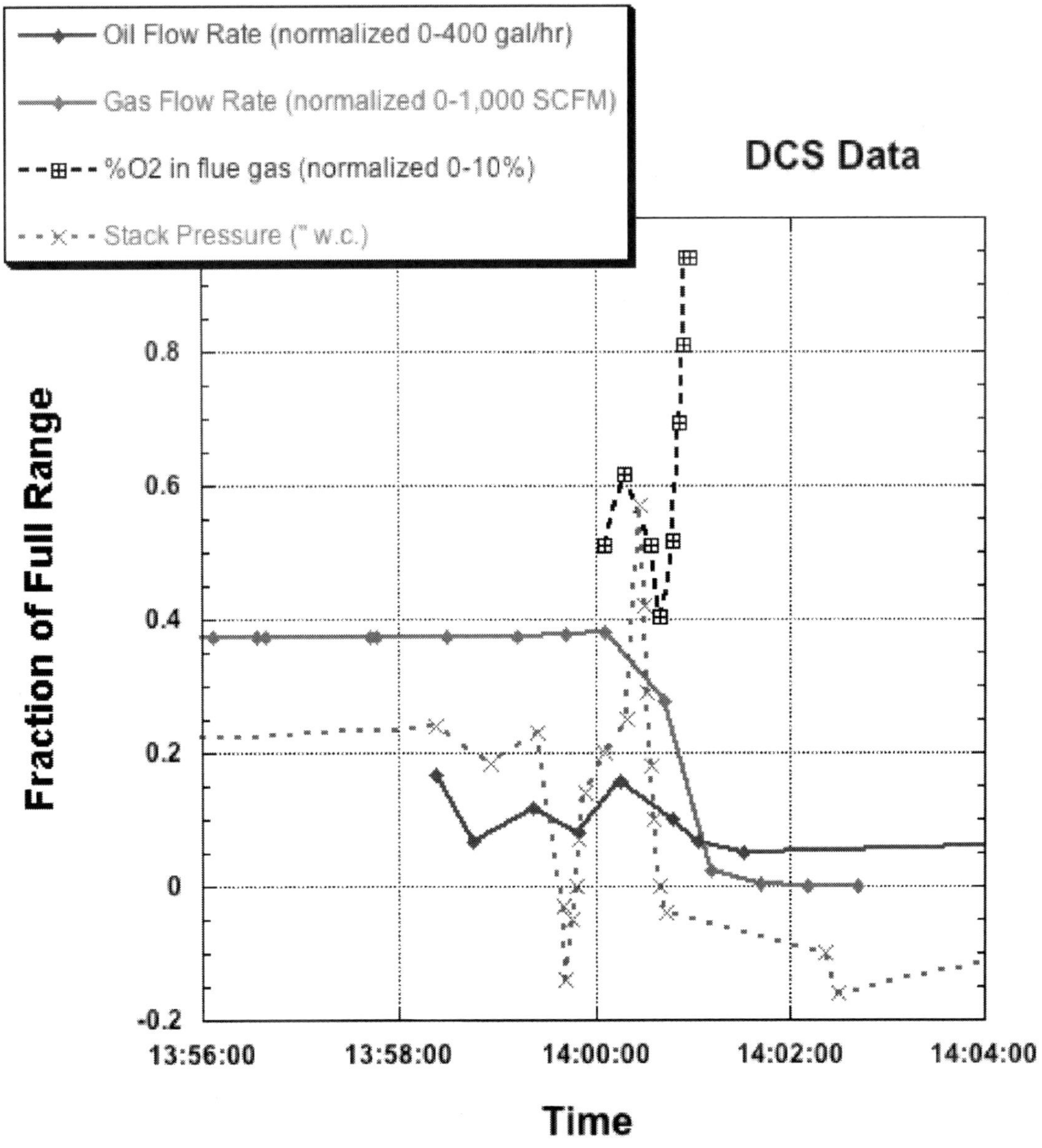

Figure 3: Process Data Recorded on the DCS System

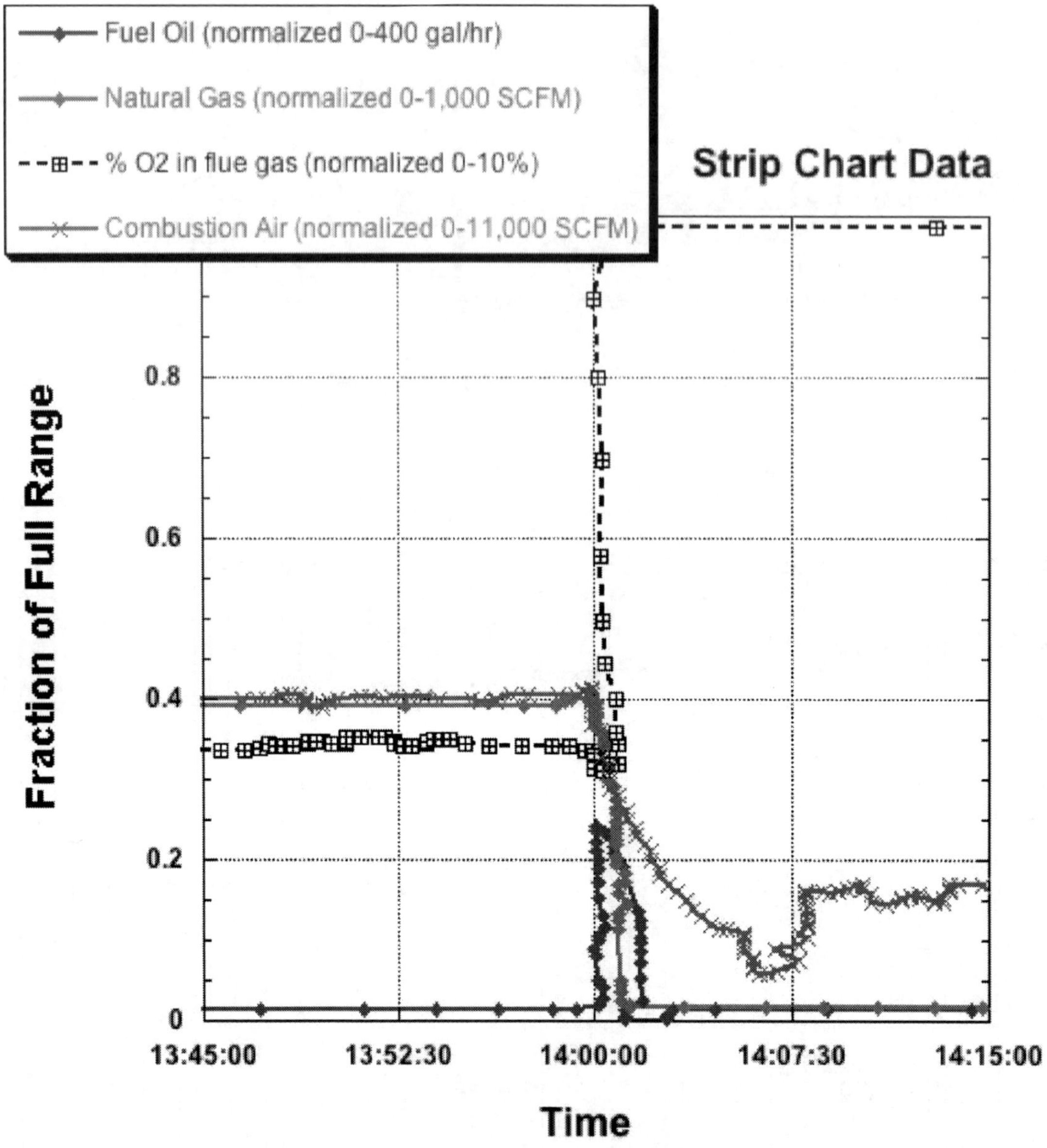

Figure 4: Process Data Recorded on the Strip Charts

Moments before 14:00, the recorded natural gas and air flow rates were approximately 372 SCFM and 4400 SCFM respectively, while the $\%O_2$ in the flue gas was recorded on the strip charts to be around 3.4%. Sometime just after 14:00, #6 fuel oil was introduced to the burner. The maximum flow rate of #6 fuel oil recorded after 14:00 by the DCS and the strip charts was approximately 1.04 gal/min (62.4 gal/hr) and 1.6 gal/min (96.9 gal/hr), respectively. The required air for ideal or stoichiometric combustion (i.e. the amount of air required to burn all the fuel and have 0% excess O_2 in the flue gas) at the moment when #6 fuel oil was introduced and both fuels were burning in the burner can be seen in Table 1.

Table 1: Required Combustion Air

Data Source	Natural Gas (SCFM)	#6 Fuel Oil (gal/min)	Air (SCFM)	Required Air (SCFM)
DCS	372	1.04	4400	5100
Strip Chart	372	1.6	4400	5900

The supplied combustion air was not sufficient for complete combustion of both fuels after the #6 fuel oil was added. Why then did the $\%O_2$ in the flue gas not drop off to zero? Based on the process data, the $\%O_2$ in the flue gas started to decrease and then rose sharply after the addition of the #6 fuel oil, while the stack pressure was initially positive and then became negative. Testing of the other process heaters at the facility showed that there was a significant time delay (on the order of 30-60 seconds) between the addition of the #6 fuel oil and the moment the $\%O_2$ in the flue gas dropped off during a transition from natural to fuel oil. This explains why the $\%O_2$ in the flue gas did not drop off to zero on the time scale associated with this incident.

2.3 Scenario

The most likely scenario is the amount of combustion air was not sufficient to accommodate the addition of the #6 fuel oil and resulted in an unstable, pulsating flame. The flame instability resulted in a partial or total extinction of the flame, followed shortly thereafter by a re-ignition of the unburned fuel/air mixture. The subsequent re-ignition created an overpressure that blew out the convection section.

3. Corrective Actions

The procedure employed at the facility for fuel transition from natural gas to fuel oil differed from that outlined in NFPA 85 in four areas: (1) the gas-firing rate could not be reduced to the low fire position for fuel transfer, (2) the manual transfer occurred at the control valves and not at the manual valves, (3) the heat input for both fuels was not maintained at a nearly constant level and (4) excess air was not maintained at all times. The two major areas that were addressed at the facility after the incident involved reducing the gas-firing rate to the low-fire position and maintaining excess air at all time during fuel transition.

The burner manufacturer was contacted to address problems associated with fuel transition while the gas-firing rate was in the low-fire position. Loss of flame was a common problem associated with fuel transitions below 40% gas-firing rate. The goal was to minimize the amount of fuel required during transition, hence minimize the risk associated with transfer. Once the minimum in the gas-firing rate was determined for a stable fuel transition, the issue of excess air could be addressed.

Knowing the firing rates of both the natural gas and #6 fuel oil during transition, the facility established a protocol that calculated the amount of combustion air needed such that $\%O_2$ in the flue gases was approximately 5%. The combustion air was maintained at this level throughout the fuel transition. After completion of the transfer, the combustion air was adjusted to the appropriate set point for the given fuel oil rate.

4. Conclusions

Burning in a fuel rich environment can have undesirable effects and should be avoided. These effects include: flame instabilities resulting in an unsteady/pulsating type combustion in the chamber, and incomplete combustion resulting in the formation of CO and other unburned hydrocarbons. Therefore, facilities should always maintain excess air during fuel transitions. One way to ensure excess air is by employing procedures similar to those outlined in NFPA 85 for automated transfers. For example, for transitions from natural gas to fuel oil this would include: (1) ensure the natural gas is set to the light-off position, (2) ensure the position of the control valve for the fuel oil is set to the light-off position, (3) gradually increase the combustion air to a preset value corresponding to the addition input of the fuel oil, (4) establish oil flow to the burner, (5) establish a stable oil flame, (6) shutoff the natural gas and (7) remove the excess combustion air.

It is expected that these issues will be addressed in a future NFPA standard.

What I Learned as an Investigator with the CSB

John F. Murphy, PE
Process Safety Services
2304 Kenya Lane
Punta Gorda, FL 33983
Phone: 941-624-0171
hamjfm@earthlink.net

Abstract

After a long career with the Dow Chemical Company and several years in process safety consulting, I had the privilege of working as an investigator with the U. S. Chemical Safety and Hazard Investigation Board. During my five years at the CSB, I led four major investigations and participated in a total of nine investigations. The investigations ranged from the Reactive Hazard Investigation which analyzed reactive chemical incidents in the U S. over a twenty year period to individual incidents that occurred in small, medium and large companies both inside and outside the chemical processing industries. From these investigations, I have learned some common lessons regarding the nature of chemical plant incidents. This paper discusses the incident investigations that I was involved in and the common lessons learned from these investigations.

1. Introduction

After a long career with the Dow Chemical Company in manufacturing and process safety roles and several years working as a process safety consultant, I had the privilege of working as an investigator with the U. S. Chemical Safety and Hazard Investigation Board. During my tenure with the CSB I formed opinions on what the common lessons learned are from the investigations that I was involved in. These opinions are my own and do not necessary agree with those of the organizations I have been a part of during my career. I will discuss the lessons learned from the investigations that I have been involved in with the hope that it will help prevent future incidents. I pass these thoughts along to those involved in incident investigation today and in the future so that they can learn from my experience.

2. Incidents Investigated

This paper is based on my experience as an investigator on the nine investigations that I participated in at the CSB. I was the lead investigator on four investigations, led the initial field investigation on two more investigations and participated as a team member on the others. The following is a list of the nine investigations with a short incident description taken from the CSB web site.

2.1. Bethlehem Steel Corporation Gas Condensate Fire: Chesterton, IN, February 2, 2001 (team member)

Incident Description:

On February 2, 2001 a fire occurred at the Bethlehem Steel Corporation's Burns Harbor mill in Chesterton, Indiana. One Bethlehem Steel millwright and one contractor supervisor died. Four Bethlehem Steel millwrights were injured, one seriously. Workers were attempting to remove a slip blind and a cracked valve from a coke oven gas line leading to a decommissioned furnace. During the removal of the valve, flammable liquid was released and ignited.[10]

2.2. Improving Reactive Hazard Management: September 17, 2002 (lead investigator)

Incident Description:

In August, 2000 following the investigation of a serious incident at Morton International, the Board initiated a comprehensive review of reactive hazards nationwide. The purpose of the investigation was to develop recommendations to reduce the number and severity of such incidents.[10]

2.3. Georgia-Pacific Corp. Hydrogen Sulfide Poisoning: Pennington, AL, January 16, 2002 (initial lead in the field, team member)

Incident Description:

On January 16, 2002 a highly toxic hydrogen sulfide gas leaked from a sewer man way at the Georgia–Pacific Naheola mill in Pennington, Alabama. Several people working near the man way were exposed to the gas. Two contractors from Burkes Construction Inc. were killed. Eight people were injured-seven employees and one employee of Davison Transport, Inc. Choctaw County paramedics who transported the victims to the hospital reported symptoms of hydrogen sulfide exposure.[10]

2.4. DPC Enterprises Chlorine Release: Festus, MO August 14, 2002 (lead investigator)

Incident Description:

On August 14, 2002 a chlorine transfer hose ruptured during a rail car unloading operation at the DPC Enterprises chlorine repackaging facility near Festus, Missouri. The hose rupture ultimately led to the release of 48,000 pounds of chlorine, causing three workers and 63 residents to seek medical treatment.[10]

2.5. Technic Inc. Ventilation System Explosion: Cranston, RI, February 7, 2003 (lead investigator)

Incident Description:

On February 7, 2003 a worker was seriously injured in an explosion at Technic Inc. in Cranston, Rhode Island. The explosion occurred during the maintenance on a ventilation system connected to multiple reactors, evidently due to an accumulation of hazardous material inside. The plant manufactures precious metal processing chemicals.[10]

2.6. Honeywell Chemical Incidents: Baton Rouge, LA, July 20, 2003 (initial field lead, team member)

Incident Description:

On July 20, 2003 there was a release of chlorine gas from the Honeywell refrigerant manufacturing plant in Baton Rouge, Louisiana. The accident resulted in the hospitalization of four plant workers and required residents within a half-mile radius to shelter in their homes. On July 29, 2003 a worker was fatally injured by exposure to antimony pentachloride when a gas cylinder released its contents to the atmosphere. On August 13, 2003 two plant workers were exposed to hydrofluoric acid, and one was hospitalized.[10]

2.7. Hayes Lemmerz Dust Explosions and Fire: Huntington, IN, October 29, 2003 (team member)

Incident Description:

On the evening of October 29, 2003 a series of explosions severely burned two workers, injured a third, and caused property damage to the Hayes Lemmerz manufacture plant in Huntington, Indiana. One of the severely burned men subsequently died. The Hayes Lemmerz plant manufactures cast aluminum automotive wheels, and the explosions were fueled by accumulated aluminum dusts, a flammable byproduct of the wheel production process.[10]

2.8. DPC Enterprises Chlorine Release: Glendale, AZ, November 17, 2003 (lead investigator)

Incident Description:

On November 17, 2003 there was a release of chlorine gas from the DPC Enterprises chlorine repackaging facility in Glendale, Arizona, near Phoenix. Fourteen people, including ten police officers, required treatment for chlorine exposure. The release occurred when chlorine vapors from a rail car unloading operation escaped from a system designed to recapture the material, known as a scrubber. Owing to the exhaustion of absorbent chemicals in the scrubber, chlorine gas was released.[10]

2.9. Acetylene Service Company Gas Explosion: Perth Amboy, NJ, January 25, 2005 (team member)

Incident Description:

On January 25, 2005 a gas explosion killed three workers at the Acetylene Service Company plant in Perth Amboy, NJ. The blast originated in a wood shed located near six large storage tanks that received liquid waste from the plant's acetylene generating system. The plant produces, repackages, and distributes acetylene used in welding.[10]

3. Lessons Learned

Every incident is, of course, unique and has its own story and ultimately its own causes, but there are surprisingly similarities. The similarities that I see include:

- The hazards involved in the incident were well known
- There were warning signs leading up to the incident
- There were management system problems
- Incidents occur in companies of all sizes
- Many chemical related incidents occur in businesses outside of chemical manufacturing
- Incidents often occur during start up, shut down, or maintenance rather than during normal operations
- Regulations are often inadequate or not enforced
- Management thought they had a good safety program
- There are often safety culture issues
- They were all preventable.

I will discuss each one of these observations. I will give examples from the investigations I participated in and show why each observation is important in our effort to eliminate or at least minimize the number and severity of future incidents.

3.1. The hazard(s) involved in the incident is well known

For all the incidents I was involved in this was the case. In the Reactive Chemicals Hazard Investigation we discovered that 90% of the 167 incidents analyzed involved reactive hazards that are documented in publicly available literature accessible to the chemical processing and handling industry.[2] In the Bethlehem Steel incident the hazards of line opening and the flammability of coke oven gas condensate were well known.[1] In the Georgia Pacific incident the reaction chemistry of sulfuric acid and sodium hydrosulfide (NaSH) to form hydrogen sulfide was also well known.[3] Table 1 lists the well known hazards for each of the incidents that I was involved in. What is the problem then if the hazards involved in incidents are already well known? The problem is that the knowledge of the hazard is not always known to the people who need to know when they need to know it. One of the keys to incident prevention then is the dissemination of the existing knowledge of the hazards to those who need to know through effective communication and training.

Table 1. Known Hazards Involved in Incidents

Incident Name	Known Hazard(s)
Bethlehem Steel	Line opening, flammability of coke oven gas condensate[1]
Improving Reactive Hazard Management	Over 90 % of the incidents analyzed by CSB involved reactive hazards that are documented in publicly available literature[2]
Georgia Pacific Corp.	Toxicity of hydrogen sulfide, reaction of sulfuric acid and NaSH to form hydrogen sulfide[3]
DPC Enterprises: Festus, MO	Wet chlorine and stainless steel corrosion, toxicity of chlorine[4]
Technic Inc.	Silver nitrate is a strong oxider reactive with organics, silver nitrate and ammonia can form silver azide and silver amide which are shock sensitive[5]
Honeywell Chlorine Release	Toxicity of chlorine, hazard of occupied buildings in chemical processing areas[6]
Honeywell Contaminated Antimony Pentachloride Exposure	Toxicity of antimony pentachloride, line and vessel opening[6]
Honeywell Hydrogen Fluoride Release	Toxicity of hydrogen fluoride, hazards of nonroutine operations[6]
Hayes Lemmerz	Combustibility of aluminum dust, hazard of combustible dust accumulation in confined spaces[7]
DPC Enterprises: Glendale, AZ	Reaction of bleach at low PH to produce chlorine, toxicity of chlorine[8]
Acetylene Service Company	Flammability of acetylene, hazard of flammable gases in confined spaces[9]

3.2. There were warning signs leading up to the incident

Many times this is the case. In the DPC Festus chlorine release, there was visible evidence that wet chlorine had corroded the piping assembly used to connect chlorine rail cars, but there was no effort to replace the pipe or properly test the emergency shut down system by verifying that the shut off valve on the rail car actually closed. When the hose failed on the day of the incident the emergency shut down valves did not close properly because of the accumulation of corrosion products in the valve seats resulting in the release of 48,000 pounds of chlorine.[4] In the Technic incident there had been problems with proper air flow in the ventilation system, but there was no effort to check the duct work for the accumulation of incompatible solids. The ventilation system was not designed to handle dust. Likely the reaction of an oxidizer and organic dust or the detonation of a shock sensitive material was the cause of the explosion which resulted in severe injuries to an employee.[5] In the Hayes Lemmerz incident there had been a history of fires in the duct work involving aluminum dusts, but there was aluminum dust accumulated throughout the building housing the chip reprocessing process. On the day of the incident the accumulated dust found an ignition source resulting in the explosion that killed one worker and injured six.[7] Table 2 lists some of the warning signs associated with the incidents that I was involved in. We need to pay attention to abnormal situations and use near miss reporting systems so we can correct small problems before they become major incidents. When abnormal situations occur, we must ask why.

Table 2. Incident Warning Signs

Incident Name	Warning Signs
Bethlehem Steel	Three prior incidents demonstrated that coke oven gas condensate was flammable[1]
Improving Reactive Hazard Management	Many of the 167 incidents had warning signs[2]
Georgia Pacific Corp.	Chlorine dioxide had previously been observed leaking from the (sealed?) man way [3]
DPC Enterprises: Festus, MO	Corrosion was easily visible on the unloading piping assembly[4]
Technic Inc.	Insufficient air flow was observed in part of the ventilation system[5]
Honeywell Chlorine Release	Chlorine odors had been detected in the control room several times in the past[6]
Honeywell Contaminated Antimony Pentachloride Exposure	Cylinders had been mislabeled and labels were difficult to read[6]
Honeywell Hydrogen Fluoride Release	Problems with the venture stick*[6]
Hayes Lemmerz	Previous history of frequent duct fires, dust flashes and smoldering fume hood[7]
DPC Enterprises: Glendale, AZ	Sampling procedures never followed[8]
Acetylene Service Company	Previous problems with the acetylene check valve[9]

* a small portable jet like device used to evacuate equipment

3.3. There were management system problems

For the incidents that I was involved in this was always the case. In the case of the Honeywell chlorine release there were several management system problems. The Honeywell mechanical integrity system failed to identify problems with the chlorine cooler prior to its failure. Annual inspections prior to the incident did not show any flaws in the cooler tubes. The Process Hazard Analysis done on the process was too general and did not identify the potential for chlorine leaking into the coolant system. The chlorine release affected the community and injured seven people. [6] In the Reactive Chemical Hazard Investigation the management systems most often cited as a cause of an incident was inadequate hazard identification and inadequate hazard evaluation.[2] If reactive hazards are not properly identified and evaluated, prevention, protection and mitigation measures cannot be implemented. In the Technic incident there were problems with the process safety review, management of change, process and equipment integrity, and emergency planning and response management systems. There was no process safety review process that should have identified the hazard of incompatible materials accumulating in the ventilation system. There was no management of change system that would call for a hazard evaluation when changes were made to the ventilation system. A proper process and equipment integrity program would include an inspection of the ventilation system. Because incompatible materials were allowed to accumulate in the duct work, the explosion resulted and severely injured an employee. The emergency planning and response system could also be improved by making the plan more detailed regarding the inventory of hazardous materials. Preplanning by the plant management and the fire department management would have improved communication during the emergency and likely would have resulted in less damage to the facility.[5] Properly designed and implemented process safety management systems are necessary to prevent chemical related incidents. Table 3 lists the management system problems found in the incident investigations that I participated in.

Table 3. Incident Related Management System Problems

Incident Name	Management System Problems
Bethlehem Steel	Management systems for supervision, planning, and execution of maintenance work were inadequate[1]
Improving Reactive Hazard Management	Most frequently cited management system problems were inadequate hazard identification and evaluation[2]
Georgia Pacific Corp.	No reactive hazard management system, process safety information was not incorporated into policies, programs and procedures, near miss investigation was inadequate, emergency response did not address decontamination adequately[3]
DPC Enterprises: Festus, MO	Inadequate mechanical integrity program, deficiencies in the emergency response program, inadequate community notification[4]
Technic Inc.	No project safety review, inadequate management of change, process and equipment integrity, and emergency planning and response[5]
Honeywell Chlorine Release	Inadequate mechanical integrity, hazard analysis, management of change[6]
Honeywell Contaminated Antimony Pentachloride Exposure	Inadequate standard operating procedures and training[6]
Honeywell Hydrogen Fluoride Release	Inadequate nonroutine job planning and hazard analysis[6]

Table 3. Incident Related Management System Problems (continued)

Incident Name	Management System Problems
Hayes Lemmerz	Inadequate management of change, hazard evaluation, engineering project controls, procedures and training[7]
DPC Enterprises: Glendale, AZ	Inadequate procedures and training[8]
Acetylene Service Company	Inadequate procedures and training[9]

3.4. Incidents take place in companies of all sizes

When I joined the CSB I had the opinion that major chemical plant accidents primarily occurred in small companies without the resources to implement strong process safety systems. During the Reactive Hazard Investigation I discovered that reactive chemical incidents occurred in companies of all size. Of the incident investigations I was involved in, four companies I would classify as large (Bethlehem Steel, Georgia Pacific, Honeywell, and Hayes Lemmerz), two as medium (DPC Enterprises and Technic), and one as small (ASCO). This tells me that companies of all sizes need to do a better job of incident prevention. It is not just having resources available for improved process safety but using the available resources wisely.

3.5. Many chemical related incidents occur in businesses outside of chemical manufacturing

I had another misconception when I joined the CSB. I thought that incidents involving chemicals took place primarily in chemical manufacturing facilities. I was wrong again. During the Reactive Hazard Investigation I discovered over 30% of the incidents we investigated took place in companies that stored or handled chemicals rather than manufacture chemicals.[2] Of the eight other investigations I was involved in only three were chemical manufacturing facilities (Technic, Honeywell, and ASCO). Bethlehem Steel manufactured steel. DPC Enterprises repackaged and distributed chlorine. Georgia Pacific manufactured paper products. Hayes Lemmerz manufactured aluminum wheel castings. What this tells us is that the process safety message and lessons learned from incidents needs to be communicated more broadly than just to the traditional chemical processing industries.

3.6. Incidents often occur during start up, shut down, or maintenance rather than during normal operations.

This statement may be a surprise to some, but my experience in the chemical processing industries has convinced me that this is true. During the Reactive Hazard Investigation I found this to be true although we did not tabulate the data to prove this statement. Of the incidents I investigated I would classify only the Honeywell chlorine release as taking place during normal operations. Three incidents took place during maintenance (Bethlehem Steel, Technic, and the Honeywell hydrogen fluoride release). Two incidents occurred during start up (Hayes Lemmerz and ASCO). Three incidents occurred during unloading or loading operations (Georgia-Pacific, DPC Festus and DPC Glendale). The Honeywell contaminated antimony pentachloride exposure took place during a cylinder cleaning operation. This tells us that we need to address these operations specifically in our process safety and incident prevention programs.

3.7. Regulations are often inadequate or not enforced

Some think that answer to improved process safety and less incidents involving chemicals is more regulation by government. Although I believe the OSHA Process Safety Management Standard has improved process safety and reduced incidents, we still have incidents in PSM and non PSM plants. The Reactive Hazard Investigation found that over 50 % of the incidents investigated involved chemicals not covered by the existing OSHA or EPA process safety regulations, although many of the processes involved were covered because they contained the threshold quantities of other listed chemicals.[2]

Of the incidents I investigated only Bethlehem Steel, Technic, and Hayes Lemmerz were not covered by OSHA PSM. Technic was a chemical manufacturing company but did not have the threshold quantities of listed substances to be covered by OSHA PSM. The others had PSM programs that were in someway inadequate that contributed to the incident. I believe that existing regulations are helpful but there are not enough enforcement resources to verify compliance with existing process safety regulations never mind new regulations. I believe we need to introduce new regulations cautiously, only when industry has failed to address its own problems. We must convince industry that process safety management can help them meet their business objectives. We need to make the business case for process safety.

3.8. Management thought they had a good safety program

In all the incidents that I was involved in, management thought they had a good safety program. Often times they were tracking personal injuries to measure the effectiveness of their safety program. Process safety performance is not easily measured by personal injury statistics. Process safety requires different measures such as the number of spills, leaks, fires, near misses, equipment failure, and compliance to company process safety management requirements. To be sure you have a good process safety program you must make and track the right measurements.

3.9. There are often safety culture issues

When there are process safety management systems in place but there are still incidents, there could be a safety culture problem. Safety culture involves the beliefs and attitudes of the organization that drive people to follow the process safety rules and programs. Safety culture is strongly influenced by the actions of top management over time. Employees look at what is done not only what is said by top management. The organizations that I think should evaluate their safety culture are DPC Enterprises and Honeywell. Both had multiple incidents investigated by the CSB. Safety culture should be evaluated as a part of any major process safety incident investigation. When existing process safety standards are not followed, it is often because of the underlying safety culture.

3.10. They were all preventable

As I look back at all these incidents, I can say they were all preventable. There were no acts of God. The CSB and the chemical processing and handling companies must continue to investigate all incidents so future incidents can be prevented. Most incidents have happened to someone else before. We need to continue to share lessons learned at this conference and others so others can learn. We need to find new effective ways to communicate the known hazards to the right people at the right time.

4. Summary

I learned a lot about the nature of incidents at the CSB. You may or may not agree with me based on your own experiences. I would encourage all of you who have an interest in reducing the number and severity of chemical related incidents to read the reports I cited and others at the CSB web site, www.csb.gov. We need to stop the same or similar incidents from reoccurring. You can form your own opinions and improve your incident prevention program.

5. References

[1] "Steel Manufacturing Incident, Bethlehem Steel Corporation," U. S. Chemical Safety and Hazard Investigation Board, Report No. 2001-02-I-IN, January 2002.

[2] "Improving Reactive Hazard Management," U. S. Chemical Safety and Hazard Investigation Board, Report No. 2001-01-H, December 2002.

[3] "Hydrogen Sulfide Poisoning, Georgia Pacific," U. S. Chemical Safety and Hazard Investigation Board, Report No. 2002-01-I-AL, January 2003.

[4] "Chlorine Release, DPC Enterprises, LP," U. S. Chemical Safety and Hazard Investigation Board, Report No. 2002-04-I-MO, May 2003.

[5] "Vent Collection System Explosion, Technic Inc.," U. S. Chemical Safety and Hazard Investigation Board, Report No. 2003-08-I-RI, August 2004.

[6] "Chlorine Release, July 20, 2003; Contaminated Antimony Pentachloride Exposure, July 29, 2003; Hydrogen Fluoride Release, August 13, 2003," U. S. Chemical Safety and Hazard Investigation Board, Report No. 2003-13-I-LA, August 2005.

[7] "Aluminum Dust Explosion, Hayes Lemmerz International, Inc.," U. S. Chemical Safety and Hazard Investigation Board, Report No. 2004-01-I-IN, September 2005.

[8] "Chlorine Release, DPC Enterprises, LP," U. S. Chemical Safety and Hazard Investigation Board, Report No. 2004-02-I-AZ, February 2007.

[9] "Explosion at ASCO: Dangers of Flammable Gas Accumulation," U. S. Chemical Safety and hazard Investigation Board, Safety Bulletin No. 2006-01-B, January 2006.

[10] http://www.csb.gov/index.cfm?folder=completed_investigations.

Recognizing In Situ Oxygen Analyzers as Ignition Sources in Fired Equipment

Sheila Van Geffen
Process Safety Technology Leader
The Dow Chemical Company
Louisiana Hwy 1, P.O. Box 150
Plaquemine, LA 70765 U.S.A.
Phone: 225-353-8783
sfvangeffen@dow.com

Bernardo Corripio
Furnace Technology Leader
The Dow Chemical Company

Andy Weber
Hydrocarbons & Energy Global Analytical Business Partner
The Dow Chemical Company

Abstract

Most modern ethylene cracking furnaces use close-coupled extractive or in situ oxygen analyzers based on a zirconium oxide sensor in the stack area to provide fast and reliable measurement of the oxygen content of the flue gas exiting the firebox. This allows for precise control of the excess combustion air in the firebox, resulting in higher energy efficiency for the furnace than is typical using conventional extractive sample-based analyzer systems where the analyzer might be several hundred feet away from the sample point. These analyzers use a zirconium oxide sensing element that operates at 700 C, which is well above the auto-ignition temperature of many flammable hydrocarbon mixtures. Although, flame arrestors can be installed on these analyzers to prevent flame propagation from the probe element into the firebox there are still situations that can exist that the analyzer can pose as an ignition source. This paper will discuss the two typical zirconium oxide oxygen analyzer designs and, how the analyzer can be a source of ignition, and review an incident in a Dow cracking furnace which resulted from an in situ oxygen analyzer igniting a flammable mixture in a furnace.

Introduction

Ethylene furnaces are the largest consumers of energy in an ethylene plant and efficient combustion of the fuel gas used in them is important for the energy efficiency and profitability of the plant. Key to efficient furnace combustion is the effective control of excess air in the firebox. This control is typically accomplished by measuring the oxygen concentration in the flue gas leaving the furnace and holding this concentration at a desired minimum value by using dampers and draft fans to manipulate the air flow into the firebox. The oxygen measurement used for this control scheme was originally accomplished using extractive sample analyzer systems where the analyzer was typically located several hundred feet away from the sample point. This design has the disadvantage of providing a slow response to changes in oxygen content and a correspondingly slow change in excess air. That in turn results in inefficient control: either the oxygen content remains low for too long and fuel is not completely combusted, or it remains too high and air is needlessly heated and leaves the stack in the flue gas.

Because of these disadvantages of the extractive sample analyzers, Dow and other ethylene producers, as well as other operators of industrial furnaces, have replaced them with either close-coupled extractive or in situ oxygen probes which provide a much faster (15 seconds – 1 minute), measurement of the oxygen concentration in the flue gas leaving the furnace. These probes contain zirconium oxide sensors which operate at high temperatures (700 C), above the auto-ignition temperatures of many light hydrocarbon mixtures. While these analyzers are designed to operate only when non-flammable mixtures are present and have safety features, sometimes offered as options, to prevent ignition of a flammable mixture, it is possible for an operating analyzer to ignite a flammable mixture, causing a fire or explosion.

This paper will describe an incident in which a flammable mixture formed in the stack area of an offline Dow ethylene furnace and was detonated by an in-service oxygen analyzer, in this case an in situ type, causing severe damage to the furnace. Some findings and corrective actions from the investigation of this incident will be described, along with Dow's plans for dealing with the hazards of these devices.

Description of Zirconium Oxide Based Oxygen Analyzers

There are a variety of zirconium oxide based analyzers on the market using two basic designs, in situ and close-coupled extractive. The in situ design places the zirconium oxide sensor, operating at 700°C, inside the fired equipment or stack. The close-coupled extractive method uses an eductor or aspirator to extract the gas to the 700°C sensor that is located outside of the fired equipment or stack. The gas is then returned to the fired equipment or stack. Both the in situ and close-coupled extractive devices are available with sample gas in-line flame arrestors; some models come standard with flame arrestors and some as optional, to prevent flame propagation via the sample gas path if flammable material is ignited.

There are several available probe configurations for the in situ analyzers, depending on whether flame arrestors and heaters are included in the probe assembly:

1. Zirconium oxide element with no heater or flame arrestor. The flue gas heats the zirconium oxide sensor to its operating temperature. This type requires the flue gas to be at least 550°C before the sensor becomes active.
2. Zirconium oxide element with element heater and no flame arrestor.
3. Zirconium oxide element with element heater and flame arrestor. The flame arrestor is designed to stop the propagation of a flame back into the fired equipment if a flammable mixture is present in the fired equipment. Since the sensor is above the auto-ignition

temperature of the flue gas, it could act as an ignition source if either the flame arrestor is broken or if the proper probe is not used. The certification of the analyzer should be checked carefully and the vendor should be consulted to make sure that the analyzer is designed not to be an ignition source inside the fired equipment (Figure 1). This was the arrangement in the Dow furnace incident.

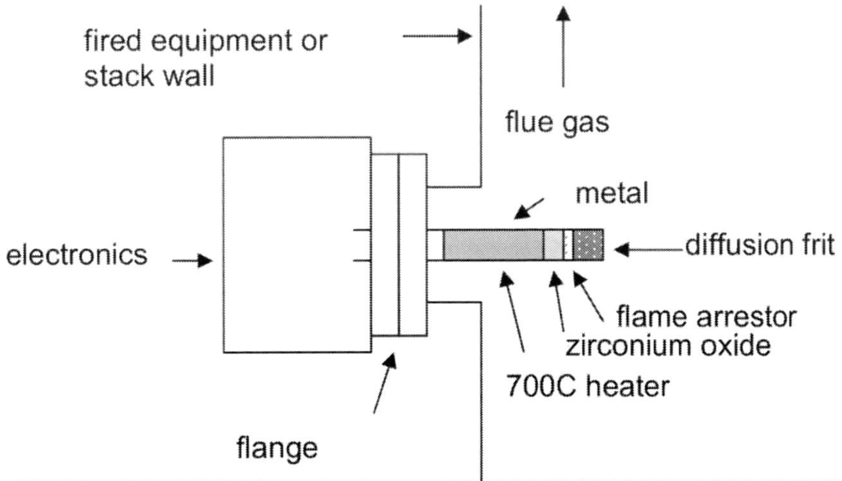

Figure 1. In Situ Zirconium Oxide Probe w/ Heater & Flame Arrestor

Dow Ethylene Furnace Incident

In 2005, a flammable hydrocarbon mixture formed in one of Dow's ethylene furnaces and ignited, severely damaging the furnace. The furnace was offline due to a plant outage caused by a power failure. The investigation of this incident determined that this mixture was ignited by an in situ zirconium oxide oxygen probe in the furnace stack. The probe was equipped with a heater to achieve the 700 C operating temperature of the element and with a flame arrestor.

The probe had been placed in service in the 1990's and had functioned well with only routine preventive maintenance. When removed from the field and inspected, it was found that the threaded housing holding the flame arrestor to the probe body was loose. It was conjectured that the loose threads formed a path for flame propagation, bypassing the flame arrestor. In addition there was evidence of flame impingement on the probe body just outside the thread area. These facts led to the conclusion that the flame arrestor function was disabled by the loose threads.

During the follow-up to this incident, it was determined that the flame arrestor alone would not be an adequate protection against this scenario, even if some means were established to ensure that the arrestor function was not compromised. Other procedural, hardware, and software measures would be taken to ensure that this probe could not act as an ignition source for a flammable gas. There were follow-up actions addressing the malfunctions that led to the formation of the flammable mixture, but they are not relevant to this paper.

There was another layer of protection in the oxygen analyzer system which could have prevented this incident but failed to do so. There is a software controller which activates the probe and turns on the probe heater. This controller is programmed to lock out a probe which has lost its signal for any reason; a technician must manually override the controller to allow the probe to be activated and heat up. The probe had lost power when the plant lost power and tripped; it was initially thought that it would not have reactivated automatically. However, the software controller had lost power at the same time as the probe. When power was restored, the controller and probe were powered up simultaneously and the controller did not recognize that the probe had failed to function for some time. It allowed the probe to heat back up without any human intervention. This behavior has been confirmed by field tests.

Alternative Oxygen Analyzer Systems

The solution to this hazard of the in situ oxygen probes is not as simple as a return to extractive analyzer systems. Close-coupled extractive analyzers can also be a source of ignition if the proper flame arrestors are not used in the installation. Even with flame arrestors, ignition can be a problem; with one close-coupled extractive model the process side of the flame arrestor can reach a temperature above 450°C, a temperature that could lead to auto-ignition of some light hydrocarbon mixtures.

Corrective Actions

Procedural and control changes have been made in the plant to force the deactivation of these types pf oxygen probes when a furnace is taken offline. Software changes to deactivate all probes when the controller loses power are underway. In addition, Dow has recently completed a study comparing the available oxygen analyzer systems and is in the process of forming a final recommendation for a standard application. This is a complicated matter as there are pros and cons associated with each alternative and there is no solution that fits all applications.

Conclusion

Because of a plant incident, Dow has learned of the hazards associated with a zirconium oxide oxygen probe in a furnace stack. As part of the company's commitment to public and industry safety, we have chosen to share this experience in this forum so that our fellow producers are aware of these hazards and can inspect their systems and perform the necessary evaluation of what if any changes they should make to assure the safety of their processes.

Case Study: Aluminum-Dust Explosion

Vahid Ebadat, Ph.D., and Richard W. Prugh, PE, CSP
Chilworth Technology, Inc.
250 Plainsboro Road, Building 7
Plainsboro, New Jersey 08536
Phone: 609-799-4449
safety@chilworth.com

Abstract

An explosion occurred in equipment that was being used to pulverize aluminum-alloy ingots, to form a powder with small particle size. In this process, air was recycled through a grinder, cyclone product separator, and blower, with a side stream to extract undesired dust to a dust collector. However, the concentration of dust in the circulating air stream was above the Minimum Explosible Concentration in most of the recycle system. Ignition of this dust/air dispersion occurred as a result of an exothermic oxidation of accumulated combustible dust in the return-air ductwork, accelerated by the relatively high temperature of the circulating air. Because the system was not protected against internal explosion, an access door on the grinder was forcibly ejected, and the door struck the LPG cylinder on a nearby forklift truck. Escaping LPG vapors – ignited by the dust fireball from the open grinder door – resulted in a flash fire that caused fatal burn injuries to the forklift driver.

This powder-producing operation initially was a batch process, and it had operated without incident for many years. When a decision was made to convert the batch process to a continuous process, some hazardous changes were introduced that were not recognized. These changes included (1) a much higher rate of dust generation, (2) high concentrations of very fine dust in the recycle stream, and (3) accumulation of heat in the recycle stream. In addition, the installer of the recycle system did not have experience with combustible powders, and no explosion prevention or protection devices or systems were provided. Further, major items of equipment (including a grinder and dust collector) were installed indoors, thus exposing operating personnel to explosion and fire hazards.

Among the lessons learned from this incident were: (1) every significant change in a process should be subjected to a Management of Change analysis; (2) producers of hazardous materials (and particularly small-company producers) need to follow Codes and Standards that apply to their processes; (3) code-enforcing authorities should become more familiar with Codes and Standards that apply to dust-explosion hazards; and (4) dust-explosion characteristics need to be included in Material Safety Data Sheets.

1.0 Introduction

During the past four years, there have been several serious dust explosions, resulting in fatal injuries and multi-million dollar property losses. The following Table summarizes some of these incidents:

LOCATION	DATE	COMBUSTIBLE MATERIAL(S)	FATALITIES	INJURIES
Des Moines, Iowa	06-01-05	Combustible Dust	0	2
Mayfield, Kentucky	02-25-05	Resin	0	1
Huntington, Indiana	10-29-03	Aluminum	1	6
Corbin, Kentucky	02-20-03	Phenolic Resin	7	37
Kinston, North Carolina	01-29-03	Polyethylene	6	38
TOTALS	-	-	14	84

The U. S. Chemical Safety and Hazards Investigation Board [CSB] has reported that there have been 281 dust explosion incidents in the United States during the period 1980 to 2005, and that 119 fatalities and 718 injuries resulted from these incidents [Reference 1]. The purpose of this Case Study is to illustrate how a process that may be common in several industries can lead to a destructive dust explosion

1.1 Process Description

In the process of interest, aluminum-nickel alloys are prepared by mixing the powdered metals in a gas-fired crucible. The aluminum melts at about 660°C, and the nickel reacts exothermally with the molten aluminum, thus increasing the temperature. The nickel essentially dissolves in the aluminum, forming a solution of nickel in aluminum, similar to the precursor for Raney nickel.

The molten alloy is poured onto a large cooling table, where it freezes, and the solidified alloy is broken into fragments for further size reduction. The fragments are loaded into a pre-grinder to cut the material into smaller particles, and these particles are loaded into a hopper that provides feed for the grinder.

A diagram that shows the subsequent steps in the preparation of the alloy powder is shown as Figure 1.

The grinder consists of two large roller-type manganese-steel cylinders which grind the material. The ground particulates drop into a pneumatic conveyor system that conveys them to a cyclone separator, where the product is collected. The process produced about one ton per hour, with an average particle size of about 36 microns.

From the top of the cyclone, dust-laden air returns to a blower that provides the motive force for the pneumatic conveyor. A side stream takes dusty air to a bag-house dust collector, and the cleaned air discharges to the atmosphere, outside the building.

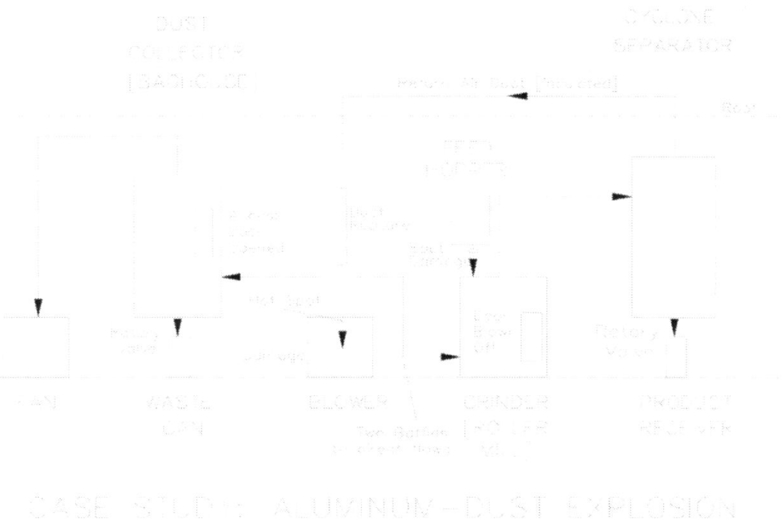

Figure 1: Case Study: Aluminum-Dust Explosion

A batch process, with "once-through" flow of air through a hammer mill and dust collector, had operated for several years, before the continuous-process equipment was installed. The continuous process involved a major change in the operation, since the conveying air would be recycled and alloy dust would be suspended in the recycled air. The grinding and recycle process operated for about four months, for a total operating time of about 240 hours, until the incident occurred.

1.2 Deficiencies In The Process

Several deficiencies in the process led to this explosion incident:

1. Finely-divided aluminum alloy is very reactive with air, exothermally forming aluminum oxide, but there was no inerting of the atmosphere in the recycle loop.

2. The supplier of a particle-size classifier, cyclone separator, and blower was not informed concerning the explosibility of dusts that would be formed in the mill.

3. The installer of the grinder and recycle system had little experience with highly air-reactive combustible and explosible powders and dusts (experience with clay, limestone, and some coal and activated carbon).

4. The throughputs of alloys (and the generation of dust) were much higher than those of the previous batch process.

5. The side stream to the bag-house dust collector took a very small percentage of the flow through the blower, grinder, and cyclone; thus, the concentration of aluminum-nickel dust increased with time, as limited only by the duration of the grinding operation.

6. There was no heat-removal apparatus in the recycle loop, and the cyclone separator (downstream of the roller mill) became too hot to touch. In addition, much of the recycle ductwork was insulated, further aggravating the accumulation of heat.

7. There was no indication or procedure for emptying the dust collector, and an excessive accumulation (high level) in the dust collector would interfere with its proper operation.

8. The dust collector was installed inside the operating area.

9. Explosion-venting was not installed on the dust collector, and no explosion-suppression system was installed.

10. The velocity of air through the recycle system was less than the recommended 4,500 feet per minute.

11. A forklift that was powered by LPG was used in the alloy-dust area, whereas a DX diesel-powered truck or an EX electrically-powered truck should have been used, provided that such use was permitted by the "authority having jurisdiction".
12. The Material Safety Data Sheet and employee training did not emphasize the combustibility and explosivity of aluminum and aluminum-alloy dust.

13. Properties of aluminum dust and the 63/33/4 Ni/Al/Mo dust that was involved in the incident are as follows:

PROPERTY	ALUMINUM	NICKEL/ALUMINUM *
Min. Explosible Conc.	30 grams/cubic meter	825 grams/cubic meter *
Stoichiometric Conc.	300 grams/cubic meter	500 grams/cubic meter
Min. Oxygen Conc.	5.0 volume-%	18.5 volume-%
Min. Ignition Energy	10 millijoules	205 millijoules
Max. Explosion Pressure	12.4 bar, gauge [180 psig]	5.7 bar, gauge [83 psig]
Deflagration Index, Kst	415 bar-meters/second	80 bar-meters/second
References	2, 3, 4	5

* It is possible that the Nickel/Aluminum dust had oxidized (prior to testing), such that it was resistant to combustion.

Data concerning thermal stability of the Nickel/Aluminum alloy – as obtained by Chilworth Technology, Inc. – are presented in Appendix B.

As shown in the preceding Table, ignition of the alloy dust could result in high explosion pressures at a high rate.

1.3 Description of the Incident

The two injured employees arrived for work at about 2:00 AM on the day of the explosion. The grinding/milling operation is begun, and all operations are normal. At 6:30 AM, most of the crew left the operating area for a breakfast break. The two employees stayed in the operating area to continue the grinding process.

A few minutes later, an explosion occurred in the processing equipment. Access doors were blown off of the dust collector and the grinder, and the most of the inlet duct to the blower was blown away. One of the doors struck the forklift that was being driven by the fatally-injured employee, causing the hose to the LPG cylinder to rupture. Despite the clouds of dust caused by the explosion, and the fact that his clothes were on fire, this employee drove the forklift out of the building. Other employees extinguished the fire in his clothing, but he succumbed to burn injuries a few days later. The second employee was not seriously injured.

1.4 Investigation Of The Causes Of The Incident

An explosion that involves a mixture of combustible dust and air requires that the concentration of the dust in air be above the Minimum Explosible Concentration [MEC] at the temperature of the mixture; thus, the MEC must have been exceeded in this incident. Since the side-stream to the dust collector was very small in relation to the circulating flow through the mill, the concentration of dust in the circulating air stream would gradually increase with time. Thus, the "fuel" leg of the fire triangle was "satisfied".

There was no inerting of the circulating air stream; thus, the "oxidant" leg of the fire triangle was "satisfied."

It was later discovered that there were several possible ignition sources in this process (not in order of likelihood):

1. In the dust collector, the cages for the filter socks may not have been grounded, allowing accumulation of static electricity, with discharges that were capable of igniting dust/air mixtures.

2. In the dust collector, there was evidence that a cage for a filter sock may have been rubbing on the interior of the collector, creating a metal-to-metal "hot spot" that was capable of igniting dust/air mixtures.

3. In the dust collector, there was evidence that the rotary valve at the bottom of the collector was rubbing on the interior surface of the valve, creating a "hot spot" that was capable of igniting dust/air mixtures.

4. In the blower, there were indications of rubbing between the metal impeller and the metal housing of the blower, creating a "hot spot" that was capable of igniting dust/air mixtures.

5. At several locations in the duct from the cyclone separator to the blower, there were indications of localized high temperature, caused by reaction of alloy with moisture, creating "hot spots" that were capable of igniting dust/air mixtures.

6. In the rotary valve at the bottom of the cyclone separator, large lumps of alloy were found, with some of the vanes deformed, and friction between these lumps and the metal housing of the rotary valve created "hot spots" that may have been capable of igniting dust/air mixtures.

7. The forklift truck was powered by liquefied petroleum gas [LPG], and exposed exhaust, fuel system, and electric systems presented ignition sources for Group E electrically-conductive combustible dusts.

8. After the incident, the alloy dust that was recovered from the grinder (and an unburned gasket at the access door) indicated that combustion had not occurred in the grinder.

9. The partly-closed position of a baffle-valve in the side-stream to the dust collector indicated that the explosion was not initiated in the dust collector.

The most-obvious ignition sources were the "hot spots" in the ductwork where it was apparent that the duct metal (and the dust/air mixture within the duct) had been exposed to very high temperatures. Although the duct at the hot spots was not deformed, acceleration of the dust/air combustion into other equipment where there was a high concentration of dust in air led to very high overpressures. As a result, the duct at the blower was destroyed, and the doors at the dust collector and grinder were blown off. The resulting fireball of burning alloy dust filled the operating area, seriously burning one employee and igniting his clothing.

2.0 LESSONS LEARNED

The National Fire Protection Association and the Center for Chemical Process Safety provide good guidance concerning the handling and processing of combustible dusts [References 6 through 16]. Also, "Lees' Loss Prevention in the Process Industries" [Reference 17] includes extensive discussions of causes and prevention of dust fires and explosions, and protection against such incidents. Additional information concerning dust explosions is given in "Dust Explosions in the Process Industries" [Reference 18].

Producers of combustible powders and dusts should follow the guidance that is given in these documents, to "provide a safe place to work" for their employees. Also, fire marshals, code enforcers, and other "authorities having jurisdiction" need to become sufficiently familiar with these documents so that they can audit their local manufacturers for conformance.

The Process Safety Management standard of the Occupational Safety and Health Act provides good guidance concerning changes to a process, as occurred prior to this incident:

1. Even though a process might not utilize a listed "Highly Hazardous Chemical", the Management of Change procedure – and particularly the "impact of change on safety and health" considerations – should be followed.

2. If the Change requires a change in the Process Safety Information – such as piping and instrument diagrams, and material and energy balances, in this case – then a Pre-Startup Safety Review, with a Process Hazard Analysis, would be required.

Quantitative data should be provided in documents such as Material Safety Data Sheets, concerning the properties of the materials that are involved in a process.

1. The intention of MSDSs is to provide employees (and customers) with information concerning their exposure to fire, explosion, toxicity, and corrosivity hazards.

2. The MSDS standard requires that there be no "blanks" in the presently-required information, but "Not Available" or "Not Known to Us" is not an appropriate entry – particularly if the material is known to be hazardous – and the required information should be obtained by tests or a literature search, prior to handling of significant quantities of the material.

3. Even though it may not be required by current regulations, if the fire and explosion properties of a dust have been determined, this information should be provided in the MSDSs provided for employees and customers, as recommended by the CSB [Reference 19].

In this modified process, all of the circulating air stream (rather than a side-stream) should have passed through a dust collector, prior to passing through the blower. The NFPA combustible-dust standard specifies that "systems shall be designed in such a manner that combustible material does not pass through an air-moving device" (such as a blower). Maintaining a low temperature of the circulating air is not addressed in this standard, but if employees had been knowledgeable concerning the hazards of dust/air explosions, their observations that the cyclone separator was "too hot to touch" could have led to management's recognition of a fire and explosion hazard.

3.0 CONCLUSION

The National Fire Protection Association and the Center for Chemical Process Safety have developed and compiled good guidance concerning the control of dust-explosion hazards. If a process produces combustible dusts, make sure that operating personnel are familiar with the fire and explosion hazards – and that adequate prevention and protection have been provided against these hazards in the process.

4.0 REFERENCES

1. U. S. Chemical Safety and Hazard Investigation Board, "Investigation Report – Combustible Dust Hazard Study," Report No. 2006-H-1, Page 30 and Table 5 (November 2006).

2. National Fire Protection Association, "Guide for Venting of Deflagrations," NFPA 68, Table E.1(d) (2002).

3. National Fire Protection Association, "Standard on Explosion Prevention Systems," NFPA 69, Table C.1(c) (2002).

4. U. S. Department of Commerce, National Materials Advisory Board, "Classification of Combustible Dusts in Accordance with the National Electrical Code," NMAB 353-3, Table 1, Section 6 (1980).

5. Chilworth Technology, Inc., [unpublished data] (2005)

6. National Fire Protection Association, "Guide for Venting of Deflagrations," NFPA 68 (2002).

7. National Fire Protection Association, "Standard on Explosion Prevention Systems," NFPA 69 (2002).

8. National Fire Protection Association, "Recommended Practice on Static Electricity," NFPA 77 (2002).

9. National Fire Protection Association, "Standard for Combustible Metals, Metal Powders, and Metal Dusts," NFPA 484 (2006).

10. National Fire Protection Association, "Standard for Purged and Pressurized Enclosures for Electrical Equipment," NFPA 496 (2003).

11. National Fire Protection Association, "Recommended Practice for the Classification of Combustible Dusts and of Hazardous (Classified) Locations for Electrical Installations in Chemical Process Areas," NFPA 499 (2004).

12. National Fire Protection Association, "Standard for Pneumatic Conveying Systems for Handling Combustible Particulate Solids," NFPA 650 (1998).

13. National Fire Protection Association, "Standard for the Prevention of Fire and Dust Explosions from the Manufacturing, Processing, and Handling of Combustible Particulate Solids," NFPA 654 (2006).

14. National Fire Protection Association, "Standard for the Prevention of Sulfur Fires and Explosions," NFPA 655 (2001).

15. National Fire Protection Association, "Standard for the Prevention of Fires and Explosions in Wood Processing and Woodworking Facilities," NFPA 664 (2002).

16. Center for Chemical Process Safety [American Institute of Chemical Engineers], "Guidelines for Safe Handling of Powders and Bulk Solids" (2005).

17. Mannan, S. [editor], "Lees' Loss Prevention in the Process Industries," 3rd Ed., Vol. 2, Pages 17/249 through 17/288 (2005)

18. Eckhoff, R. K., "Dust Explosions in the Process Industries," 3rd Ed. (2003).

19. U. S. Chemical Safety and Hazard Investigation Board, "Investigation Report – Combustible Dust Hazard Study," Report No. 2006-H-1, Pages 37 to 41, 43, 44, 69, and 70, and Appendix B (November 2006).

5.0 APPENDIX A

NICKEL–ALUMINUM ALLOY POWDER

Material Safety Data Sheet

(as prepared by the client for its employees)

Section III – Physical Data

Boiling Point: >2600 °C

Solubility in Water: Insoluble

Evaporation Rate: N. A.

Specific Gravity: > 1.0

Section VI – Fire and Explosion Data

Flash Point: N/A

Extinguishing Media: This material is considered non-flammable. If a fire occurs in the area where this material is used, use sand or dry chemicals, avoid water.

Unusual Fire and Explosion Hazards: None known at this time.

6.0 APPENDIX B

SUMMARY OF TEST RESULTS

DUST-CLOUD EXPLOSIBILITY

[ASTM E-2019]

65/35 (approx.) Nickel/Aluminum Powder

[97% <75 microns; 0% Moisture]

SAMPLE	CONDITIONS	RESULT
No. 1	Glass Tube; 100°C; 10-kv arc 10,000 grams/cubic meter	Low-pressure explosion;
	100°C; 500-millijoule spark; 200 to 40,000 g./cu.m.	No ignitions
No. 1	Glass Tube; 25°C; 10-kv arc 15,000 grams/cubic meter	Low-pressure explosion; Bright white flame, with sparks
	25°C; 500-millijoule spark; 200 to 15,000 g./cu.m.	No ignitions
No. 2	Acrylic Tube; 25°C; 10-kv arc 200 and 400 g./cu.m.	Low-pressure explosion; Bright white flame, with sparks
	25°C; 500-millijoule spark; 200 to 60,000 g./cu.m.	No ignitions
No. 2	Acrylic Tube; 100°C; 10-kv arc 200 and 400 g./cu.m.	Low-pressure explosion
	100°C; 500-millijoule spark; 200 to 40,000 g./cu.m.	No ignitions

Note: g./cu.m. = grams per cubic meter, in a 1.2-liter test cylinder.

7.0 DIFFERENTIAL SCANNING CALORIMETER

[ASTM E-698-99]

65/35 (approx.) Nickel/Aluminum Powder

[97% <75 microns]

(4°C per minute)

PERCENT MOISTURE	TOTAL WEIGHT	ONSET TEMP.	HEAT OF REACTION	ADIABATIC TEMP.	MAX. RATE
10%	8.18 mg	63°C	166 J/g.	287°C	980 mW/g.
15%	10.43 mg	60°C	335 J/g.	512°C	1,175 mW/g.
20%	7.75 mg	65°C	296 J/g.	465°C	1,220 mW/g.

Note: The specific heat of the alloy is about 0.74 J/g.-°C (about 0.18 cal./g.-°C)

8.0 CARIUS TUBE

65/35 (approx.) Nickel/Aluminum Powder

[97% <75 microns]

(0.5°C per minute; 35 cu.cm.)

PERCENT MOISTURE	TOTAL WEIGHT	ONSET TEMP.	MAX. PRESSURE	MAX. TEMP.	MAX. H_2 AMOUNT
19%	17.11 g	49°C	145+ psig	85+ °C	4g./100g.

Note: The Carius Tube was vented periodically, as the pressure approached 10 barg (145 psig).

Lessons Learned From an Unusual Hydrogen Reformer Furnace Failure

T. Mike Rogers
Syncrude Canada Ltd.
Fort McMurray Alberta, Canada T9H-3L1
Phone: 780-790-5419
rogers.mike@syncrude.com

Abstract

On January 30, 2005 a high-energy pressure impulse occurred within several radiant tubes of a steam/ methane reformer furnace during startup after a partial shutdown. This impulse caused the simultaneous rupture of 5 tubes with sufficient force to eject portions of the tubes out through the roof of the furnace. All of the failures occurred near the inlet of each tube in the headspace above the catalyst and the 5 tubes were clustered toward the centre of the furnace in four rows.

Approximately 41 additional tubes were destroyed as a result of impacts from ejected tube segments and the subsequent pressure wave. Extensive damage to the refractory also occurred.

As a result, the furnace internal structure was condemned and a complete retube required. This paper will describe the investigation process, outcomes, and lessons learned.

Introduction

At the time of the incident, Syncrude Canada Ltd. operated a large oil sands facility in northern Alberta, Canada that produces approximately 250,000 barrels per day of light sweet blend crude termed Syncrude Sweet Blend. The facility is comprised of large mine, extraction, utilities, and, upgrading plants. The upgrading plant is similar to a large oil refinery with two separate trains that each include a crude unit, fluid coker, hydrotreaters, hydrogen reformation, H_2S scrubbing, sulfur extraction and process water treatment.

Syncrude produces 300 MSCFD of hydrogen in three hydrogen reformer units for use in the various upgrading units. The hydrogen is produced by steam reforming of methane.

Steam Reforming

The steam methane mixture enters the catalyst filled tubes at a temperature of 572°C and a pressure of 362 PSIG. The methane is reformed with the steam on the nickel alumina (NiAl) catalyst to produce:

$$\underset{\textbf{steam}}{H_2O} + \underset{\textbf{methane}}{CH_4} \xrightarrow[815^{\circ}C\ (1500^{\circ}F)]{\textbf{nickel catalyst}} \underset{\textbf{carbon monoxide}}{CO} + \underset{\textbf{hydrogen}}{3H_2}$$

Two of the reformer units were designed by GKN Birwelco and constructed in 1979.

The third Steam/Methane Reformer, which suffered this incident, was designed by KTI Corporation. The furnace that was constructed in 1987 and put on line in 1988 has a maximum production rate of 81 MSCFD.

The radiant section of the furnace contains eight (8) banks of reforming tubes. Each bank or row contains forty-six (46) tubes for a total of 368 tubes composed of spun cast HP-45 Niobium Modified Alloy with a design wall tube temperature of 935°C (1715°F).

The tubes are mounted vertically with the flow entering in the top and exiting at the bottom. Each tube has a 10.2cm (4") ID and is 12 meters (39'10") long containing 102 cubic decimeters (3.6 ft^3) of nickel alumina (NiAl) catalyst. Figure 1 provides the general layout of a radiant tube.

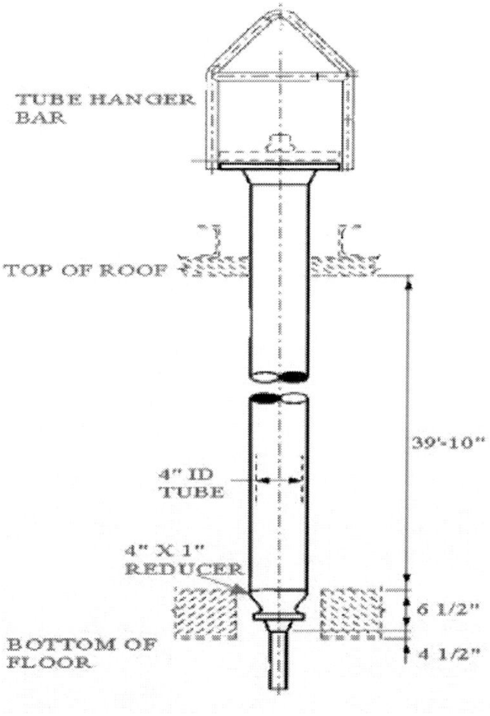

Figure 1. General tube arrangement

The furnace also has two convection sections containing boiler feed water and mixed feed (steam/gas) preheat coils.

Heat for the reformation process is supplied by nine (9) rows of natural gas burners installed on the top of the furnace. Each row consists of twelve (12) burners. The hot gases from the burners travel downward through the radiant section and are directed to the convection section via flue gas tunnels running along the furnace floor.

Investigation and Findings

A formal investigation was conducted to determine the cause of the incident. A lead investigator was assigned to conduct the investigation with support from a multidisciplinary team (process, maintenance and technical). Eyewitness interviews were conducted immediately after the incident. An inspection of the incident area was conducted as soon as the area was secured for entry. The investigation team established a sequence of events and conducted a root cause analysis. Due to the extent of the damage access was not granted into the furnace for several days. Much of the initial evidence was gathered using photography and videography.

Under normal operation, the loss of the hydrogen from this reformer is estimated to restrict production by 4.5 million barrels of Syncrude Sweet Blend over a 71-day outage. The cost of retubing the furnace was approximately $20 million, which does not include other associated shutdown costs. The cost of a new catalyst charge was approximately $1million.

The Incident

Isolated hot spots occasionally develop on these radiant tubes during operation. Rather than shut the furnace down to replace these isolated tubes, the tubes are externally pinched at their inlet and outlet without shutting down the furnace. Since these tubes are isolated with no process gas flow to keep them cool, they tend to grow. On line insitu evaluation of 6 previously pinched tubes revealed that they had encroached onto the flue gas tunnels and other, in service, tubes as a result of severe bowing. The decision was made to shut down this unit to remove these tubes and cap the inlet and outlet headers before they damaged adjacent in service tubes or the flue gas tunnels. There was a planned shutdown scheduled for August 2005, however due to the rapid bowing it was deemed necessary to remove these pinched tubes as soon as possible. A window of opportunity presented itself and the plant outage was planned for January 21, 2005. This was also a minimum scope shutdown involving only the reformer furnace. Since this was to be winter shutdown, the decision was made to leave steam in the steam generating circuit to prevent freezing.

The plant was successfully shutdown and the reformer furnace isolated. The 6 bowed tubes were removed from the furnace and preparations for startup commenced on January 28, 2005.

The startup procedure required the furnace to be heated up to 350°C (662°F) prior to introducing 4136 kPa (600 psig) steam into the radiant tubes. Once steam flow is established the furnace temperature is increased until 500°C (932°F) is reached, at which time the methane feed is introduced and the unit is brought on line producing hydrogen.

At 22:39 January 30, 2005 just after 4136 kPa (600 psig) startup steam was introduced into the reformer furnace inlet the control room alarm journal reported an extreme positive pressure spike at the same time a single loud bang was reported by the operations personnel on the furnace structure. An emergency shutdown of the furnace was initiated and the start up aborted.

At the time of the incident there were 7 people on the furnace. Four were operations staff plus three maintenance personnel. One serious injury was sustained to an operator by flying shrapnel.

The investigation revealed that as the steam flow was increased it picked up some remnant water and carried it into the furnace. Under the conditions present, the water underwent an instantaneous phase change and the energy release was sufficient to rupture the tubes. The entire event was calculated to have taken only a few milliseconds but involved extremely high-energy release rates. The power generated by a small amount of water entering a single tube at 10 lbm/sec would be 25,500 HP – 28,300 HP[1]. Subsequent shock calculations indicate that the dynamic pressure to fail the reformer tubes would be 41.3mPa (6000 psi) and the yield strength for this material would be no greater than 21.4 mPa (3100 psi). The volume of water to create this pressure would be approximately 0.4kg (0.9lb.)[2]. This was sufficient energy to catastrophically fail 5 tubes.

Extent of Damage

A total of 46 reformer tubes were destroyed in the incident. The inlet portions of 5 tube assemblies were ejected from the furnace with sufficient force to severely injure one worker and launch debris beyond the furnace structure. The other 41 tubes were subsequently destroyed either by shrapnel strikes or by the impulse released into the firebox after the initial tube ruptures. Five of the flue gas tunnels were destroyed and the firebox walls were penetrated in several locations by tube shrapnel.

Broken light fixtures, insulation and walkway gratings damaged by shrapnel impacts were found above the furnace and on a west-side furnace platform.

The investigation team condemned the entire reformer radiant tube section as there was no method to prove the integrity of the surviving tubes and there was too much damage to simply conduct repairs. The furnace was subsequently retubed.

Photographs 1 through 6 detail some of the damage to the reformer furnace.

Photograph 1 General arrangement radiant tube inlets

Photograph 2 Pre incident general radiant tube arrangement

Photograph 3 Post incident damage. Note displaced hangers and counter weights

Photograph 4 External damage

Photograph 5 Tube 161 lifted approximately 16 ft (view looking up)

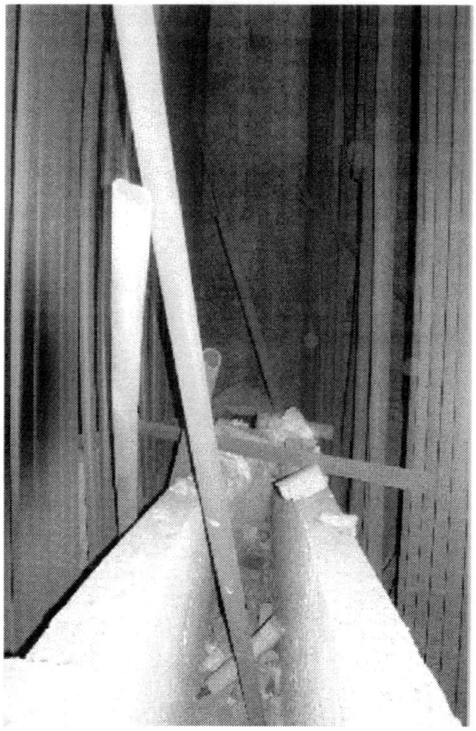

Photograph 6 Internal condition of radiant section

Upper Section Failure Mechanism

The steam flow enters the hot catalyst filled tube through the inlet pigtail and would be normally passed down through the catalyst and exit via the outlet header. Entrained water impinges on hot catalyst and possibly on the pipe wall and "instantly" flashes. This produced an extreme local pressure excursion and shock wave. The shock wave reached the pipe wall and initiated longitudinal cracking due to a hoop stress shock. It then progressed upward at sonic speed (~24,000 in/sec in 320°C steam at 75 psia) and downward (although impeded by random reflection at catalyst surfaces). Longitudinal cracking further extended due to the advancing hoop stress shock.

The shock wave reached the flange about one millisecond after water flashed at the catalyst and imparted an axial shock load on the tube. This resulted in sudden lateral brittle fractures progressing from the inside out and the upper and lower tube sections separate.

The now unconstrained upper tube end is launched out of the furnace by the extreme internal pressure. This also causes radial displacement of the free tube end segments. In areas of normally brittle material (hot areas within the furnace), high bending loads result in horizontal brittle fractures and the generation of shrapnel. Ductile material near the flange is able to yield without fracturing and produces a "banana peel" effect immediately below the flange assembly. Photographs 7 and 8 detail one of the 5 tube failures

Photograph 7 Tube # 116

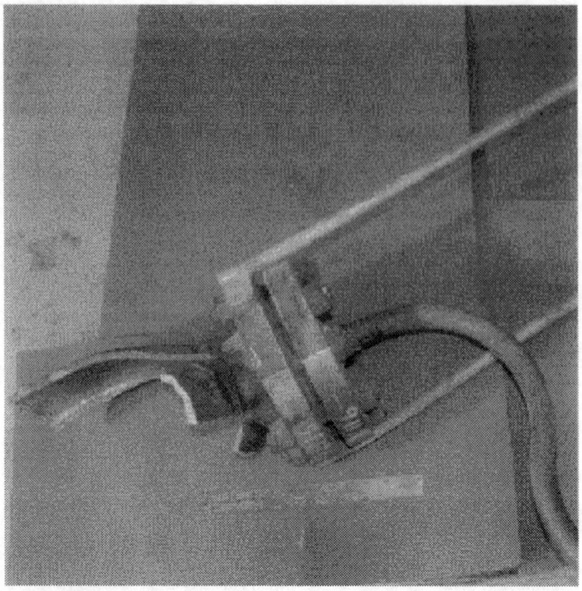

Photograph 8 Tube #116 removed from furnace

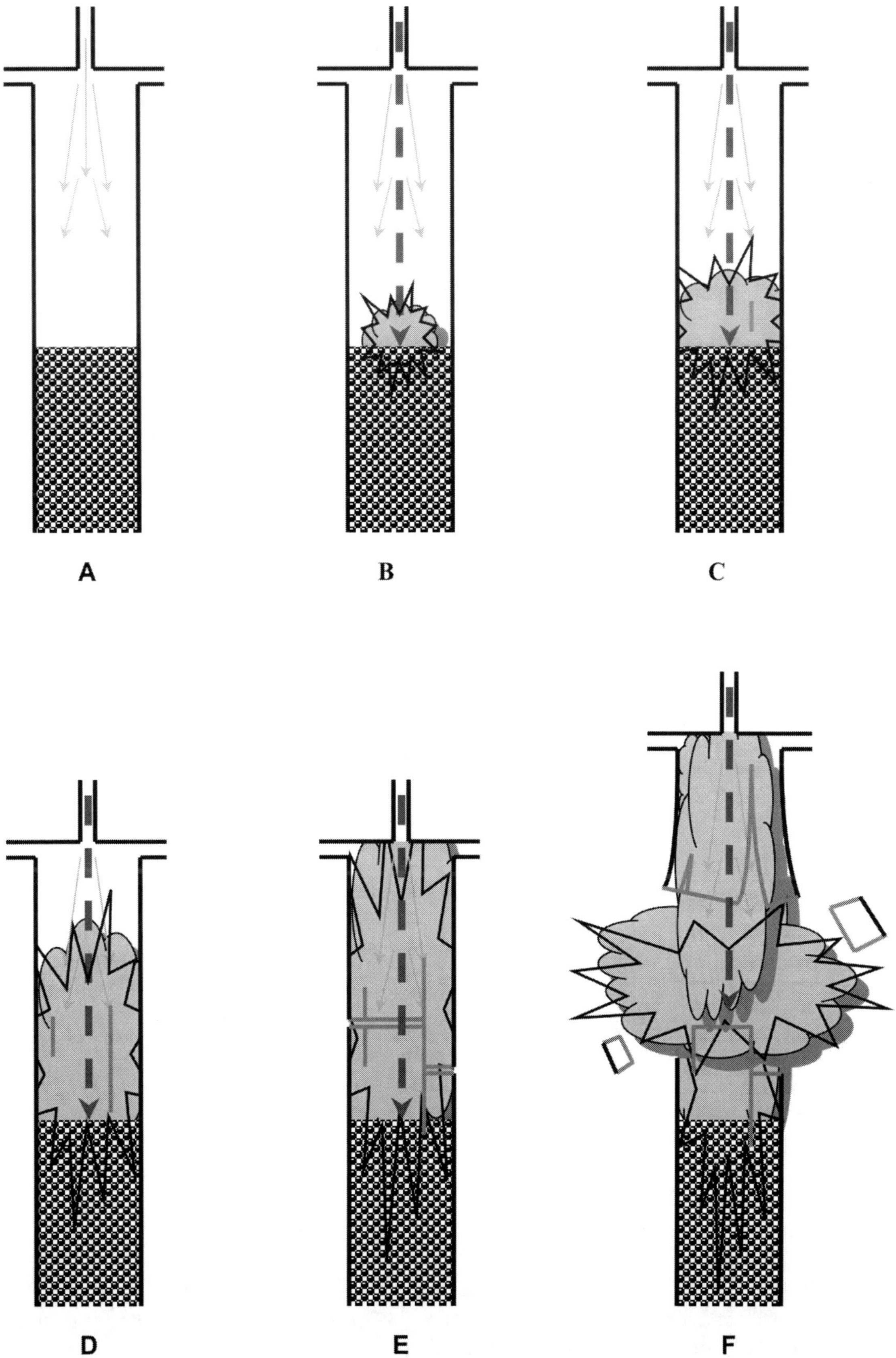

Diagram 1. Sequential Representation of Impulse

Radiant Tubes

Two distinct modes of failure were noted on the radiant tubes. The categories were arbitrarily picked as a means of distinction.

Category A Tubes

The five tubes noted above as having had their upper ends grossly lifted exhibited unique fracture features.

All of these tubes showed complete mechanically brittle circumferential separation on multiple planes of the tube material between ~14 cm to 45 cm (~0.5 ft to ~1.5 ft) below the flange. The different planes of circumferential cracking were bounded by mechanically brittle (no local distortion) longitudinal cracks running toward the flange. This indicates that longitudinal cracking preceded circumferential cracking. In many cases, the remaining material between the longitudinal cracks was highly deformed radially. Longitudinal cracking generally arrested in or near the flange hub and was sometimes accompanied by circumferential tearing at this point. Some large pieces of shrapnel found on the furnace roof, believed to have come from this region of these tube assemblies, also exhibited significant distortion in some areas and very little in others. Whether this distortion occurred before or after separating from the tube is unknown. Hence, the extent of distortion and longitudinal cracking below the separation is unknown. Although these sections would have been subjected to similar loads, the same extent of gross radial distortion (yielding) as described above is not likely due to the anticipated brittleness of tube sections exposed to radiant heating within the firebox.

The final resting place of these five tubes, the reported single bang, the full thickness of the fracture faces especially in regions of gross radial deformation and the distortion direction conclusively indicated that failure was due to "simultaneous," sudden, high internal pressure loads applied near the upper end of each assembly.

Category B Tubes

The remaining 41 damaged tubes had fractured at various lower elevations and exhibited quite different macro features than were shown by the Category A tubes.

The only macro feature these tubes had in common with those described above was the unquestionably mechanically brittle nature of the fracture faces (no local distortion). Nothing suggested that failure was primarily due to internal pressure excursions. Rather, evidence such as tube flattening with associated longitudinal cracks as well as angled cracks is indicative of externally applied bending loads. Several of these tubes exhibited varying degrees of impingement damage, due to impacting shrapnel or falling tube sections.

Process Evidence of a Wet System

The startup procedure did take into consideration the need to have the reformer furnace feed system water free and steps are included to ensure that any water would be either drained form the system or evaporated as steam during heat up.

Shortly after gas burner lighting commenced, the reformer mixed-feed temperatures began a rapid ascent before settling at ~143°C (290°F).

At the same time, but at a slower rate, the reformer outlet header temperature began to increase and eventually settled at approximately the same temp as the reformer inlet. However, no flow was going through the feed line and reformer radiant tubes. The process piping system had been purged with nitrogen and the process down stream of the reformer furnace was still isolated. The system pressure also began to increase to ~40 psig. (Saturation temperature of 40-psig steam is 141°C.)

The temperature of the reformer inlet and outlet remained constant at ~143-146°C, even though the portions of the radiant tubes in firebox were approximately 350°C. A constant temperature profile indicated that water was still present since steam generation occurs in equilibrium with water at a constant temperature, until all water is vaporized.

The investigation team calculated that after the Induced and forced draft fans were restarted, the mixed feed and superheater coils gave up approximately 56 MBtu of heat from 0900 to 1630 hrs, as evidenced by the temperature rise. This corresponds with the production of approximately 30,000 lbs of condensate over the 7-hr period. Further, from 19:00 to 22:00 hrs, the mixed feed preheat coil absorbed approximately 14.6 MBtu of heat. This would be equivalent to producing approximately 18,000 lbs of steam over this 3-hour period.

The following graph depicts these key process conditions prior to the incident.

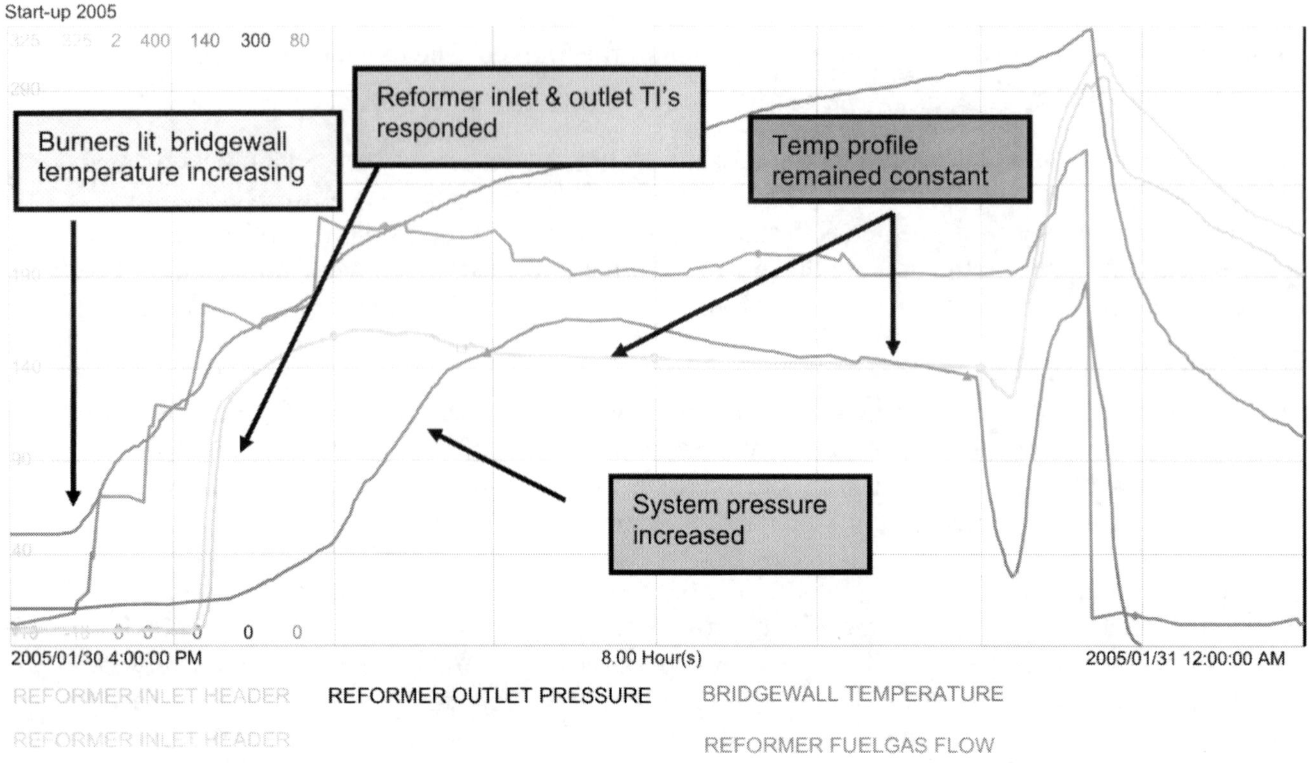

Sources of water

The entire furnace feed system was analyzed. Piping configurations and witness testimony identified several ways water could enter the feed system and get trapped.

During the shutdown sequence, the steam system will become wet as pressure is reduced and temperature falls. This is normal and expected so the design of steam and feed systems incorporate vents and drains to facilitate water removal.

Steam was left in the steam generating system and throttled to keep approximately 344kPa (50 psig) in the system to prevent freezing.

There is an argument that there is always an opportunity for water to accumulate in the steam system upstream of the reformer furnace as, at some point, the steam is commissioned to a "cold" system so the fact that steam was left on should not be a factor. However, this change is a contributor to the overall accumulation of water since the steam was commissioned for days rather than hours as per the startup procedure. This would have provided an extended time frame for water to accumulate in areas that would not drain.

The most likely source of water accumulation would be in the mixed feed preheat coil for two reasons. The inlet header to this coil will not fully self-drain and the tubes had become slightly bowed over time.

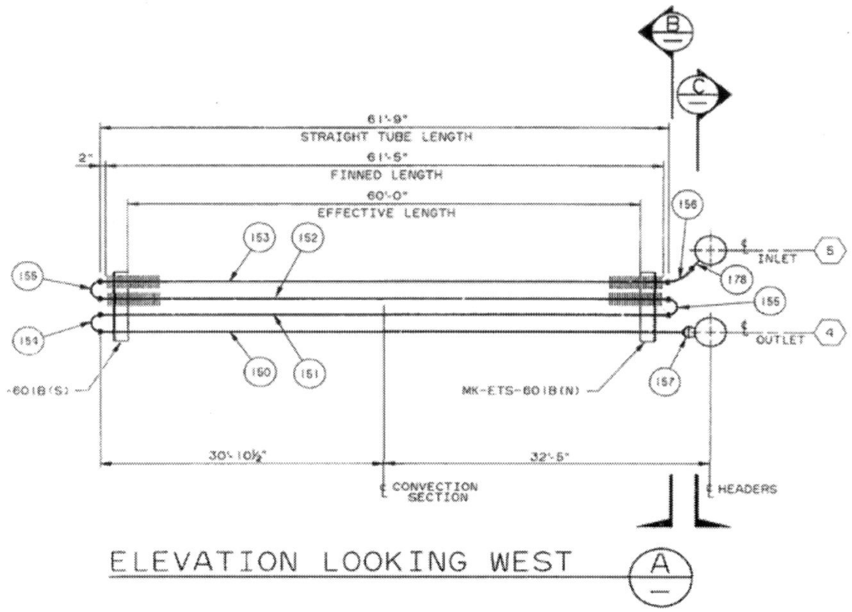

Diagram 2. Detail of Mixed Feed Preheat Coil

Since there was no accurate means to positively identify the source of trapped water for this incident, the investigation team recommendations were focused on addressing the most likely water sources and points of accumulation and procedural changes to deal with ensuring a dry feed system exists prior to adding the startup steam.

Comparison to Previous Start-ups (2000 & 1995)

The process data from the previous two shutdowns was reviewed in order to verify the conclusions regarding the source of the water. In previous startups, the reformer inlet temperature response was much slower to the heat input through the convection section and did not plateau during the startup sequence. Hence, there was no indication of boiling in the mixed feed preheat coil. While the temperatures did increase, this is attributed to conduction and general elevated temperatures in the "penthouse" region at the top of the furnace.

The reformer outlet temperature instruments hardly responded at all, and remained below 50°C (122°F) until steam flow was introduced to the tubes. No pressure increase was observed on the system.

Lessons Learned

Inadequate Safeguards

The system was designed with low point drains to facilitate water removal, however, these were found to be inadequate in both location and size. The fact that the mixed feed pre-heat coil was not self-draining was unknown prior to the incident.

After a thorough review of the entire reformer furnace feed system was completed, existing drains were increased in size and others added to ensure the entire feed system could be drained.

A HAZOP revalidation had not been performed on this plant, but had been scheduled for 2006.

Inadequate Procedure

The startup procedure did not account for a startup of a cold furnace with no hold points for catalyst reduction or refractory dry out. As a result, the time to reach the critical "steam in" temperature of 350°C (662°F) was short as compared to previous startups. Also the procedure provides little direction for confirming that the reformer furnace feed system is dry.

Modifications to the procedure were completed that included a longer heat up period, the addition of more detailed guidance for verifying the feed system is dry and a formal sign off by both operations and engineering personnel.

Also a separate cold eyes review by external experts was completed as part of the pre start up safety review.

Lack of Management of Change

The startup procedure had two hold points for refractory dry out and new catalyst reduction, during the heat up phase prior to introducing the 4.1mPa (600psig) startup steam. These hold points were not utilized since it appeared neither was required. Consequently the heat up cycle was artificially shortened. It became apparent that this alteration to the startup sequence was not viewed as a change by operations. Several sections of the procedure were not performed since they did not apply to this startup.

When the conditions were met the 4136 kPa (600psig) startup steam it was introduced as per the procedure.

Shutdown and startup procedures are designed to take a unit from safe operation to a zero energy state and then return it to safe operation. Changing these sequences by an intentional omission is a change and must be properly assessed for risk.

The decision to leave some steam flow in the steam generating system for this winter shutdown was made to keep the system warm and prevent freezing. However, no formal risk assessment was performed and no MOC was generated. A risk assessment was performed prior to the startup but the change in status of the steam system was not evaluated. In fact the decision to leave steam in was seen as a safeguard from the risk of freezing. This provided an opportunity for water to accumulate upstream of the reformer furnace.

Non-essential Personnel

At the time of the incident there were, as previously mentioned, seven people on the furnace structure. Only the operations personnel were essential. Changes have been made to ensure non-essential personnel are cleared from the areas during startup activities.

References

1. Emberley D. "Fracture Power Calculations" Syncrude Canada Ltd. 2005-0037, 2005-04-05.

2. Pierorazio A., Ketchum D. Wesevich "Pressure Estimate Analysis – Hydrogen Reformer Incident" Baker Risk Engineering and Risk Consultants Inc., Project 01-1076-001-05, 2005-03-15.

Vinyl Chloride Monomer Explosion

Lisa A. Long (Speaker)
James Lay, PE; Katherine Leskin; Randy McClure, CSP; Allen Smith

U.S. Chemical Safety Board
2175 K St, NW
Washington, DC 20037
Phone: 202-261-7635
Lisa.long@csb.gov

Prepared for Presentation at
American Institute of Chemical Engineers
2007 Spring National Meeting
41st Annual Loss Prevention Symposium
Houston, Texas
April 22–26, 2007

UNPUBLISHED

Vinyl Chloride Monomer Explosion

Lisa A. Long (Speaker)
James Lay, PE; Katherine Leskin; Randy McClure, CSP; Allen Smith

U.S. Chemical Safety Board
2175 K St, NW
Washington, DC 20037
Phone: 202-261-7635

Abstract

On April 23, 2004, an explosion and fire at the Formosa Plastics Corporation, Illiopolis, Illinois (Formosa-IL) polyvinyl chloride (PVC) manufacturing facility killed five and severely injured three workers. The explosion and fire destroyed most of the reactor facility and adjacent warehouse and ignited PVC resins stored in the warehouse. The plume of smoke from the smoldering fire drifted over the community and resulted in an evacuation that lasted 48 hours. The facility has been closed down since the incident and no actions to rebuild have been taken.

Vinyl chloride (VCM), a highly flammable chemical and known carcinogen, and the major component in the facility's manufacturing process, was the primary fuel for the initial fire and explosion. Formosa-IL used VCM to manufacture a variety of PVC resins in twenty-four heated and pressurized reactors. Formosa-IL, a wholly owned subsidiary of Formosa Plastics Corporation, USA (FPC USA), purchased the Illiopolis facility from Borden Chemical and operated it for approximately two years before the incident.

This paper will include discussion of the incident's root and contributing causes, and CSB recommendations to prevent recurrence.

Background

The Formosa-IL PVC plant was purchased from Borden Chemical, and included a commodity PVC resin process (PVC1) and a specialty PVC resin process (called "Paste"). The reactor building (Figure 1) housed the PVC1 and Paste production areas.

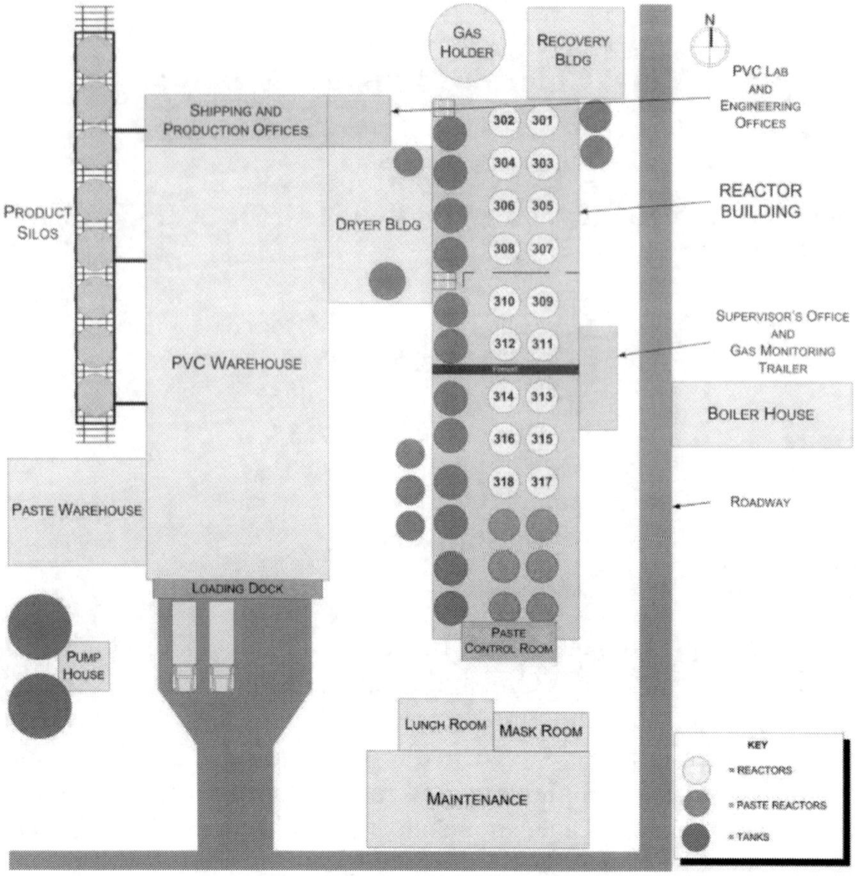

Figure 1. Formosa-IL plant layout

Formosa-IL produced PVC by the polymerization of VCM.[1] VCM is a colorless gas with a mild, sweet odor, which the average person cannot smell at concentrations below hazardous levels.[2] VCM can ignite when its concentration in air is between 3.6 and 33 percent by volume. It has serious short and long-term health effects, and is a known human carcinogen.

Two operators, a poly operator and a blaster operator, were responsible for the reactor involved in this incident. The poly operator worked exclusively on the upper level of the building where the reactor controls and indicators were located, while the blaster operator worked on both the upper level and the lower level, where additional valve controls were located.

---

[1] The Formosa-IL PVC1 suspension unit produced both co-polymer and homopolymer resins. Because a homopolymer reactor was involved in the incident, only the homopolymer process is described.

[2] The VCM odor threshold for most people is approximately 3,000 parts per million (ppm), 600 times higher than the OSHA Short-Term Exposure Limit (STEL) (a 15-minute time-weighted average concentration that should not be exceeded).

Figure 2. Reactor building elevation view

To make a batch of PVC, the poly operator:

- readied the reactor,
- added the raw materials (liquefied VCM, water, suspending agents, and reaction initiators),
- heated the reactor,[3]
- monitored the reactor temperature and pressure until the batch was complete, then
- vented pressure from the reactor, and told the blaster operator to transfer the batch to the stripper.

To transfer the batch to the stripper, the blaster operator opened the transfer and reactor bottom valves from the lower level.

---

[3] Temperature was automatically controlled by adding steam or cooling water into the reactor jacket.

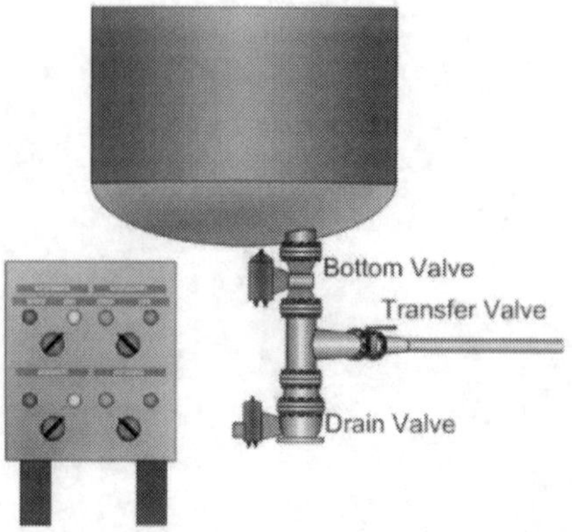

Figure 3. Reactor bottom, transfer, and drain valves

When the transfer was complete, the blaster operator closed the transfer valve, and the poly operator purged the reactor of hazardous gases to prepare the reactor for cleaning. The blaster operator cleaned the reactor by

- opening the reactor manway on the upper level,
- power washing PVC residue from the reactor walls, and
- opening the reactor bottom valve (if not left open from the transfer) and drain valve to empty the cleaning water to floor drains.

Once the blaster operator finished cleaning the reactor, he closed the reactor bottom and drain valves, and gave the poly operator a completed checklist indicating that the reactor was ready for the next batch of PVC.

Reactor Emergency Procedures

The reaction of VCM to form PVC generates heat. If this heat is not removed, a runaway reaction may occur, potentially increasing reactor pressure above the pressure relief device set point, venting the reactor contents, including VCM, to the atmosphere.[4] The U.S. Environmental Protection Agency (EPA) Clean Air Act (CAA) regulations require companies to report VCM releases to the atmosphere.

Formosa-IL had procedures to prevent reactor overpressure and minimize the possibility of an atmospheric release. However, if these procedures failed to control the reactor pressure, the operators followed an emergency transfer procedure that required them to manually open the reactor bottom and transfer valves and connect to another reactor.

---

[4] A relief device installed on the reactor would open and relieve pressure to the atmosphere to protect the reactor. Each reactor was protected by a rupture disk and a relief valve, in series, that relieved to the atmosphere.

Borden Chemical created this emergency transfer procedure, concluding that connecting two reactors would promote reaction inhibitor mixing, maximize cooling, and provide extra volume so that the pressure would not exceed the reactor relief device set point and thereby avoid a reportable VCM release.

Reactor Bottom Valve Interlock and Bypass

The reactor bottom valve had a safety interlock that prevented opening it if reactor pressure exceeded 10 psi. This interlock was installed to prevent the bottom valve on an on-cycle reactor from being inadvertently opened.

During an emergency transfer the reactor pressure was greater than 10 psi, so Borden developed a manual interlock bypass to open the valve. The bypass consisted of quick-connect fittings on the air hoses so operators could disconnect the valve actuator from its controller and open the valve by connecting an emergency air hose directly to the actuator.

The emergency transfer procedure required a supervisor to authorize, but not necessarily witness, bypassing the interlock.

VCM Alarm System and Deluge System

The plant had a detection system to monitor airborne VCM concentrations and alert personnel of harmful levels. When the system detected airborne concentrations of one part per million (ppm) or greater, horns sounded and red lights flashed above all area entryways.

Additionally, the plant had a deluge, or water spray, system for fire protection and loss control. The deluge system could be activated by heat, high VCM concentration, loss of instrument air, or manual pull stations. When activated, water spray nozzles blanketed an area with water.

Incident Description

Operations at the plant were normal in the hours before the incident. At approximately 10:30 p.m., just moments before the incident, all PVC1 reactors were on-cycle except for reactor D306, which was being cleaned.

A few minutes after 10:30 p.m., workers throughout the plant heard a very loud rumbling[5] and some smelled VCM. The shift supervisor and operators in the south (Paste) end of the reactor building heard the Paste section deluge alarm, indicating deluge system activation in that area.

The shift supervisor left the Paste control room to check the VCM gas detection system reading, and found that two areas had levels above the instrument's measurable limit, suggesting a large release. He said that on his way to investigate the release, he walked past an open doorway in the PVC1 area near reactor D310 and saw material spraying from the bottom of D310 creating a

---

[5] Some described it as the sound of a jet engine.

foaming mixture on the floor about 1.5 feet deep. He immediately climbed the stairs to the upper level.

According to the shift supervisor, operators on the upper level of PVC1 reported that reactor D310 pressure was rapidly falling. He told two operators that he had seen material spraying from the bottom of D310, so they immediately began checking the D310 valves and controls. The supervisor and one operator tried to go to the lower level through an interior stairwell, but high VCM concentrations forced them to retreat.

The shift supervisor instructed operators to open the reactor D310 vent valves to relieve pressure and slow the release. At this point, he saw that the reactor pressure had already dropped to 10 psi, further indicating a large release. Attempting again to go to the lower level, the shift supervisor was descending an exterior stairway when a series of explosions occurred.[6]

Four operators were killed by the explosions; two working near the top of the reactor and two working on the lower level. A fifth operator died in the hospital two weeks later. The shift supervisor and two workers were hospitalized, and four workers were treated at the scene.

The explosions destroyed much of the reactor building. The PVC laboratory, the safety and engineering offices, and an adjacent building housing safety equipment- all located hundreds of feet away- were also destroyed. The fire spread to the PVC warehouse and burned for hours, sending a plume of smoke into the community.

In response, emergency responders evacuated approximately 150 residents living within one mile of the plant and the Illinois State Police closed the major roadways in the area.

Most Likely Incident Scenario

The CSB used physical evidence and survivor interviews to determine the most likely incident scenario.

Employee interviews and day shift records indicate that at the time of the incident, reactor D310 was full and heating up to temperature. After the incident, D310 was found empty with the reactor bottom and drain valves open, and both valve switches in the "Open" position. The normal air hose to the bottom valve actuator had been removed and replaced with the emergency air hose.

The as-found condition of Reactor D306 indicated that it was being cleaned: the top manway was open, the pressure washer was inside, the bottom valve was open, and the drain and transfer valves were closed. The next cleaning step that the blaster operator would have performed would be to open the drain valve and empty the reactor.

Because of the as-found condition of reactors D306 and D310, the CSB concluded that the blaster operator cleaning reactor D306 likely went to D310 by mistake and tried to open the

---

[6] According to witnesses, several explosions followed the initial explosion.

bottom valve to empty the reactor. The D310 valve would not open, however, because the reactor was on-cycle with a pressure greater than 10 psi.[7] Because he likely thought he was at the correct reactor (D306), the blaster operator may have presumed the bottom valve was not functioning properly. The CSB investigators concluded that the blaster operator, who likely believed the reactor contained only cleaning water, used the emergency air hose to bypass the bottom valve pressure interlock and open the reactor bottom valve releasing the contents.

According to the shift supervisor, the blaster operator did not request permission to bypass the interlock (required by procedure) or inform anyone that he had bypassed it. Consequently, operators attempting to control the release likely believed that a failure or other malfunction had occurred and tried to relieve the pressure inside the reactor to slow the release.[8] The VCM vapor cloud ignited and exploded while the operators were working at the top of the reactor.

Incident Analysis

Minimizing Human Error

Studies show that human error is a large contributor to fatalities, injuries, and property damage in the chemical industry [1]. At Formosa-IL, several factors combined to make the human error in this incident more likely.

Reactor Layou

The PVC1 reactors were arranged in groups of four, with bottom valve and drain valve control panels for every two reactors. Both reactors (D306 and D310) involved in this incident occupied the same relative position among groups of four reactors (Figure 2).

[7] Plant personnel estimate that the pressure inside D310 was approximately 70 psig when the interlock was overridden.

[8] Approximately 15,000 pounds of VCM was released, along with water, PVC, and other reaction ingredients.

Figure 2. Cutaway of reactor building

Reactor bottom and drain valve control panels were on the lower level. Each reactor was labeled with the reactor number. The control panels, also labeled with the reactor number, had clearly labeled indicator lamps showing the bottom and drain valve positions.

Similar equipment layouts are common in the chemical industry and can create a possibility for human error. Companies need to be aware of this and ensure appropriate safeguards are in place.

Communications

Operators working on the lower level had no means to communicate with operators on the upper level where the reactor status information was displayed.[9] Consequently, an operator at the valve control panel on the lower level, who questioned why a bottom valve would not open, would have to climb the stairs to the upper level to determine reactor status. This "inconvenience" may have contributed to a scenario in which the blaster operator might guess reactor status instead of climbing the stairs to get the status.

Staffing Changes

When FPC USA bought the Borden facility in 2002, management reorganized and reduced staff. Interviews and documents obtained by the CSB investigators indicate that when considering the staffing changes, FPC USA compared similar Formosa plant sites and Formosa operating philosophy, but did not conduct a formal analysis of the safety impact of these changes.[10]

Under Borden Chemical, each area (such as PVC1) had assigned operators and a group leader (a working supervisor). The group leader was an hourly employee responsible for supporting the operators in his/her area. Employees interviewed indicated that the group leader, a respected position at the Illiopolis facility, had an elevated level of responsibility. If an operator had a problem, the group leader for the area was skilled and readily available. For example, if a problem occurred that required that the reactor bottom valve safety interlock to be bypassed, the PVC1 group leader had the knowledge of how, and more important, under what circumstances it could be done. Furthermore, the group leader could usually obtain the shift supervisor's permission to bypass the interlock.

FPC USA management eliminated the group leader positions when it bought the facility. Consequently, instead of each operational area having a group leader, a single shift supervisor was responsible for the entire plant. In an emergency, or if an operator needed advice or assistance, the shift supervisor might not be as available as the previous group leaders.

---

[9]At Formosa-IL, operators did not normally carry radios.

[10]Management of Change (MOC) programs evaluate the safety consequences of changes to facilities that affect a covered process. The OSHA PSM standard requires companies that handle highly hazardous chemical processes to have MOC programs. Since 1989, CCPS has advocated applying MOC to organizational and staffing changes.

This lack of availability, combined with communication difficulties and that the use of the safety interlock bypass would be undetectable, increased the likelihood that an operator might act independently and may have contributed to the unauthorized safety interlock bypass use.

Learning from Incidents

Prior to the April 23, 2004, Formosa-IL incident, operators mistakenly opened a reactor bottom valve at both the FPC Baton Rouge and Formosa-IL facilities. FPC USA had received reports of both incidents but did not recognize a key similarity: operators could mistakenly go to the wrong reactor and bypass safeguards to open a reactor bottom valve.

2003 Baton Rouge Incident

In June 2003, about 8,000 pounds of VCM was released at FPC's Baton Rouge facility when an operator cleaning a reactor mistakenly opened the bottom valve and blind flange on an on-cycle reactor.[11]

This incident was similar to the fatal incident at Formosa-IL, as the operator opened the top manway on the reactor to be cleaned and then went to the lower level, where he mistakenly went to an adjacent reactor, removed a bolted flange cover, and opened a motor-operated drain valve, emptying the reactor contents onto the floor. The operator tried unsuccessfully to close the valve, then evacuated and notified the supervisor. The supervisor immediately switched the valve operation to "Remote" and closed the valve from the control room.[12] In this incident the VCM dispersed without igniting.

The FPC Baton Rouge investigation team could not determine why the operator went to the wrong reactor, as both reactors and the drain valve controls were labeled and the reactor agitator could be heard. Since this incident, all reactor drain valves at the Baton Rouge facility are locked when not in use, and only supervisors have access to the keys.

February 2004 Illiopolis Incident

At Formosa-IL, about 60 days prior to the April 23, 2004, fatal incident, an operator mistakenly transferred the contents of an on-cycle reactor to a stripper tank. The pressure in the stripper tank exceeded the rupture disk set point and VCM was released to the atmosphere through a vent pipe.

Formosa-IL investigated this incident. As in the fatal incident, an operator used the emergency air hose to bypass the reactor bottom valve interlock without supervisor authorization. In response, Formosa-IL retrained the operators on emergency reactor transfer procedures, including the requirement for supervisory approval prior to using the emergency air hose.[13] In

[11] The operation at Baton Rouge was different than at Illiopolis: reactors were not cleaned after each batch and the reactor bottom drain was closed with an 8-inch motor-operated locally operated drain valve and a bolted cover.

[12] At Formosa-IL the reactor bottom valve can only be operated on the lower level under the reactor.

[13] Records were unavailable to determine if the blaster operator on duty the night of April 23 received this training.

addition, the investigation report recommended redesigning the interlock bypass to prevent unauthorized use and established a deadline of April 1, 2004, for completion.

Employee interviews indicated that Formosa-IL personnel met several times to discuss a new bypass design, but were unable to agree on one. Formosa-IL personnel continued discussions after the April 1 deadline, but had not implemented any design changes before the April 23, 2004, incident.

2005 Delaware City Incident

In May 2005, an operator mistakenly transferred the contents of an on-cycle reactor at Formosa's Delaware City facility, resulting in a VCM release of about 2,500 pounds. The reactor bottom valve on all reactors had been chain-locked with two locks to prevent inadvertent operation; one key was held by the cleaning operator, and the other by the control room operator. On the day of the incident, the control room operator gave his/her key to the cleaning operator in the control room[14] who went to the wrong reactor, unlocked both valves, and transferred the contents of the operating reactor.

PVC Industry Incidents

The Encyclopedia of PVC [2], lists several PVC manufacturing incidents similar to the April 23 Formosa-IL incident in that operators went to wrong reactors and opened bottom valves. These incidents demonstrate that inadvertently draining a reactor is a serious hazard in the PVC manufacturing process.

The 2003 Baton Rouge incident, February 2004 Formosa-IL incident, and several incidents listed in *The Encyclopedia of PVC* illustrate that an operator can mistakenly go to the wrong reactor and open a bottom valve. Moreover, the February 2004 incident illustrates that easily defeated safeguards may do little to prevent operator error.

The CSB concluded that FPC USA corporate safety personnel and Formosa-IL managers should have recognized the similarities in these incidents, specifically the possibility that an operator might mistakenly open the wrong reactor bottom valve, and made the installation of a more effective interlock an urgent priority.

Companies with a learning culture purposefully study problems such as those revealed in incident investigations, and quickly make changes to improve systems [3]. This includes reviewing incidents across the corporation, recognizing similar patterns, and ensuring that all facilities take appropriate corrective actions

---

[14] The policy required the control room operator to hand over his/her key in the control room so that he/she did not have to leave the control room unattended.

Analyzing High-Consequence Hazards

Borden Chemical completed a process hazard analysis (PHA) on the PVC1 unit in 1992 and a partial revalidation of the PHA in 1999.[15] These PHAs were missed opportunities to identify the hazard of dumping an on-cycle reactor and take action to reduce the associated risk.

1992 Borden PHA

The April 23, 2004, fire destroyed much of the 1992 Borden PHA documentation. What remained verified that the PHA team analyzed the reactor emergency transfer procedure and recommended

- modifying the plant-wide philosophy "to require that no plant interlock be defeated without direct hands-on approval and witnessing of operating supervision"[16], and
- replacing the emergency air hose system with one that required a supervisor's key to override it.

These recommendations, which addressed some key deficiencies that the CSB concluded contributed to the unauthorized use of the bottom valve interlock bypass on April 23, 2004, were never implemented. In addition, site management continued to require that employees obtain only the supervisor's verbal authorization to bypass the reactor bottom valve control.

1999 PHA Revalidation

In 1999, Borden Chemical conducted the partial PHA revalidation of the PVC1 process. The PHA team addressed a previous incident where the contents of an on-cycle reactor were inadvertently transferred, a scenario very similar to what occurred during the February 2004 incident. The PHA team identified potentially severe consequences of fire, injury, and environmental and/or offsite impact, but accepted the existing bottom valve safety interlock, procedures, and training as suitable safeguards.

Inadvertently transferring the contents of an operating reactor was identified as a high-consequence event with a significant likelihood of occurrence (it had occurred at least once). Operating procedures are typically viewed as less reliable safeguards for high-consequence events [4], yet the safeguards accepted for this high-consequence event were written procedures and an interlock with a bypass controlled only by procedure (the emergency transfer procedure). Such administrative protections did not adequately prevent the severe consequences i.e., fire, injury, and environmental release of a toxic substance.

The CSB investigators noted that neither PHA addressed reactor cleaning procedures. Addressing the reactor cleaning procedure in a formal PHA[17] may have identified the potential

---

[15] Formosa-IL had not revalidated the 1999 PHA or conducted a new PHA prior to the April 23, 2004, incident.

[16] Formosa's policy at the time of the incident required that a supervisor authorize the override instead of being present and directly involved.

for an operator to go to the wrong reactor during the cleaning process, and might have prompted Borden Chemical to implement more rigorous safeguards, like a key lock system, to ensure operators could not accidentally open the bottom valve on an operating reactor.

Risk Assessment

After identifying likelihood and severity of identified hazards, risk should be assessed to determine if existing safeguards are adequate. Risk, a function of likelihood and severity, is highest when the most severe consequences have a likelihood of occurring frequently. These risks require a higher level of safeguard.

Borden Chemical and Formosa-IL used a qualitative approach to evaluate hazards and assess risk for PHAs. [18] A qualitative PHA is conducted by a team of knowledgeable and experienced individuals. The team uses experience and judgment to assess a hazard's severity and likelihood and makes recommendations for additional safeguards when the risk associated with the hazard is perceived to be too high. Relying solely on judgment without standards or written guidance can lead to accepting or specifying an inadequate safeguard for an unacceptable risk. For example, in the 1999 Borden Chemical PHA revalidation, the PHA team identified the high-risk scenario of transferring the contents of an operating reactor, but accepted weak administrative safeguards as adequate to control the risk.

While qualitative approaches to hazard analyses are common in the chemical industry, the Center for Chemical Process Safety (CCPS) has recently promoted a semi-quantitative method called Layer of Protection Analysis (LOPA)[5]. LOPA is a more rigorous approach to determine if safeguards are sufficient to adequately control risks in hazardous systems such as chemical reactors.

In this case, LOPA could be used to determine

- if the reactor bottom valve interlock qualifies as an independent protection layer (IPL), a safeguard whose effectiveness can be measured and audited; and
- if the effectiveness of all combined IPLs is adequate to control risk.

A reactor bottom valve interlock LOPA would likely identify that an easily used, poorly controlled, untraceable bypass mechanism would disqualify the interlock as an IPL. Additionally, since LOPA does not accept procedures as IPLs, the procedural requirement that a supervisor authorize the use of the interlock bypass would not likely qualify.[19] The absence of any other IPL on the reactor bottom valve would likely lead to the conclusion that the risk of

---

[17] The CCPS *Guidelines for Hazard Evaluation Procedures* suggests performing PHAs on procedures for batch processes like that used at Formosa-IL and include a worked example for a PVC production facility [6].

[18] Formosa PHA procedures include guidelines for risk assessment. However, instead of using risk assessment to determine if a recommendation should be made, Formosa used risk assessment to prioritize recommendations once they have been made.

[19] Procedures help improve human performance, but because they are not independent of that performance, they are not considered to be IPLs.

inadvertently opening the valve is inadequately controlled and that more robust safeguards are necessary.

Interlocks

Formosa-IL relied on the reactor bottom valve interlock as a key safeguard to protect against accidental transfer or dumping of the contents of an operating reactor, even though the interlock could be easily bypassed. In addition to the reactor bottom valve interlock, the CSB identified other safety equipment that could be easily bypassed. [20]

The design of an interlock must be highly reliable, effective, and secure. The Formosa-IL reactor bottom valve interlock lacked the physical controls needed to make it secure, in that anyone could access and use it. In addition, failure to provide indication of the bypass condition meant that bypassing could be undetected, compromising the effectiveness of the safety equipment.

Although Formosa-IL had a specific procedure for using the emergency air supply to bypass the bottom valve interlock, the procedure did not include recognized and generally accepted good practices for controlling safety interlocks. The 1992 Borden PHA recommended a "plant-wide" philosophy for controlling interlocks; however, none was implemented at the time of the April 23, 2004, incident.

CCPS offers best practice guidance for bypassing interlocks:

- Implement a formal facility policy for bypassing interlocks; and
- Write and implement procedures that require
    - process monitoring while the interlock is bypassed,
    - analysis of the appropriateness of the bypass method,
    - a time limit for bypass, and
    - documentation for and communication to all affected personnel [7].

In addition, *Loss Prevention in the Process Industries* [8] recommends supplementing bypass procedures with additional security measures such as keyed switches or passwords.

Following the April 23, 2004, incident, Formosa's Baton Rouge, Delaware City, and Point Comfort management instituted site-wide policies for controlling interlock bypasses. [21] These

---

[20] Wooden blocks used to prevent inadvertent activation of the deluge system during maintenance hung from each riser valve. Facility management intended that authorized personnel use the blocks only during maintenance; however, because the use of the blocks was uncontrolled, they were vulnerable to misuse.

A toggle switch installed to disable the VCM gas detection system and prevent automatic activation of the deluge system during maintenance. Other than the switch position, no other indication existed to alert personnel that the system was disabled.

[21] Prior to the April 2004 incident, FPC USA only required its PVC facilities to institute policies for interlock override as part of a work permit process used for maintenance. Furthermore, some of the FPC USA facilities did not have interlocks on the reactor bottom valve.

policies require a manager's signature on a written permit, identification of alternate safeguards to ensure ongoing operational safety, and re-evaluation of bypassed interlocks every 24 hours. A site-wide interlock bypass policy like this, combined with appropriate physical or engineering control over the emergency air hoses, would likely have prevented the April 2004 fatal incident.

Emergency Preparedness

Worker actions during this incident demonstrate that they were poorly prepared for a catastrophic release of VCM. For example, facility emergency response procedures contained conflicting requirements, workers tasked with critical response duties had not received requisite training, and the workers had not rehearsed a response to a large VCM release in more than 10 years.

The supervisor joined the two operators who were trying to control the release rather than call for an evacuation. The CSB estimates that the operators continued their work until the explosion occurred, up to five minutes after the release started.[22] Despite knowingly working directly over a toxic VCM cloud, the operators did not put on protective breathing apparatus, activate emergency alarms, or evacuate the facility, contrary to emergency response actions outlined in Formosa-IL procedures addressing emergency response to hazardous material releases.

The Formosa-IL facility had six different procedures addressing VCM release emergencies, which made determining Formosa's plan for responding to a large VCM release difficult. For example, one procedure stated that a trained emergency brigade would respond to releases of less than 55 gallons. Another directed the shift supervisor and unit operators (not necessarily trained in emergency response) to put on protective equipment and take necessary actions to stop a release, yet it did not define the size of the release that employees were to respond to. These ambiguities created uncertainties that increased stress and compounded the potential for mistakes. Given that the procedures outlined employee response actions and that employees did respond to control the VCM release, the CSB concluded that the employees would have been better prepared to respond to this emergency if Formosa-IL had fully implemented the emergency response requirements outlined in OSHA's Hazardous Waste Operations and Emergency Response Standard (HAZWOPER, 29 CFR 1910.120).

CCPS recommends that emergency response plans specify the type and frequency of drills and exercises, and contain procedures for organizing, conducting, and evaluating them. Furthermore, CCPS suggests annually conducting a "full scale" exercise and periodic credible incident scenario drills for each process [9]. Formosa-IL had no written requirements or procedures for conducting drills or exercises; instead of planning exercises or drills, the facility only documented actual occurrences (false alarms, minor system upsets and releases, and employee evacuations) and after such occurrences, conducted "lessons learned" sessions with management and employee representatives. The CSB found that the Illiopolis facility had not conducted a drill of a large release scenario in more than 10 years.

---

[22] Estimates range from 2.5 to 5 minutes as the time from the start of the release until ignition.

Designing Deluge Systems to Prevent Fires and Explosions

Deluge systems can sometimes prevent vapor cloud fires and explosions through several mechanisms:

- Dispersing vapor and inducing air flow into the release, ultimately reducing the vapor concentration below the lower flammable limit [10].
- Absorbing the substance released; this, however, depends on the properties of the substance, primarily its solubility in water [10].
- Cooling and condensing vapors from materials with high boiling point temperatures, which may also reduce the vapor concentration below its lower flammability limit [11].
- Preventing ignition by reducing the potential for static electricity-generated sparks, and by cooling hot surfaces [11].[23]

Formosa-IL employees believed that the reactor building water deluge system would prevent or mitigate an explosion or fire from a major VCM release.[24]

The PVC1 deluge system did not activate during the April 2004 incident. The CSB was unable to determine a cause for the malfunction. However, the CSB concluded that the PVC1 deluge system would probably not have prevented the fire and explosion by reducing the vapor concentration below the lower flammable limit because

- the PVC1 building was enclosed, thereby preventing effective dispersion of VCM vapors;
- VCM is only slightly soluble in water, making water sprays ineffective for absorbing VCM releases; and
- VCM vapors cannot be condensed by water sprays due to the low VCM atmospheric boiling point.[25]

CCPS [13] provides some general guidelines for the design of deluge (water spray) systems for air entrainment and suggests a strategy for designing a system to mitigate flammable material releases that requires concise definition of the protection scenario. For specific critical design features of fixed water spray systems, CCPS refers to NFPA 15, *Standard for Water Spray Fixed Systems for Fire Protection* [11].[26]

NFPA 15 states that preventing fire is a legitimate design objective for water spray systems. The standard also provides specific, quantitative guidance on designing fixed water spray systems to

---

[23] If electrical components are not sealed properly, water ingress can actually create ignition sources by causing sparks from electrical short circuits [12].

[24] The CSB collected information on the PVC industry through an informal sampling of PVC manufacturers and learned that other PVC industry personnel have similar expectations.

[25] The boiling point of VCM is -13.6°C.

[26] NFPA 15 Chapter 7, Section 7.5 contains design objectives for fixed water spray systems (NFPA, 2001).

extinguish and control burning and for exposure protection. However, NFPA 15 does not provide a clear strategy or detailed guidelines on how to design systems to prevent flammable vapor fires, nor does it include information on, or discussion of, the limitations of such systems. The CSB concluded that more comprehensive and detailed guidance would help manufacturers specify and understand the limitations of deluge systems to prevent flammable vapor fires and explosion.

Root and Contributing Causes

The investigation identified the following root causes:

1. Borden Chemical did not adequately address the potential for human error:
   a) Borden Chemical did not implement 1992 process hazard analysis (PHA) recommendations to change the reactor bottom valve interlock bypass to reduce potential misuse.
   b) In a 1999 PHA, Borden identified severe consequences for opening the reactor bottom valve on an operating reactor, but accepted the interlock, controlled by procedures and training, as a suitable safeguard.
2. Formosa-IL did not adequately address the potential for human error:
   a) After a 2003 incident at FPC USA's Baton Rouge facility, Formosa-IL did not recognize that a similar incident could occur at the Illiopolis facility or take action to prevent it.
   b) Formosa-IL site management did not implement corrective actions identified in the investigation of a similar incident in February 2004 at Formosa-IL.
3. Formosa-IL relied on a written procedure to control a hazard with potentially catastrophic consequences.

The investigation identified the following contributing causes:

1. FPC USA did not have written guidelines for matching safeguards with risk.
2. FPC USA did not have comprehensive written standards managing interlocks at its PVC facilities.
3. FPC USA did not recognize and address common elements among several serious incidents at its PVC facilities.
4. Formosa-IL employees were unprepared for a major VCM release.

Recommendations

As a result of this investigation, the CSB made recommendations to Formosa Plastics Corporation; the Vinyl Institute; the National Fire Protection Association; the U.S. Environmental Protection Agency; and the American Institute of Chemical Engineers Center for Chemical Process Safety. The recommendations are listed in the CSB report, which can be found on the CSB website, www.csb.gov.

References

[1] Center for Chemical Process Safety (CCPS), *Guidelines for Preventing Human Error in Process Safety*, AIChE, 1994.

[2] Nash, L.I., ed. *Encyclopedia of PVC*, Marcel Dekker, Inc., 1976.

[3] Hopkins, A., *Safety, Culture and Risk The Organisational Causes of Disasters*, Sydney, Australia. CCH Australia Limited, 2005.

[4] Bird, F. E. and Germain, G. L., *Practical Loss Control Leadership*, International Loss Control Institute, Inc., 1985.

[5] CCPS, *Layer of Protection Analysis*, AIChE, 2001.

[6] CCPS, *Guidelines for Hazard Evaluation Procedures*, AIChE, 1992.

[7] CCPS, *Guidelines for Safe Automation of Chemical Processes*, AIChE, 1993.

[8] Lees, F. P., "Pressure System Design" in *Loss Prevention in the Process Industries: Hazard Identification, Assessment and Control*, Vol. 1, Butterworth-Heinemann, 1996.

[9] CCPS, *Guidelines for Technical Planning for On-site Emergencies* AIChE, 1995.

[10] CCPS, *Guidelines for Fire Protection in Chemical, Petrochemical, and Hydrocarbon Processing Facilities*, American Institute of Chemical Engineers (AIChE), 2003.

[11] National Fire Protection Association (NFPA), Standard *for the Inspection, Testing, and Maintenance of Water-Based Fire Protection Systems*, NFPA 25, 2002.

[12] Thomas, G.O., "On the Conditions Required for Explosion Mitigation by Water Sprays," *Process Safety and Environmental Protection*, Vol. 78, Part B, September 2000.

[13] CCPS, *Guidelines for Postrelease Mitigation Technology in the Chemical Process Industry*, AIChE, 1997.

AUTHOR INDEX

Agreda, Anita Garcia 81
Bailey, James R. 121
Baker, Quentin A. 41
Bennett, Raymond H. 41
Berger, Scott 325, 355
Blair, Angela S. 331
Childs, Matt 261
Clark, David G. 23
Corripio, Bernardo 385
Davis, Scott 367
De Maaijer, M. 1
Deetjen, Caroline 13
Dharmavaram, S. 279
Di Benedetto, Almerinda 177
Di Carluccio, Antonio 81
Ebadat, Vahid 389
Fabbrocino, Giovanni 81
Fitzgerald, Gary 217
Francis, Christina 119
Geng, Jihui 167
Gilmour, Sandy 349
Gutierrez, Anthony 217
Hansen, Olav 193, 295
Harris, Steven P. 99
Hendershot, Dennis 339
Herrmann, David 135
Horowitz, Daniel 349
Jillings, Lee Anne 325
Klein, A. 1
Knight, Greg 217
Lay, James 414
Leblanc, John 249
Leprette, E. 151
Leskin, Katherine 414
Levitan, Marc L. 121
Liserio, Frank F. 53
Liu, Lijun 315
Long, Lisa 414
Louvar, Joseph 339
Manfredi, Gaetano 81
Mannan, Sam 261, 315
McClure, Randy 414
McPhate, Jeff 261
Middha, Prankul 193, 295
Murphy, John 375
Ozbey, M. Cem 69
Papadaki, Maria 315
Person, Karen 355
Pertuz, David 13
Polcyn, Michael A. 41
Proust, Ch. 151
Prugh, Richard 237
Prugh, Richard W. 389
Qiao, Anna 13
Rogers, T. Mike 400
Rogers, William 315
Russo, Paola 177
Salzano, Ernesto 81, 177
Sari, Ali 69
Shochet, I. 151
Sipkema, Anthony 261
Smith, Allen 414
Suardin, Jaffee 261
Svoboda, Alan 53
Thomas, Kelly 167
Van Geffen, Sheila 385
Versloot, N. 1
Wang, Yanjun 315
Weber, Andy 385
Wieczorek, Christopher 249
Wilson, Duane O. 99
Zhang, Yingchun 315